Solvent-Dependent Flexibility of Proteins
and Principles of Their Function

Advances in Inclusion Science

Alex I. Käiväräinen

Institute of Biology, U.S.S.R. Academy of Sciences

Solvent-Dependent Flexibility of Proteins and Principles of Their Function

D. Reidel Publishing Company

A MEMBER OF THE KLUWER ACADEMIC PUBLISHERS GROUP

Dordrecht / Boston / Lancaster

7149-2902

CHEMISTRY

Library of Congress Cataloging in Publication Data

Käiväräinen, A. I. (Aleksandr Ivanovich)
    Solvent-dependent flexibility of proteins and principles of their function.

    (Advances in inclusion science)
    Translation of: Dinamicheskoe povedenie belkov v vodnoi srede i ikh funktsii.
    Bibliography: pp.
    Includes index.
    1. Proteins. 2. Water of hydration. I. Title. II. Series.
QP551.K4913   1983   574.19′245   83-19121
ISBN 90-277-1534-3

Published by D. Reidel Publishing Company,
P. O. Box 17, 3300 AA Dordrecht, Holland

Sold and distributed in the U.S.A. and Canada
by Kluwer Academic Publishers,
190 Old Derby Street, Hingham, MA 02043, U.S.A.

In all other countries, sold and distributed
by Kluwer Academic Publishers Group,
P. O. Box 322, 3300 AH Dordrecht, Holland.

All Rights Reserved

© 1985 by D. Reidel Publishing Company, Dordrecht, Holland
No part of the material protected by this copyright notice may be reproduced or utilized in any form
or by any means, electronic or mechanical including photocopying, recording or by any information
storage and retrieval system, without written permission from the copyright owner

Printed in The Netherlands

Table of Contents

Q P551
K4913
1985
CHEM

Foreword

Molecular biology has now advanced to the point where it is no longer possible to give a complete review of the available data on the conformational features of proteins. New data keep streaming in, and there is obviously an urgent need for some sort of general treatment of the subject. A systematic treatment of the large amount of data obtained by a great variety of methods on a great variety of objects must be based on the use of models; these should be as simple as possible, should conform to well-established scientific laws, and at the same time be sufficiently flexible. The validity of the models finally arrived at is then confirmed or otherwise by testing the conclusions arrived at with their aid by means of experiment.

After a suitable model has been adopted, it can be used in analyzing the experimental data. Such an analysis may result in one of three possible situations: neither the experimental results nor their interpretation contradict the proposed model; the experimental results do not contradict the proposed model but their interpretation by the authors does; finally, both the experimental results and their interpretation may be found to be incompatible with the fundamental principles underlying the proposed model.

The first situation is clearly the most desirable, and presents no difficulties. If the second situation obtains, we would do well to recall the wise words of Leonardo da Vinci: "Experiments are never wrong; it is only opinions which may be..." The third situation is the most serious: if the experimental setup does not give rise to any systematic errors, the model must be modified or even discarded altogether. Here, one firm 'against' outweighs any number of 'fors'.

This book will consider three models, each succeeding model more complex than the previous one: the dynamic equilibrium model of behavior of proteins in water; the dynamic association-dissociation model of specific complex compounds; and the dynamic model of enzymatic activity.

The principles of these models derive from relatively simple systems, which are taken as examples. The problems involved in the interaction of proteins with water form one of the principal subjects of this book. A number of realistic examples are employed to demonstrate the validity of the model, and in order to show that different experimental data can be reduced to a common denominator by its use. As a result, an altogether novel interpretation of familiar facts becomes possible.

It is my pleasant duty to thank L.A. Blumenfeld, G.I. Likhtenstein, Yu. I. Khurgin, V.I. Lobyshev, S.I. Aksenov and A.V. Filatov for their assistance and for the comments they offered on the manuscript.

Introduction

The properties of proteins, in the form of results obtained by biochemical and physico-chemical studies, form the subject of a very large number of publications. Interest in the connection between the structure and the biological functions of proteins is steadily growing. The concept of the active protein as a cooperative system with definite degrees of freedom has now become firmly established, and most experts now believe that there are certain general principles which govern protein activity.

The major advances of molecular biology are now apparent to all; nevertheless, the mechanism of functional activity is not yet fully understood, not even for a single protein. Attempts to solve this problem, including the application of the powerful tool of X-ray diffraction analysis, have so far failed, even though detailed information about the initial (ligand-free) and final (ligand-bonded) states of the macromolecule was available. Even if it is assumed that the conformational states of the molecule in solution and in the crystal are identical — an assumption which is itself open to doubt — we are still unable to effect an experimental follow-up of the complicated pathways of the structural rearrangements involved in the conversion from the initial to the final state or to grasp the meaning of their particular sequence. Likely methods of approach to this problem in future will most probably be dynamic such as NMR, EPR, fluorescence methods, deuterium exchange techniques, stopped flow method, temperature jump method, etc.

During the past few years the principal scientific interest has shifted to the dynamic aspects of protein physics. These questions are intimately connected with the problems involved in the hydration of biopolymers, which are now being intensively studied.

A detailed treatment of the nature of the interaction between proteins and water, including correlation times and percentage contents of the different fractions is given in numerous studies, with the main stress on the role of water in the mobility of the protein matrix. However, practically nothing has been published on the conformational changes taking place in the solvate hull as a result of conformational rearrangements of the protein, except possibly about its changed extent of hydration. Only very few workers consider the sorbed water to be an active participant in the structural changes taking place in the macromolecule.

The principal subject of this book is the treatment of the dynamic properties of proteins as related to the properties of their solvate hulls, and of the general relationships governing the interaction of proteins with water. It will be shown that the available data can be readily fitted into our dynamic model of behavior of proteins in water, without resulting in any inconsistencies. The model offers a systematic explanation for the results of studies conducted on a given object by several different methods, and makes it possible to consider a variety of different phenomena from a uniform point of view. Several non-trivial implications of the model were confirmed by series of specially conducted experiments. Conversely, no facts could be discovered which would contradict any fundamental conception, even though every effort was made at an objective approach to this problem.

Our model introduces one novel concept — the idea of 'clusterophilic' interaction between protein cavities and water. The concept follows from the hypothesis that protein cavities with their active sites assume, in one of their states, a special structure (geometry, amino acid composition, etc.), in which the water molecules which have entered the cavity arrange themselves into an ordered, cooperative 'cluster', since this is thermodynamically favored. The lifetime of such 'clusters' is not more than about 10^{-8} second; it varies with the properties of the protein and with external conditions, and is several orders longer than that of hydrogen bonds in homogeneous liquid water (10^{-11} - 10^{-12} sec). If, for any reason, the water has lost its capacity to form clusters, its interaction with the cavity is thermodynamically less favored; as a result, the stability of the protein cavity decreases, and the cavity tends to assume another structure which is stable by virtue of its hydrophobic properties. The former state will be referred to as the 'open' B-state, while the latter will be denoted as the 'closed' A-state. It is probable that this adaptability of the properties of proteins to those of water is the result of a prolonged biological evolution, in the course of which the macromolecules have 'learned' to utilize the phase transitions of water to regulate their own conformation and stability (Käivär-äinen, 1975b).

The idea that water may have affected the evolution of protein properties was also ·put forward at an earlier date, though in a less specific form : "Proteins are a machine, which has been programmed by evolution to undergo chemical reactions with liquid water" (Lamri and Biltonen, 1973). According to these workers, this is realized by way of fine changes, which involve both the geometry of the proteins and the state of the solvent, including the adaptation of the conformations of the protein and the water to each other. They identify this process as the swelling of proteins, which is accompanied by free exchange of water. At $25°C$ the enthalpils required to produce a hole of the approximate size of one H_2O molecule in water, in short-chain aliphatic alcohols and in pure aliphatic hydrocarbons are about 84, 29 and 8.5 kilo-joules respectively*. It is thought, accordingly, that cavities in proteins will appear mostly at the concentration sites of hydrophobic residues, and at the sites of contact between the hydrophobic and the polar groups.

The problem of the effect of water on biological systems was actively studied by Drost-Hansen (1970), who explained the observed anomalies in the variations of various biological processes in terms of polymorphic transitions between the different structures of water. A study of a large number of data on temperature anomalies led this worker to specify four typical transition temperatures: 15, 30, 45 and $60°C$. However, the attempt by O'Nell and Adami (1969) at confirming temperature-dependent isomorphism of water was unsuccessful.

Following the work of Frank and Evans (1945), who pointed out that nonpolar compounds tend to stabilize the adjacent water molecules, many hypotheses on the role of 'ice-like' water in biological systems have been advanced. Klotz (1964) suggested that water reacts with nonpolar side chains of the protein to form crystalline hydrates and

---

* All physical units in this book are SI units.

that this is accompanied by the stabilization of the macromolecule; that thermal denaturation follows the fusion of these crystalline hydrates; and that urea, which destroys the hydrate structure, destabilizes proteins in this way. He compared the effect of nonpolar amino acid groups on water to that produced by model compounds such as tetrabutylammonium ion and by certain gases (argon, krypton, chlorine and methane) which form crystalline hydrates. In all these cases the enthalpy of hydration is about the same, viz., 67±8 KJ. It would appear that the stability of hydrates does not depend on specific interactions between nonpolar molecules and water, but rather on the stability of the water itself in the presence of these molecules. Crystalline hydrates have a pentagonal rather than a hexagonal structure, such as is typical of ordinary ice; they are very stable and melt only at about 30°C.

We consider this theory to be inadequate, since it leaves out of account the dynamic properties of protein-water systems and the special role played by protein cavities. Short-lived ordered structures may admittedly be formed around the nonpolar surface groups; however, experimental studies performed on hydrocarbons indicate that this type of protein-water interaction is not thermodynamically favored. We think, accordingly, unlike Klotz, that the sorption of water on non-polar surface zones results in their destabilization and is responsible for the appearance of the mobility in the tertiary domain (or block) structure of the proteins.

Klotz interprets the properties of membranes by postulating that major changes in the permeability of the nerve membrane to ions are due to the transition of the water inside the membrane from the solid to the liquid state. Here, again, the dynamic and the structural aspects of the process are left out of account. This notwithstanding, Klotz' ideas on the capacity of water to form cooperative systems in the hydrate hull and in the protein membrane are very important in themselves.

Fisher (1964), who adopted a diametrically opposite approach to the problem of the interaction between nonpolar protein zones and water, proposed a model of the protein globule, in which all nonpolar amino acid residues are located inside the globule and are inaccessible to water, while the surface of the globule consists of hydrophilic residues only. If this is accepted, it follows that the role of hydrophobic interactions in the formation of tertiary protein structure has been greatly overrated. However, it is evident from the results of X-ray diffraction analysis that a large proportion of nonpolar zones are in fact accessible to the solvent, while in the interior of the proteins polar and nonpolar residues are about equally distributed. The structure of protein cavities is mostly nonpolar.

That the reaction of proteins with water may be accompanied by transitions resembling phase transitions of the first kind was pointed out by Likhtenshtein as early as 1967; such transitions were subsequently used (Lumry and Rajender, 1970) to explain the enthalpy-entropy compensation which accompanies conformational changes in proteins. Unlike these workers, we believe that such transitions, which in actual fact take place only under certain circumstances, affect only the water contained in protein cavities.

According to the literature data which are now available three principal types of mobility may be distinguished in protein structures.

1. Thermal displacements of atoms and reorientations of amino acid residues, the amplitude and frequency of which increase from the center of the globule towards its

periphery. The motion of polar and of charged side groups is characterized by correlation times of between $10^{-9} - 10^{-11}$ sec. The mobility of nonpolar aromatic residues (histidine, tyrosine, phenylalanine and tryptophan) is much lower than that of the polar residues. An increase in the temperature results in a thermally induced activation of the rotation of these residues (Gurd and Rothgeb, 1979).

Such motions can be studied by the following methods: determination of electron density distribution in diffraction patterns of crystalline proteins (Frauenfelder *et al.*, 1979; Artymiuk *et al.*, 1979); computerized modeling of molecular dynamics of protein structures (Karplus *et al.*, 1980; McCammon and Karplus, 1980); NMR relaxation and high-resolution NMR (Nelson *et al.*, 1976; Aksenov, 1979; Williams, 1979; Levine *et al.*, 1979).

2. Micro-openings of the structure, migration of defects and other processes, which are responsible for the diffusion of small-sized molecules (D_2O, I^-, NO_3^-, CNS^-, O_2, acrylamide, etc.) into the interior of the globule. They are detected by hydrogen exchange (Linderström-Lang and Shellman, 1959; Hvidt, 1973; Wagner and Wütric, 1979) and from phosphorescence and fluorescence quenching (Saviotti and Galley, 1974; Eftink and Chiron, 1976; Burshtein, 1977; Lakovicz, 1980).

3. Relative displacement of protein domains or subunits in oligomeric proteins, which is accompanied by changes in the geometry of protein cavities. This type of motion is detected directly by X-ray diffraction analysis (Steitz *et al.*, 1979; MacDonald *et al.*, 1979; Pickover *et al.*, 1979). Information about such motion in solution is obtained by dynamic methods; fluorescence polarization (Zagyansky *et al.*, 1969; Glotov *et al.*, 1976; Bishop and Ryan, 1975), nanosecond fluorescence polarization (Hanson *et al.*, 1981; Cathou, 1978) and by a modified spin label method (Käiväräinen, 1975a). This effect is known as the conformational flexibility of proteins.

A view which has recently gained wide acceptance is that it is this third type of motion which plays a determining role in the functional mechanism of proteins (Käiväräinen, 1975b; Ptitsyn, 1978; Beece *et al.*, 1980; Huber, 1979; Careri *et al.*, 1979; Bennet *et al.*, 1980). This is also the type of motion about which least is known — indeed, until a short time ago, doubt was cast upon its very existence. Nothing is known at all about the part played by water in the phenomenon of conformational flexibility of proteins. It would seem that this is due, on one hand, to the as yet insufficient awareness of the significance of the flexibility of proteins as a major problem and, on the other, to the fact that the physical methods by which this effect may be studied are rather complex in handling.

We shall attempt in this book to fill this gap in molecular biophysics and to outline the pathways for future investigations. Three, successively more complex models served as theoretical basis in the analysis and systematization of the experimental data: the dynamic equilibrium model of protein behavior in water; the dynamic model of association and dissociation of specific complexes; and the dynamic model of enzymatic action.

Before we formulate the basic assumptions which underlie our dynamic model of behavior of proteins in water, we shall consider certain properties of pure water and the effect of model compounds on these properties. This is important since, according to the ideas which will be developed in this book, it is the properties of water which largely determine conformational mobility of proteins, the structure of their cavities, and the thermodynamics of their structural conversions.

Chapter 1

The Properties of Water

That physicochemical factors affect biological evolution has been repeatedly confirmed and is now considered as certain (see, e.g., Shnol', 1979). One such highly important factor is the 'habitat' of the systems undergoing evolution, whether they be macromolecules or complex organisms. This is why a discussion of the properties of water in this context is of particular interest.

1.1 Pure Water

X-ray investigations of liquid water indicate that its average structure resembles that of ice *I*. The fusion of ice *I* at 0°C since only 15% of the hydrogen bonds, and is accompanied by an increase in the packing density of H_2O molecules, and by a decrease in the molar volume from 19.65 to 18.02 cm^3/mole. Ice *I* crystallizes in the hexagonal system; each oxygen atom is located in the center of a tetrahedron constituted by four oxygen atoms at a distance of 0.276 nm from one another. Each water molecule is linked by hydrogen bonds to four other water molecules located in its immediate vicinity. In ice, any molecule can assume six possible orientations, so that hydrogen atoms in an ice crystal may assume a large number of possible orientations. The unit lattice cell (a = 4.52 Å, c = 7.37 Å, or a = 0.452 nm, c = 0.737 nm) contains four water molecules. The volume of a single water molecule is 32 $Å^3$ or 0.032 nm^3.

Frozen solutions of certain biopolymers contain other species of ice as well – vitreous ice and cubic ice (*I*c) (Dowell et al., 1962), which are converted to ice *I* at higher temperatures.

According to Lobyshev and Kalinichenko (1978), ordered water structures other than the hexagonal ice *I* may also be formed in heterogeneous biological systems, owing to the polymorphic nature of ice. It should be noted that the thermodynamic parameters of phase transitions between the various modifications of ice are not very different from one another.

The properties of water and ice are determined, to a large extent, by the hydrogen bonds O-H...O, which interlink oxygen with hydrogen atoms. Hydrogen bonds in ice are rectilinear. According to Zatsepina (1974), the melting of ice is accompanied by the activation of hydrogen atom vibrations across the direction of the hydrogen bond and as a result the hydrogen bonds in liquid water become deformed and its structure becomes denser. The appearance of these additional degrees of freedom is also manifested by the increase in the heat capacity which occurs during the fusion of ice. The energy of the hydrogen bonds is the resultant of the following contributory factors: electrostatic forces

1

generated by electron delocalization, and the difference between the dispersive interactions and the energy of repulsion. Calculations of the hydrogen bond energy in ice *I*, based on different models of water, yielded values ranging from 19.7 to 34.36 KJ/mole, the experimental value being 25.56 KJ/mole (Eizenberg and Kauzman, 1975).

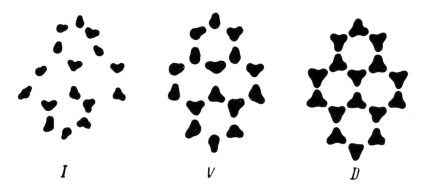

I   *V*   *D*

Figure 1. Schematic representation of three structures of ice I. After Eizenberg and Kauzman (1975).

The last-named workers introduced the concept of three ice "structures": I, V and D (Fig. 1). The water molecules in the crystal oscillate about their average locations, whose distribution in the sample constitute its crystalline network. Obviously, the average locations of the molecules at different times t within the order of their oscillation period $\tau_k = 2 \cdot 10^{-8}$ sec will correspond to different structures. If $t \ll \tau_k$, we shall obtain the instantaneous or the I-structure of the crystal, in which the distance between the nearest neighbors may deviate by as much as 15% from the equilibrium position.

In addition to the oscillatory motion, the crystal molecules also undergo rotational and translational displacements which, in the case of ice I at 0°C, may be described by the time $\tau_p \cong 10^{-5}$ sec. If the locations of the molecules are averaged over the time $\tau_k < t < \tau_p$, we shall obtain the vibrationally averaged V-structure. Finally, if $t \gg \tau_p$, we obtain the D-averaged or the diffusional structure. In the case of water $\tau_k = 2 \cdot 10^{-13}$ sec, while $\tau_p \cong 10^{-11}$ sec. This means that water, just like ice, has three types of structure: I, V and D. If the temperature of water is raised, the molecular displacements become faster and their frequency approaches the frequency of the oscillations, when the difference between the structures I and V disappears. At the time of writing experimental methods of obtaining information about the I-structure of water are as yet nonexistent. The D-structure of water is characterized by its thermodynamic properties and by other parameters such as the dielectric constant, refractive index and chemical NMR shift. Information about the V-structure is obtained by methods responding to fast processes, e.g., IR and Raman spectroscopy or inelastic neutron scattering (Fig. 2).

According to Samoilov (1957), an ice-like network is preserved in liquid water except that, unlike in ice, the molecules are found not only at the nodes of the lattice, but also at a number of locations between the nodes, as a result of which the density of water is

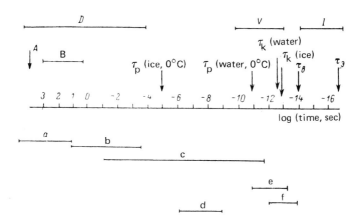

Figure 2. Time scale of molecular processes taking place in ice and in liquid water. After Eizenberg and Kautsman (1975).

A — periods corresponding to the various molecular processes: molecular displacements τ_p and oscillations τ_k, O-H stretching vibrations with a period of τ_v, and the time τ_e required to complete one electron rotation in the first Bohr orbit.

B — time intervals determined by various experimental methods: a — X-ray diffraction, thermodynamic properties, light scattering and refractive index; b — chemical NMR shift; c — dielectric relaxation; d — ultrasonic absorption; e - inelastic neutron scattering; f — IR and Raman spectroscopy.

higher than that of ice. Light scattering and other experimental studies indicate that water is a medium without any pronounced structural inhomogeneities (Vuks, 1977).

The molecular mechanism of autodiffusion in ice has not yet been fully clarified, but is undoubtedly related to the imperfections of the crystal lattice. The activation energy of autodiffusion, the energy of dielectric relaxation and the energy of viscous flow of water are all about 19.3 KJ/mole at 25°C. It has been shown that the average length of a diffusional jump is equal to the distance between the nearest molecules, and that the unit of the liquid undergoing orientational and positional changes is one water molecule (H_2O) (Wang *et al.*, 1953).

If deuterium is substituted for protium, the geometry of the free water molecule, which is determined by its electron structure, remains unchanged. The dipole moments of H_2O and D_2O are practically equal. D_2O is 1.2 times more viscous at 30°C than H_2O. The rate of molecular reorientations and translations in liquid D_2O is slower than in ordinary water, because heavy water is more viscous, has a longer relaxation time and a lower autodiffusion coefficient.

The approximate estimates of Sokolov (1964) indicate that the hydrogen bond energy is 25.85 KJ/mole in liquid water, and is slightly higher — 27.65 KJ/mole — in heavy water. The properties of heavy water change at a faster rate than those of ordinary water at increased temperatures and pressures. This is because D_2O molecules are more highly ordered and their cooperative properties are more pronounced; accordingly, the formation of ordered structures by D_2O molecules is thermodynamically more favored than in the case of H_2O molecules (Lobyshev and Kalinichenko, 1978). This is an important consideration, which must be borne in mind when studying the effects of solvation of bio-polymers in H_2O - D_2O mixed solvents.

The equilibrium $H_2O + D_2O \rightleftharpoons 2HOD$ becomes established very soon after the H_2O and D_2O are intermixed. The relative concentrations of all three forms of water in the mixture varies nonlinearly with the concentration of deuterium (Fig. 3).

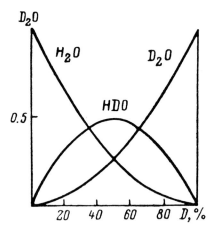

Figure 3. Relative proportions of the three forms of water as a function of the variation in the concentration of deuterium between 0% and 100%. After Gold (1969).

Numerous results indicate that both H^+ and OH^- ions are strongly hydrated in water. The heat of hydration of H^+ is about 1150 KJ/mole at 25°C, i.e., is higher by 420 KJ/mole than the heat of hydration of any other monovalent ion. As a result, oxonium ions H_3O^+, as well as more complex ions such as $H_9O_4^+$, are formed. The average time period elapsing between two successive associations of a given water molecule with a proton is about $5 \cdot 10^{-4}$ sec, while the average lifetime of a protonized water molecule is 10^{-12} sec.

The heat capacity of ice tends to zero at very low temperatures. As the temperature increases, the C_p-value increases to 9 Kcal/mole or 38 KJ/mole (Fig. 4). As a result of fusion, the C_p-value is nearly doubled and remains approximately constant thereafter.

The changes in enthalpy, entropy and free energy at any temperature T relative to 0°K are found by numerical integration of C_p:

$$H_T - H_0 = \int_0^T C_p dT + \Delta H \text{ p.t.} \tag{1.1}$$

$$S_T - S_0 = \int_0^T \frac{C_p}{T} dT + \Delta S \text{ p.t.} \tag{1.2}$$

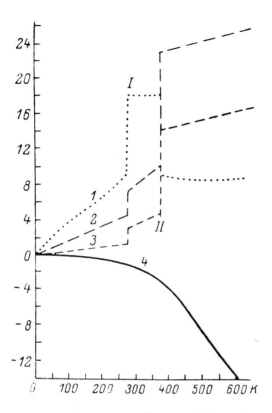

Figure 4. Variation of thermodynamic parameters of water with increasing temperature, and during fusion (I) and vaporization (II) under a pressure of 1 atm. 1 — heat capacity, cal/mole·°K; 2 — entropy, cal/mole·°K; 3 — enthalpy, Kcal/mole; 4 — Gibbs' free energy, Kcal/mole.

where ΔHp.t. and ΔSp.t. are changes of enthalpy and entropy during phase transitions taking place between $0°$K and T; and

$$G_T - G_0 = H_T - H_0 - TS_T, \tag{1.3}$$

since S_0 is assumed to be zero.

Ice I displays sudden changes of thermodynamic parameters at its melting and boiling points(Table I). The essential difference between the properties of liquid and crystalline water are to be attributed to the different velocities of molecular motions. Liquid water molecules undergo 10^{11}-10^{12} reorientations and translational shifts in one second, while the corresponding figure for ice is only 10^5-10^6.

At temperatures above $0°$C the enthalpy and entropy of the quasi-crystalline water in protein cavities may be expected to be higher than in ice, but they may be lower than in free liquid water.

One of the important characteristics of liquid water is the fluctuation of its parameters such as density, concentration, pressure, etc.

Strictly speaking, the conclusions of the thermodynamic theory of fluctuations are only applicable if the zones, v, in which the fluctuations are observed, are infinitely large. In practice, theoretical results are found to be in satisfactory agreement with the experimental data if these zones contain about 10^4 molecules or fewer. The theory is not yet sufficiently developed to be applicable to small zones containing only a few dozen molecules. Experimental data on micro-fluctuations are still very scanty, but there is reason to believe that fluctuations in small liquid volumes are much like those observed in the vicinity of the critical point liquid-vapor (Shakhparonov, 1976).

TABLE 1. Thermodynamic phase transition constants at $P = 1$ atm, or $1.01 \cdot 10^5$ Pa. After Eizenberg and Kauzmann, 1975.

Thermodynamic transition constants	Fusion		Vaporization	
	H_2O	D_2O	H_2O	D_2O
T, K	273.15	276.97	373.15	374.95
ΔC_p, J/mole·°K	33.98	39.72	41.98	
ΔH, KJ/mole	6.018	6.289	40.713	41.590
ΔS, J/mole·°K	22.13	22.7	109.10	
ΔV, cm³/mole	-1.621	–	$3.01 \cdot 10^4$	

Consider the fluctuation of a parameter χ in a sufficiently large zone v, which interacts weakly with the environment. As a result of the fluctuations, there is a difference between χ and its statistical average value $\langle\chi\rangle$. In the course of time a closed system tends to assume its equilibrium state, and χ asymptotically tends to $\langle\chi\rangle$. This process is known as relaxation. Let $\Delta\chi = \chi - \langle\chi\rangle$; then the rate at which χ tends to assume the equilibrium value $\langle\chi\rangle$ is given by the equation

$$\partial (\Delta x)/\partial t = - (x - \langle x \rangle)/\tau, \qquad (1.4)$$

where τ is the relaxation time, which describes the average duration of the fluctuations.

It is a requirement of the thermodynamic theory of fluctuations that τ be greater by at least one order than the time period required for the Maxwell velocity distribution to be attained in the zone v. In such a situation dynamic quantum effects need not be considered. Evaluations based on the uncertainty principle indicate that the average "lifetime" of thermodynamic fluctuations may be described by the inequality

$$\tau \gg \hbar/kT \simeq 10^{-11}/T, \qquad (1.5)$$

where k is the Boltzmann constant.

It follows from the above that at room temperatures the frequency of the fluctuations $v \simeq 1/\tau$ may be expected to be less than 10^{12} sec^{-1}. The time of dielectric relaxation at 25°C for water is about 10^{-11} sec (Eizenberg and Kauzman, 1975), which means that the condition (1.5) is satisfied.

In accordance with the Boltzmann principle, the probability density χ of fluctuations in an isolated system is given by the equation

$$w_x = \text{const } e^{\Delta S(x)/k}, \tag{1.6}$$

(Shakhparonov, 1976), where ΔS is the difference between the entropies of the equilibrium state and the state under study. The probability of occurrence of density fluctuations $\Delta \rho$ in zone v of a liquid is

$$dW = w_{\rho, v} \Delta \rho, \tag{1.7}$$

where ρ is the average density of the liquid.

If the state of the liquid differs from that of the immediate surroundings of the critical point liquid-vapor, the probability density has a normal distribution (Leontovich, 1944):

$$w_{\rho, v} = \frac{1}{\sqrt{2\pi \langle (\Delta \rho)^2 \rangle}} \exp\left[-\frac{(\Delta \rho)^2}{2 \langle (\Delta \rho)^2 \rangle}\right]. \tag{1.8}$$

The statistical mean square of density fluctuations is directly proportional to the isothermal compressibility β_t, and is inversely proportional to the volume v in which the fluctuation is considered:

$$\langle (\Delta \rho)^2 \rangle = \frac{kT \beta_t \rho^2}{v}. \tag{1.9}$$

In the vicinity of the critical point liquid-vapor β_t increases by more than two orders, and the probability of density fluctuations increases correspondingly. These equations are valid provided the fluctuations $\Delta \rho_i$ and $\Delta \rho_j$ in any two zones v_i and v_j of the liquid are statistically independent, i.e., if

$$\langle \Delta \rho_i \Delta \rho_j \rangle = 0. \tag{1.10}$$

In the vicinity of the critical point this condition is no longer satisfied. The radius of the liquid concentration around any arbitrarily selected atom increases as well.

The density fluctuations in the liquid phase are of two types: adiabatic ($S = $ const) and isobaric ($P = $ const). For this reason the density fluctuation is related to pressure fluctuations ΔP and entropy fluctuations ΔS by the equations:

$$\Delta \rho = (\partial \rho / \partial P)_S \Delta P + (\partial \rho / \partial S)_P \Delta S. \tag{1.11}$$

S and P are independent variables, and for this reason their fluctuations are statistically independent: $\langle \Delta P \Delta S \rangle = 0$.

As a result of adiabatic fluctuations, sonic waves produced by thermal motion are propagated through the liquid. As distinct from adiabatic fluctuations, isobaric density fluctuations do not become displaced with time. They are mainly caused by local changes in temperature, and are connected with the changing concentration of the associates in single-component liquids – for example, with the increase or decrease in the number of hydrogen bonds in a given micro-volume of water, whose radius is larger than about 5 nm.

In water, unlike in other liquids, the isothermal and adiabatic compressibilities are practically identical between 0°C and physiological temperatures (Vuks, 1977).

Thus, the functional environment of biopolymers is distinguished by a complicated dynamic behavior. It is obvious that this fact could not fail to affect the evolutional pathways of biological macromolecules. However, the properties of water in heterogeneous systems differ somewhat from those of free water. In what follows, we shall attempt to deduce the nature of the changes which are to be expected in the properties of water as a result of its interaction with biopolymers, using various model systems for this purpose.

1.2 Water in Model Systems

The concept of hydrophobic interactions, which are believed to play an important part in the stabilization of biopolymers, had its origin in an experimental study of the solubility of nonpolar hydrocarbons in water. It was shown (Kauzmann, 1954, 1959) that the transfer of hydrocarbons from a nonpolar to a polar solvent (water) is accompanied by an increase in the free energy of the system:

$$\Delta G = \Delta H - T\Delta S > 0. \tag{1.12}$$

It was found in the course of these studies that both H and S decrease, but the decrease in H is smaller than in S, so that ΔG is positive. The dissolution of nonpolar compounds in water is thermodynamically disfavored, since interactions between two nonpolar compounds are more favored thermodynamically than interactions between water and a nonpolar compound. Hence the term 'hydrophoby'.

Amino acid residues show considerable variations in their hydrophobic properties, and as a result may display both polar and nonpolar characteristics (Table 2). The degree of hydrophoby is defined as the free energy change per side group of the amino acid during its transfer from ethanol to water (Tanford, 1967). However, the distribution of polar and nonpolar properties along these molecules is usually non-uniform, and for this reason the molecules may display different properties, depending on their location in the protein and on their environment.

The commonly accepted interpretation of thermodynamic properties of nonpolar groups in water is that given by Frank and Evans (1945). According to these workers, negative enthalpy and entropy are caused by the formation of ordered structures of water molecules around the nonpolar groups. These structures gradually melt as the temperature is increased. Since this is accompanied by a decrease in solubility, we must conclude that the interaction between nonpolar compounds and ordered water is thermodynamically favored over the interaction with non-ordered water. The mechanism by which water becomes stabilized by nonpolar substances has not yet been fully clarified.

The abnormally high partial molar heat capacity of nonpolar groups in water is probably caused by the fusion of the ordered water structures — clusters. Denaturation may be accompanied by a major change in the heat capacity of the solution, since nonpolar residues are now exposed to the solvent and are solvated. This conclusion follows from the experimental studies of spiral-to-coil transitions in aqueous solutions of synthetic

TABLE 2. Hydrophoby of amino acid residues, J/mole.

Residue	ΔG	Contributions		
		Hydrophobic	Polar	By aromatic group
Tryptophan	12570	24511	−2514	−13408
Isoleucine	12444	10894		
Tyrosine	12025	18855	−2514	−6704
Phenylalanine	11103	18855		−6704
Proline	10894	8170		
Leucine	10140	10894		
Valine	7081	8170		
Lysine	6285	10894	−2514	
Histidine	5866	10894	−5028	
Methionine	5447	8170		
Alanine	3059	2723		
Arginine	3059	8170	−7542	
Cysteine	2723			
Glutamic acid	2304	5447	−5028	
Aspartic acid	2263	2723	−5028	
Threonine	1844	5447	−2514	
Serine	168	2723	−2514	
Glycine	0			
Asparagine	− 41.9	2723	−5028	
Glutamine	−419			

polypeptides of the type of weakly hydrophobic polyglutamic acid, and of the copolymer of this acid with the hydrophobic leucine. In the latter case the heat capacity increased strongly; in the former, it remained practically unchanged (Fasman *et al.*, 1964, as quoted by Volkenstein, 1975).

According to Brands, clusters are highly cooperative systems, which are disrupted by the slightest disturbance, including the addition of salts. The presence of electrolytes reduces the excess heat capacities in solutions of nonpolar substances. This may be the result of destabilization of the water clusters by the stronger effect of ionic fields, which impose on water poles an orientation which is unfavorable to the preservation of the cooperative system of hydrogen bonds.

Ions with negative enthalpy and entropy of solution are known as structure-forming, while ions with a positive enthalpy of solution are known as structure-disrupting. It would appear that both these types are capable of destroying water clusters in the vicinity of nonpolar compounds. According to the Frank-Ven model, three forms of water exist in electrolyte solutions: a form produced by Coulomb type interaction with the dissolved ion; an intermediate type with a disrupted structure; and ordinary water.

In general, the effects of solvation by D_2O ions are less strong than those by H_2O — i.e., the interaction of ions with heavy water is weaker. This may be expected to enhance association into ion pairs in solutions of biopolymers in D_2O, thus strengthening the salt bridges. Both salts and non-electrolytes have lower solubilities in heavy water. Several workers reported that hydrophobic interactions were stronger in D_2O solutions (Oakenfull and Fenwick, 1975; Lobyshev and Kalinichenko, 1978).

The analysis of X-ray patterns obtained during a study of X-ray scattering in aqueous methanol solutions led to the conclusion that the presence of small amounts of methanol reinforces the structure of water without introducing significant changes in it (Radchenko and Shestakovskii, 1955). The correlation time of rotation of water molecules in alcoholic solutions, as determined by NMR (Hertz and Zeidler, 1964), was longer than in pure water. That the mobility of water molecules decreases in solutions of substances with nonpolar alkyl groups is also indicated by the increase in the time of dielectric relaxation (Hasted and Elsabeh, 1953).

When water is intermixed with alcohols, the bands of the vibration (IR) spectrum undergo a shift, indicating that the bonds between water molecules are stronger at low alcohol concentrations, while the bonds between water and alcohol molecules are stronger at high alcohol concentrations (Belousov *et al.*, 1974). These workers, who studied the thermodynamic and spectroscopic properties of aqueous alcohol solutions, came to the conclusion that the negative enthalpy effect in the presence of low alcohol concentrations is largely due to the stabilizing effect of the nonpolar groups of the alcohol on the water molecules. This effect became considerably weaker when the temperature was raised from $0°C$ to $60°C$.

In the view of Vuks (1977), the variation of ultrasonic absorption, and of the concentration fluctuation function with the content of ethanol in aqueous solutions are qualitatively similar. This worker had in fact previously suggested that the absorption of ultrasonic waves by aqueous alcohol solutions is caused by fluctuations in the concentration of the solute.

It was shown by Lisnyanskii (1974) that ultrasonic wave absorption by aqueous solutions of *tert*-butanol may take place in two frequency ranges — $5 \cdot 10^7$ to 10^8 sec^{-1} and $0.6 \cdot 10^7$ to $2 \cdot 10^7$ sec^{-1}. This absorption is due, in the former case, to the attentuated fluctuations of pressure, while in the latter it is due to the fluctuations of entropy. Both attenuated fluctuations and entropy fluctuations may exist in liquids, the latter being related with the propagation of shear deformations.

Dowell *et al.* (1962) carried out an interesting study on the formation of ice structures in gelatin gel, conducted by the X-ray diffraction method. In 'dilute' gels, containing less than 20% of protein below $0°C$, the bulk water is present as ordinary hexagonal ice *I*. When the gelatin content increases, there are indications of incipient appearance of other ice varieties — cubic and vitreous ice. These varieties were metastable above $-70°C$ and probably underwent exothermal conversion to hexagonal ice.

Frank and Evans (1945) described a highly interesting intensification effect: when a nonpolar compound (ethanol) was added to a system of water with another nonpolar substance, such as argon, the free energy of the system decreased. This effect, which was also studied by Brands (1973), takes place when two hydrophobic molecules approach one another to a distance which is sufficiently small for their ordered solvate hulls to become superposed; as a result, the cooperativity zone increases, and the stability of the ordered structures increases as a result. A similar, but stronger effect may well take place in protein cavities.

It is instructive to compare the sorption of water on biopolymers with its sorption on structurally nonmobile disperse systems. Sorbents of non-biological origin such as silica gel, rutile, etc., if sufficiently moist, contain several types of water molecules with different mobilities (Kvlividze, 1970). The range of their correlation times is 10^{-6} to 10^{-9} sec. The relative contents of water types with different mobilities vary with the magnitude of the adsorption.

The study of the sorption of water on zeolites constitutes an approach to the problem of the relationship between the size of the voids and the organization and the properties of the water filling these voids. The structure and geometry of zeolite voids containing water molecules can be revealed by X-ray diffraction results. The size and the shape of the voids vary with the type of the zeolite. When heated, zeolites lose water while maintaining the integrity of their alumosilicate matrix. In other words, zeolites are rigid, and their structure is not effected by their interaction with water. Zeolite voids contain ion-exchange cations, and these compensate for the negative charge on the AlO_4 tetrahedra, which constitute the zeolite matrix. In protein cavities, the charged groups responsible for the polarization of H_2O molecules may be neutralized by H^+, OH^- and H_3O^+.

The state of water in zeolites is closely connected with the geometry of the voids in the matrix. Zeolites may be subdivided into three main groups: those containing only a few water molecules, which are 'firmly' bound at room temperature, and which are contained in tiny pores; those with pores $0.4 - 0.7$ nm in diameter, in which water molecules have two rotational degrees of freedom; and those with pores of up to 1.3 nm in diameter, which contain water molecules with all rotational and translational degrees of freedom.

The NMR spectra of water contained in zeolites belonging to the first group are doublets. In zeolites in the second group the NMR spectra are also doublets, but the intercomponent distance is small, and the state of the water then depends not only on the interaction with matrix, but also on the interaction of water molecules with each other. In zeolites in the third group the NMR spectrum at room temperature is a single narrow line, which means that the water molecules have not only rotational but also translational degrees of freedom. If the voids are amply filled, the average number of bonds per water molecule is nearly four. In this case the organization of the water is tetrahedral, such as is typical of ice, at least during 10^{-6} to 10^{-8} sec; this is confirmed by studies of the temperature variation of the width, of the second moment and of the shape of the NMR signals given by the water sorbed in the pores (Kvlividze, 1970).

The experimentally determined second moment S_2 can be expressed in terms of nuclear coordinates:

$$S_2 = 6/5 I \, (I+1) \, g^2 \mu_0^2 N^{-1} \sum_{j>k} r_{jk}^{-6} + 4/15 \mu_0^2 N^{-1} \sum_{j,f} I_f \, (I_f+1) \, g_f^2 r_{jf}^{-6}, \quad (1.13)$$

where I is the spin of the nucleus whose resonance is being observed, g is the split factor, μ_0 is the nuclear magneton, N is the number of the nuclei in the system, r_{jk} is the length of the vector interconnecting nuclei j and k, and I_f is the nuclear spin of the second species.

If the pores in the zeolite are filled to a sufficient extent, the value of S_2 at $90°K$

(-183°C) is close to that obtained for pure polycrystalline ice. This indicates that the structure of the water frozen in the pores is close to that of hexagonal ice. When the temperature is raised to a certain value, a narrow line, typical of liquid water, appears against the background of the broad line of the NMR spectrum, which is a characteristic effect given by ice. The temperature at which the second phase appears in the spectra depends on the size of the sorbent pores and on the amount of the sorbed water. In zeolites with 1.3 nm pore diameter containing 5, 12 and 30 water molecules, the fusion begins at -153, -133 and -103°C and is complete at -90, -80 and -60°C respectively. The respective acvivation energies of fusion for the pores filled as above are 14.6, 18.8 and 23.9 KJ/mole. In a silica gel with 7 nm pore diameter, with the pores completely filled with water, the fusion began at -78°C and terminated at -8°C, the activation energy being 31.4 KJ/mole.

These data indicate that, as the pores become filled with water molecules, the cooperative system of hydrogen bonds becomes larger and its stability increases. Consequently, filling the pores with water is a thermodynamically favored process. The larger-sized the water cluster in the void, the more does its structure resemble that of free water, as indicated by the rise in its point of fusion.

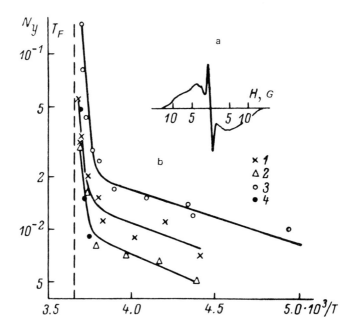

Figure 5. Variation of the relative proportions of sharp and wide components of the NMR spectrum as a function of the temperature. After Kurzaev *et al.* (1975). *a* — signal of frozen protein solution (concentration: 25 mg in 1 ml of water at 248°K); *b* -relative content N_y of the mobile phase as a function of $1/T$ for protein solutions (1 — native protein; 2 — denatured protein) and dispersions (3 — silica powder, 150 mg/ml; 4 — Teflon). T_{pl} — melting point of ice.

Physically different surfaces — polar and nonpolar surfaces — may be expected to affect the properties of the bound water in different manners. Kurzaev *et al.* (1975) compared temperature variations of NMR spectra of water protons for a number of various frozen objects: solutions of native and heat-denatured human serum albumin; water-disperse system of nonporous hydrophilic particles of Aerosil [silica powder], radius $\cong 10$ nm; and a water-disperse system of hydrophobic Teflon particles, radius $\cong 17.5$ nm (Fig. 5). They found that the water next to the nonpolar surface begins to melt at much higher temperatures (above -10°C) than the water interacting with polar groups on surfaces of silica powder and of proteins (between -90 and -60°C), the fusion itself taking place over a rather narrow interval. Thus, nonpolar surfaces distort the structure of water to a smaller extent than do polar surfaces, and the fusion of water is then more cooperative in nature.

Denatured albumin showed a decreased concentration of the mobile phase as compared with the native protein below the fusion points of water in the vicinity of nonpolar zones. This may be due to an increase in the number of hydrophobic residues available to water upon the denaturation of the protein.

*   *   *

Water molecules tend to become stabilized around hydrophobic compounds. This type of solvation is thermodynamically more favored than interaction with non-ordered water. It seems that water structures around nonpolar compounds are cooperative, and the overlapping of ordered solvate hulls increases the size of the cooperative zone and enhances the stability of the hulls.

The pore size is interrelated with the organization and properties of the water contained in the pores. If zeolite pores not smaller than 1.3 nm are filled with water to a sufficient extent, the water inside them tends to become stabilized in the tetrahedral form, which is typical of ice I, at temperatures below 0°C; above that temperature, the correlation time of water molecules in the pores is between 10^{-6} and 10^{-8} sec. Filling the pores with water is a thermodynamically favored process, since, as a result, the cooperativity between the hydrogen bonds in the system increases and, judging by the activation energy of fusion, the stability of the system is enhanced.

Polar and nonpolar surfaces affect the properties of the adsorbed water in different manners. The nature and the temperature intervals of water fusion prove that the structure of the water is distorted to a lesser extent by nonpolar than by polar surfaces. The structure of the ice is more stable, and its melting point is higher (above -10°C) near a nonpolar than near a polar surface:

We shall now discuss the main principles underlying the dynamic model of behavior of proteins in water; this model will subsequently be applied to the interpretation of the data on the dynamics and the hydration of biopolymers.

Chapter 2

Principles and Implications of the Dynamic Model of Behavior of Proteins in Aqueous Media

In order to facilitate analysis and systematic treatment of the large amount of experimental data on the dynamic properties of proteins and their hydrate hulls, it is first necessary to lay down certain fundamental principles; such principles must not be inconsistent with the accumulated experience and must permit prediction of properties, by means of which the validity of the model can be established.

2.1 Description of the Properties of Proteins in Terms of the Dynamic Model

The proposed model of dynamic behavior of proteins in aqueous media is based on the following general principles (Käiväräinen, 1975a,b, 1979a, c).

1. The protein molecule contains one or more cavities (clefts), which are capable of fluctuating between state B which is more accessible to water and state A which is less accessible to water; the frequencies ν of such fluctuations vary between 10^4 and 10^7 sec^{-1}. The transitions between these states may take place as a result of shifts and rotations of protein domains (blocks) relative to one another, but may also result from sudden reorientations of individual amino acid residues. In view of the cooperative properties of proteins, cavity fluctuations are interrelated.

2. The water interacting with the protein consists of two main fractions. The larger fraction forms a disordered hydrate hull, which is incapable of undergoing phase transitions of the first kind, while the smaller fraction may, under certain conditions, undergo transitions of a similar type during the fluctuations of protein clefts. Exchange is possible between these fractions and the free water of the solution. In some cases the protein may contain a small amount of its own internal water, which does not participate in phase transitions.

3. On interacting with a cavity in the 'open' state B, water forms a cooperative system – cluster – the lifetime of which depends on the geometry, on the surface stability of the cavity and on external conditions (temperature, pH, ionic strength, etc.). If, when in this state, the cavity contains charged groups, their polarizing, anisotropic effect of the water may be neutralized by H_3O^+, H^+ and OH^- ions.

X-ray diffraction data seem to indicate that the protein cavities – the active site, the clefts between the domains of multi-globular proteins, the vacant spaces between the subunits of oligo-proteins, etc. – are mainly constituted by nonpolar amino acid residues.

The adaptation of the properties of proteins to those of water may have taken place as

14

a stage in the natural evolution since water, as a thermodynamic system, always tends to assume the state corresponding to the lowest free energy – either a crystalline or a liquid form. Since the properties of liquid water become strongly distorted upon adsorption on any surface, nature has chosen the former alternative: water is capable of coexisting with biopolymers in their nonpolar cavities in a quasi-crystalline form (Käiväräinen, 1975b). It has been shown (para 1.2) that the structure properties of water are less disturbed by interaction with nonpolar than with polar surfaces. The thermal fluctuations of biological macromolecules, which reflect their polymeric properties, set a limit on the lifetime of the quasi-crystalline water clusters in the cavities. Moreover, a cluster comprising a sufficiently large number of water molecules will spontaneously dissociate at physiological temperatures, even if the geometry of the cavity is ideally adapted to the crystal and the cavity is perfectly stable. The dissociation of the crystal in the protein cavity disturbs the condition of minimum free energy of the water and of the system protein-water as a whole. The decomposition of crystalline hydrates heated to about 30°C illustrates the effect of dissociation of the water cluster. It is thus seen that water is capable of producing specific changes in its environment.

Shnol (1979) also believes that the evolution of biopolymers may have been affected by the properties of the water. It was suggested by Käiväräinen (1975b, 1978b) that the interaction between an ordered water cluster and the protein cavity be named 'clusterophilic' since it is intermediate between a hydrophobic and a hydrophilic reaction.

The energy of clusterophilic interaction G^{cl} is the difference between the free energies of the 'open' B-state of the cavity containing water in the form of dissociated ($G_B^{H_2O}$) and quasi-crystalline ($G_B^{H_2O*}$) cluster:

$$G^{cl} \equiv G_B^{H_2O} - G_B^{[H_2O]*}. \qquad (2.1)$$

The closer the approach of the structure of the water cluster in the cavity to an ideal crystalline lattice, the more pronounced will be its cooperative properties, the sharper will be its phase transition during the A ⇌ B fluctuations, and the more sensitive will the equilibrium A ⇌ B itself be to the changes in the geometry and stability of the cavity. In such a case the interaction between the active site and the globule, which must take place for the protein to be able to fulfill its function, will be accompanied by the smallest loss in energy. Such an optimum situation may have become permanently established in the course of natural selection, and we believe it to be realized at the physiological temperature and under conditions specific to the given protein. The equilibrium constant $K_{A \rightleftharpoons B}$ is given by the Boltzmann Law:

$$K_{A \rightleftharpoons B} = \exp[-(G_A - G_B)/RT], \qquad (2.2)$$

where G_A and G_B are the respective free energies of the A- and B-states of the cavity, the value of G_B varying with the strength of the clusterophilic interaction.

The frequent A ⇌ B transitions may form the background for the smooth relaxation vibrations $K_{A \rightleftharpoons B}$, whose specific period is determined by the rate of the change G_A-G_B. Such a change may result, for example, from the stabilization of active site fluctuations due to bonding a ligand, or as a result of sudden changes in the external conditions (pH, ionic strength, temperature, etc.). Relatively fast $K_{A \rightleftharpoons B}$ changes are connected with reorientation and rotation of the cavity-constituting domains with respect to each other; slow changes are caused by changes in the conformation and stability of these domains.

Let us consider the changes in the thermodynamic parameters of free water on its penetration into 'open' (B) and 'closed' (A) cavities. In the former case the change of state of the water may resemble a phase transition of the first kind, i.e., the decrease $\Delta S_B^{H_2O}$ in the entropy of the water is fully compensated for by the decrease $\Delta H_B^{H_2O}$ in its enthalpy. A situation may also be envisaged in which the decrease in the enthalpy is larger than the decrease in entropy. Thus, in the former case, we may write

$$\Delta G_B^{H_2O} = \Delta H_B^{H_2O} - T\Delta S_B^{H_2O} \leqslant 0. \tag{2.3}$$

The interaction between the cavity and the water is thermodynamically favored ($\Delta G_B^{H_2O} < 0$), or else is without effect on the free energy of the B-state ($\Delta G_B^{H_2O} \simeq 0$). The free activation energy of the transition B $\rightarrow$ A increases by an amount which is determined by the energy of clusterophilic interactions. In the latter case, under conditions which make phase transition of the first kind possible, the energy of clusterophilic interactions will be equal to the energy of hydrophobic interactions if $K_{A \rightleftharpoons B} \simeq 1$.

If the structure of the 'closed' cavity is mostly nonpolar, its interaction with water is typically hydrophobic. In such a case the decrease in entropy is larger than the decrease in enthalpy, and

$$\Delta G_A^{H_2O} = \Delta H_A^{H_2O} - T\Delta S_A^{H_2O} > 0. \tag{2.4}$$

The interaction with water will then be thermodynamically disfavored, and the water is displaced from the cavity, or else its A-state becomes destabilized. If the 'open' cavity contains n_1 water molecules, while the 'closed' cavity contains n_2 water molecules, only

$$\Delta n^{H_2O} = n_1^{H_2O} - n_2^{H_2O} \tag{2.5}$$

water molecules may in fact participate in the phase transition induced by the cavity fluctuations A $\rightleftharpoons$ B. It is clear that

$$\Delta n^{H_2O} \simeq (V^B - V^A)/v^{H_2O}, \tag{2.6}$$

where V^B and V^A are the volumes of the cavity in the 'open' and 'closed' states respectively, while v^{H_2O} is the volume of a single water molecule.

It is natural to expect that the larger the size of the cavity, the larger the difference $V^B - V^A$, and the more pronounced will be the cooperative properties of the water cluster and of the macromolecule. This may be manifested as the allosteric properties of oligo-proteins with a large central cavity. The change in the state of this cavity and in the structure of the water contained in it may bring about an interaction between the active sites of individual sub-units. This case will be discussed in detail for hemoglobin.

If, as a result of cavity deformation, the cluster structure is disturbed, the clusterophilic interaction weakens, since $G_B^{H_2O*}$ increases (cf. equation (2.1)), so that G_B and

$K_{A \rightleftharpoons B}$ increase as well (cf. equation (2.2)). In such a case, the change in the water properties during $A \rightleftharpoons B$ fluctuations may no longer constitute a phase transition or may constitute it only in part.

The interaction between the cluster and the interior of the 'open' cavity may involve the hydrogen bonds of the water molecules interacting with the groups of the amino acid residues constituting the cavity. In the ideal case, the tridimensional bond distribution would fit the crystalline structure of the water cluster, being its natural extension. The protein environment of the cluster here acts as a screen shielding the cluster from the effect of solvent fluctuations. Such thermal insulation may tend to prolong the average lifetime of the hydrogen bonds of the cluster, thus making it more nearly equal to the lifetime τ of the cavity in the 'open' state:

$$10^{-4} \text{ sec} \geqslant \tau \geqslant 10^{-7} \text{ sec.} \qquad (2.7)$$

τ is a magnitude which is inversely proportional to the rate constant $k^{1 \rightarrow 2}$ of the transition of the cavity from state 1 to state 2. The Eyring-Polanyi equation may be used to calculate the free energy of activity $G^{1 \rightarrow 2}$ of the rate-determining stage of $A \rightleftharpoons B$ fluctuations:

$$1/\tau \sim k^{1 \rightarrow 2} = \frac{kT}{h} e^{-G^{1 \rightarrow 2}/RT}, \qquad (2.8)$$

where k, T and h have the same meanings as before.

The inequality (2.7) may then be written as

$$50.3 \text{ KJ/mole} \geqslant G^{1 \rightarrow 2} \geqslant 33 \text{ KJ/mole} \qquad (2.9)$$

The free activation energy of $A \rightarrow B$ transition is the sum of the contributions of all non-covalent interactions which stabilize the cavity in the A-state – hydrophobic, van der Waals, electrostatic, hydrogen bonds, etc. effects. The height of the activation barrier of the $B \rightarrow A$ transition may also depend on these types of interactions but, as has been seen, the B-state becomes stabilized by clusterophilic rather than by hydrophobic interactions.

Since the rate of exchange between the cluster water and the free water in solution is fast, anisotropic fluctuations may arise, resulting in a decrease in the average strength of the hydrogen bonds of the solvent. It will be shown in Chapters 5 and 6 that this effect is actually observed.

If we assume that the amount of elementary quasi-crystalline water cells is an integral number at maximum stability of the cluster, then the linear dimensions of the 'open' cavities may be correlated with the dimensions of elementary ice cells. Such a correlation has in fact been observed in a number of cases (cf. sec. 4.2).

Special studies of general relationships governing the structure and geometry of protein cavities with the aid of computerized and X-ray diffraction techniques would be of major interest. It must be borne in mind, however, that it is the 'closed' state of the cavities which is preferentially stabilized during the crystallization of proteins, since water clusters in 'open' cavities are readily disrupted by salts or other additives which are introduced into the mother liquor to induce crystallization. The crystallization temperature is also an important factor in the stabilization of a given form of the protein.

In our view, the saturation of the hydration centers of a protein molecule by water probably takes place in the following sequence. The first to be saturated are the charged and the polar groups which are accessible to water; this followed by the hydration of nonpolar surface segments and cavities. As the cavities become filled, and the clustero-philic interactions are intensified, the filling of the cavities becomes thermodynamically more favored, i.e., the bonding of the water is a cooperative process.

The motion of individual zones with respect to each other can only occur after the inter-domain voids have been filled with water. It could be expected, by analogy with zeolites (Chapter 1), that the stability of the water cluster would increase with increasing size of the cavity, and with increasing proportion of the filled cavity volume. Since this results in an increase in the free activation energy of the transition $B \rightarrow A$, an inverse relationship between the cavity volume and its fluctuation frequency should be observed. Moreover, the stability, i.e., the dissociation rate of the cluster, may be expected to be determined by the following main factors.

1. The geometrical correspondence between the shape of the cavity and the shape of the cluster. The shape of the water cluster depends on the size, amount and location pattern of elementary cells of the quasi-crystalline water. This is an illustration of the fact that structure and physicochemical properties are mutually complementary — a principle which is frequently encountered in molecular biology, and which is a consequence of evolutionary processes.

2. Dynamic correspondence between the structure of the 'open' cavity and the cluster. The water molecules in the nodes of the quasi-crystalline lattice execute vibrations, whose frequency and amplitude are a function of the temperature. It is evident that the cluster and its protein environment can coexist as a single system only if the nature of the vibrations executed by the functional groups of the cavity — which is determined not only by the temperature, but also by its amino acid composition — is the same as that of the vibrations of the water molecules constituting the cluster. Such a correspondence may also have developed in the course of the evolution, and should be realized at the physiological temperatures specific to the given protein. This last conclusion follows from our view, according to which phase transitions of the cluster water are an important factor in the functional mechanism of biopolymers.

3. The size of the protein domains (blocks) forming the cavity, and the flexibility of their interconnecting polypeptide segments. These factors determine the diffusional hydrodynamic parameters of intramolecular mobility of proteins, i.e., the frequency of $A \rightleftharpoons B$ transitions, and hence also the maximum lifetime of the cluster.

4. The number of defects in the cooperative water-protein system (cluster + cavity). This number depends both on the effect of nonspecific perturbing agents (ions, temperature, etc.), which affect both water and proteins, and on the specific changes of the cavity induced by the bonding of ligands.

5. The degree of isolation of the cluster molecules from the ordinary hydrate hull of the protein and from the free water in the solution. This factor is relevant, since exchange processes between cluster molecules and other water fractions may destabilize the cluster.

Factors 2, 4 and 5 are responsible for conformational transitions in proteins which are not accompanied by denaturation, and which are induced by various perturbants, including the temperature. The transitions may result from changes in clusterophilic interactions, which are most sensitive to all kinds of external effects, and from the consequent shift in $A \rightleftharpoons B$ equilibrium.

We think it possible that the structure of the water in the cavities is close to the hexagonal ice I type but, should it in fact prove to be that of any other ice modification, the model would not have to be modified in any way as a result. The only important factors are the cooperative nature of the water cluster, and the variation of its stability with the structure and dynamic properties of protein cavities.

A direct verification of the validity of the model is difficult, since experimental methods of observation of rapid changes in the properties of proteins and of the water component of solutions are still lacking. Accordingly, the validity of the model can only be verified by comparing the direct implications of the conceptions which underlie the model with the available experimental results. This is in fact the course adopted throughout this book.

We shall discuss, in general outline, the application of the dynamic model to the description of the reaction taking place between the protein and the ligand (Käiväräinen, 1978a, c, d, 1979a). We shall assume that the active site has passed from the 'open' b-state to the closed a-state, while the other cavities, which will be denoted as 'auxiliary', are still in the open B-state. Owing to the existence of discrete mechanical and physico-chemical degrees of freedom, such protein conformation is 'strained', and is thermodynamically less favored than when all the cavities are in the same state – all cavities 'open' or all cavities 'closed'. Intra-protein cooperativity may also be of a different kind – interrelation between cavities of opposite types, when a 'closed' state of one cavity corresponds to the 'open' state of another.

If the association of a ligand with an active site is accompanied by the stabilization of its 'closed' state and destabilization of its 'open' state, then the corresponding dissociation must represent the reverse process. If the transition of the cavity with an active site, brought about by the presence of a ligand is mechanically related to a change of state of another protein cavity (e.g., both cavities are formed by a pair of domains, which become reoriented by rotating around a common axis), the interaction is of mechanical type. In this particular case, the states of the two cavities vary in counterphase.

In addition to the shift in equilibrium between 'open' and 'closed' states, the ligand may also induce events on a smaller scale in the active site – such as reorientation of individual residues or changes in its electronic state. The two most familiar examples of such changes are changes in the spin and in the valency state of heme iron in hemoglobin and in cytochrome C during the bonding of ligands. Perturbations of active centers by ligands include the stabilization of the thermal mobility of its functional groups which are in contact with the ligand. This effect may be expected to vary with the number of points of such contact, i.e. with the size of the ligand. It may, for instance, manifest itself in the reaction between antibodies and antigenic determinants of various sizes.

The stabilization of the internal structural mobility of active site-forming domains by ligands may result in a fine adjustment of the interaction of the active site with the auxi-

liary cavities. The disturbances located around the active site spread out towards the auxiliary cavity, perturb its geometry and the mobilities of its amino acid residues, whereby
the stability of the water cluster is altered; both this effect and the mechanical effects
may alter the $A \rightleftharpoons B$ equilibrium. This is the so-called perturbational type of interaction
between cavities.

The mechanical and perturbational interactions between the active site and the auxiliary cavity are schematically represented in Fig. 6. The figure is a schematic representa-

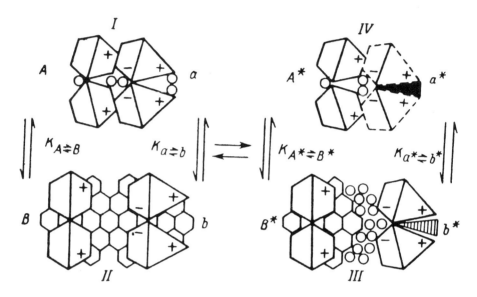

Figure 6. Schematic representation of protein association (Fab-subunits of an antibody) with the
ligands, according to the dynamic model.
I — protein conformer in which the active site cavity a and the cavity between the pair of domains are
in the 'closed' state; the water in the cavity is unable to form cooperative clusters. II — conformer
with 'open' cavities; water in the cavities forms cooperative clusters; $K_{a \rightleftharpoons b}$ and $K_{A \rightleftharpoons B}$ are equilibrium
constants between the states of the active site and the auxiliary cavity in the absence of the ligand.
III — conformer corresponding to cavities in their intermediate states: b^* — state in which water is displaced by the ligand from the active site, but the site is in a metastable form, corresponding to the
B^*-state of the auxiliary cavity, in which the structure of the water cluster is disturbed, but nevertheless continues to affect the mutual orientation of the domains of the active sites. IV — conformer
corresponding to the most stable condition of the protein-ligand complex; both spaces, a^* and A^*,
are in the 'closed' state. $K_{a^* \rightleftharpoons b}$ and $K_{A^* \rightleftharpoons B}$ are the new values of equilibrium constants between the
active site and the auxiliary cavity, induced by the ligand. The rotation of the active site-generating
domains around the rotation axis, as a result of which the geometry of the active site and of the
auxiliary cavity changes, represents the mechanical interaction between the cavities. The disturbance
of the cooperative water cluster in the auxiliary cavity, the impaired balance of electrostatic and other
types of noncovalent interactions between the domain pairs, as well as the deformations and alterations
in the internal domain mobilities (delineated by a dotted line) are manifestations of perturbational
interactions between cavities.

tion of the structural features of the Fab sub-units of immunoglobulins. The flexible segments of the two polypeptide chains, interconnecting the domain pairs, are not shown in the diagram. It must be remembered, however, that the fluctuations of the auxiliary cavity result in their conformational changes which are not random, but which are governed by their structural stereochemistry. This is then yet another contribution to the mechanical interaction between the cavities, consisting in a reorientation of the domains in each of the two pairs which constitute the molecule.

The stabilization of the 'closed' *a*- state of the cavity containing the active site, and the rotation of the cavity-forming domains are directly caused by hapten. The other pair of domains 'adapts' its structure to the domains with active sites, and assume the configuration which is thermodynamically most favored under the given conditions. This is the relaxational shift of the equilibrium $A \rightleftharpoons B$, which is accompanied by a decrease in the free energy of the complex, and by a change in the linear dimensions of the macromolecule (Chapter 6).

The determinist nature of such a process is not determined solely by mechanical degrees of freedom, but by the entire complex of effects, including clusterophilic, hydrophobic, electrostatic, van der Waals and other effects. We may assume that following complex formation, the interaction between pairs of domains is realized through certain degrees of freedom of the protein, which are conveniently denoted as 'physicochemical' (as distinct from the ordinary 'mechanical') degrees of freedom. Since such interactions disturb the non-covalent bonds, which determined the previous equilibrium state of the protein, they must be considered as perturbations. Perturbational interactions between cavities, which involve deformation or stabilization of internal domain surfaces, may be a manifestation of yet another, independent, kinetic stage of complex formation. This stage will be observed in reality, if the corresponding structural domain changes are connected with overcoming activation barriers of a kind which differs from those involved in relative domain reorientation and in the consequent shift in the $A \rightleftharpoons B$ equilibrium.

The neutralization of the charged groups in the protein, i.e., the change in the pK induced by the ligand directly or as a result of conformational rearrangements, affects the balance of intramolecular electrostatic forces, with consequent changes in the equilibrium constants between the 'open' and 'closed' states of protein cavities. It may be expected that such changes will alter the state of hydration and the effective Stokes radius of the protein. If the protein contains only a single cavity, or if all its cavities alter their state synchronously, the resulting changes in the state of the protein may be legitimately considered as transitions between two conformations (the mechanism of association and dissociation of specific complexes will be considered in detail in Chapter 6).

The specific degrees of freedom of the protein responsible for cavity fluctuations represent only some of the many degrees of freedom which determine the dynamic properties of macromolecules. The changes in the enthalpy H and in the entropy S of the system protein + water resulting from any effects, specific or nonspecific, which alter the $A \rightleftharpoons B$ equilibrium may be described in a simplified manner. A shift of this equilibrium towards the 'open' state is the principal reason for loosening the structure of the protein and for the increase in its degrees of freedom. It is natural to assume that this is accompanied by an increase in both H and S of the protein. However, it also results in an increased accessibility of protein cavities to water, which assumes the form of clusters, thus in turn reducing the H and S of the water. If the equilibrium A $\rightleftharpoons$ B is shifted toward the 'closed' state, an opposite process takes place. It is seen that the thermodynamic parameters of the protein and of the water which interacts with its cavities participate in the process as an overall resultant, and may vary in counter-phase and compensate each other to a certain extent (Käiväräinen, 1975b). This is why proteins in solution may undergo conformational changes which do not respond to calorimetric determination methods.

In a further development, the dynamic model of the behavior of proteins in water was applied to describe the mechanism of enzymatic catalysis (Chapter 7). It is assumed that the 'closed' state of the active site is stabilized by the substrate. Interaction between cavities may bring about strained, non-equilibrium states of the enzyme-substrate complex. The duration of such states will depend on the lifetime of the auxiliary protein cavities in a state which is thermodynamically unfavorable to this complex. This also applies to intermediate complexes formed during multi-stage reactions. During this period of time the activation barrier of the respective stage of the chemical reaction is lower.

The time required for the reaction [ES] $\xrightarrow{k_2}$ E + P comprises the time required for the dynamic adaptation of the active site and of the substrate, and the time needed to overcome the activation barriers of the reaction. We succeeded in obtaining expressions interrelating the rate of enzymatic catalysis according to Michaelis-Menten with the free activation energies of the fluctuations of the active site and of the auxiliary cavities of the enzyme (Chapter 7). Non-competing inhibitors which form complexes with an auxiliary cavity stabilize its fluctuations, which means that the period of dynamic adaptation is increased, and the contribution of the auxiliary cavity to overcoming the activation barrier of the reaction is reduced.

The typical features of our model include the small energy changes involved in the transglobular changes in proteins, and the possibility of transmitting signals from the active site over large distances in the macromolecule. These features may well underlie processes which result in major reversible changes in the structure of proteins and their aggregates, such as contraction of muscle proteins, cooperative transitions in membranes, etc.

2.2 A Description of the Properties of Membranes in Terms of the Dynamic Model

According to the ideas developed in this book, the water contained in the internal non-polar zones of biological membranes is also capable of forming ordered structures which determine to a large extent the cooperative properties, stability and permeability of membranes (Käiväräinen, 1980). The phase transitions of the lipid component, which take place at sufficiently high temperatures may be directly related to the fusion of such water structures.

According to the results obtained by NMR (Chan *et al.*, 1971) and by EPR (McFarland and McConnel, 1971), the correlation times of CH_2-groups in liquid-crystalline lipids are of the order of 10^{-6} to 10^{-8} sec. After fusion, these times decrease to 10^{-9} sec. Saturated hydrocarbons crystallize out as *trans*-rotamers. As a result of phase transition, the coiled *gosh*-rotamers appear along with the *trans*-rotamers (Volkenstein, 1978).

Cavities may be formed inside membranes by the nonpolar lipid 'tails'. As in the case of proteins, geometric and dynamic relationships may be expected to exist between the water clusters of membranes and the lipid cavities — the so-called 'cells'. The tridimensional distribution and the relative symmetry of the lipid tails of lipid cells in the B-state may be expected to show a correlation with the dimensions of the unit cell in the water cluster.

It would appear that the frequency of the fluctuations of lipid cells between state B in which they contain crystalline water and state A in which they contain liquid water is of the order of 10^6 to 10^8 transitions per second. In the B-state, which predominates below the fusion temperature, the distribution pattern of the lipid tails constituting the cell becomes stabilized by interaction with the cluster, i.e., by a clusterophilic interaction. In such a case the membrane is more rigid and less permeable.

If this is admitted, it is natural to conclude that the dissociation of the water cluster resulting from its own fluctuations or from a change in the geometry of the lipid environment, strongly enhances the mobility of the lipid and increases the number of its degrees of freedom. The lipids are now capable of sliding against each other. This situation corresponds to the distorted state A of the cell, when the symmetry of the lipid tails is impaired. During the $A \rightleftharpoons B$ fluctuations lateral diffusion of lipids and of membrane proteins may take place.

Specific (e.g., conformational change of the proteinic cell receptor) or non-specific (temperature, surfactants, etc.) disturbance factors affect the distribution of lipids and their dynamics, and destabilize the B-state of the cells, similarly to the situation described above for proteins. As a result, the lifetime of the B-state of membrane cells becomes shorter, the $A \rightleftharpoons B$ equilibrium becomes shifted to the left, and the frequency of these transitions increases.

Since water clusters are highly cooperative, the shift in the $A \rightleftharpoons B$ equilibrium is sharp, and may result in an inversion of the polarity and dynamic characteristics of the environment of intra-membrane enzymes, and influence their activity in this manner. The reverse may also be expected to be true: changes in enzymatic activity due to a changed concentration of the substrate or to allosteric effects alter the properties of the membrane. By virtue of this relationship the cell may adjust its surface properties in accordance with the internal metabolic processes.

If the $A \rightleftharpoons B$ fluctuations of the lipid cells induce pressure fluctuations in the intracellular medium, sonic fields in the frequency range of 10^6 to 10^9 sec^{-1} may affect the morphogenesis of the cell. In such a case the lengths of the sonic waves are of the same order as the dimensions of the cells (1 - 100 μm).

The lateral cooperativity of membranes, which is a highly important factor in the response of the cells to external effects, probably results from the fact that the outer walls of a lipid cell also act at the same time as the internal walls of the neighboring cell, etc. Accordingly, disturbances of structure or dynamics of the behavior of a single cell or a small group of cells are immediately propagated over large distances and affect a large surface of the membrane. In particular, this is how a nervous impulse may be transmitted along with the axon, with accompanying change in the ionic conductance. The myelin film protects the axon membrane from nonspecific disturbances.

Changes in the state of aqueous clusters in the membrane may be one of the factors which regulate the permeability of membranes to protons, since the rate of diffusion of H^+ ions in ice is 100 times higher than in liquid water.

Franks (1976) used the method of X-ray diffraction analysis on oriented lipid-cholesterol lamellar structures in the capacity of membrane models, and showed that the area of minimum electron density is located in the center of the bilayer. This is the contact area between the tails of oppositely oriented lipids. The zone of fluctuations of lipid cells and water clusters, which are responsible for the cooperative properties of membranes, may well be formed at this very location.

Worcester and Franks (1976) studied the distribution of water in the system by neutron diffraction by multi-layers of a cholesterol-lecithin mixture. The results thus obtained could be described by assuming two different distributions of electron density of water across the membrane, depending on whether the phosphoryl-choline group of the lecithin was oriented in the plane of the bilayer or at right angles to it.

It is interesting to note that neutron-scattering densities of systems containing H_2O and D_2O are qualitatively different.

The differing rate of passive diffusion of particles across the membrane may be due to a different behavior of both crystal and liquid water towards different particles. The considerable differences in the interaction of K^+, Cl^- and Na^+ with water are noteworthy in this connection. Thus, for instance, Na^+ ions almost double the correlation time τ of the neighboring molecules as compared with pure water. The stabilizing effect of Na^+ on water is also manifested by the fact that the activation energy ΔE of reorientation of these water molecules is positive: 1.580 KJ/mole. K^+ and Cl^- exert an opposite, though much weaker effect on water (Table 3).

TABLE 3 Parameters which determine the interaction between ions and water, and their concentration in the internal and external media of Loligo axons and in sea water. After Hodgkin (1965).

Ion	τ^*/τ^0	$\Delta E.$ J/mole	Concentration, mmoles/kg H_2O		
			Axoplasm	Blood	Sea water
K^+	0.94	−167	400	20	10
Na^+	1.89	1580	50	440	460
Cl^-	0.94	−146	40–150	560	540

Note: τ^* and τ^0 are correlation times of water protons in the presence and in the absence of ions respectively, as determined by the NMR method. After Fabricand *et al.*; quoted from Hippel and Scheich (1973).

K^+ and Cl^- interact with water in the same manner, but have opposite charges. It may be expected, accordingly, that these ions will be symmetrically distributed relative to the membrane, as is in fact observed. Since the interactions of K^+ and Cl^- with water are weak, the permeability of the membrane should vary in the same manner with the change in the phase conditions of the water in the membranes, and to a lesser extent than the permeability to Na^+.

In fact, the state of the membrane at rest is characterized by the following ratio of ionic permeabilities: $P_K:P_{Na}:P_{Cl}$ = 1:0.04:0.45 (Volkenstein, 1978). In the excited state the relative permeability of Na^+ increases sharply: $P_K:P_{Na}:P_{Cl}$ = 1:20:0.45. This may be due not only to the disturbed structure of the quasti-crystalline water, to the shift of the equilibrium A ⇌ B to the left, and to the acceleration of the passive ion transfer, but also to the formation of specific conductance channels in the protein by Na^+.

In accordance with the ideas presented in this book, specific ionic channels, which are internal cavities in membrane proteins or in their associates, may be formed during the shift of A* ⇌ B* equilibrium between the states of the cavity to the left or to the right. The 'open' conduction channels for one type of particles may constitute 'closed' channels for particles of another type. According to the dynamic model, the relaxational shifts of the A ⇌ B equilibrium of the channel are determined by the state of the active site of the receptor or the enzyme (Käiväräinen, 1979a, 1980).

It may be expected that there is a direct connection between the dynamic properties of membrane proteins and their lipid environment. Accordingly, if the conformational changes of receptors produce a shift of the equilibrium A ⇌ B of the lipid cells, the internal polarity of the membranes and the solubility of the specific ion carriers – ionophores – in the membranes are altered.

If the orientation of translocator enzymes in the membrane is determined by the different polarities of the environment on the internal and external sides, it may be concluded that temperature, and other disturbing effects such as salts in high (about 1 M) concentrations, surfactants, etc., which reduce the gradient of $K_{A⇌B}$ values across the membrane, will tend to reduce the anisotropy of its ionic permeability. Disturbing agents, including local anaesthetics, shift the A ⇌ B equilibrium to the left and labilize the membrane, thus enhancing the rate of passive transfer and reducing the effectiveness of active ion transport.

No attempt will be made here to prove the validity of the ideas presented above on the properties of membranes. We shall merely report some results which are in conformity with these conceptions: the sudden nature of conformational transitions in membranes under the effect of perturbing agents (inorganic ions: Konev *et al.,* 1972), and the variation of the mobility of membranes with their degree of hydration (Kol'tover *et al.,* 1968). The rotational and translational diffusion of rhodopsin, which is a component of the visual receptor membrane (Cone, 1972; Poo and Cone, 1974), illustrates the dynamic properties of this membrane.

Levitt (1980) recently proposed a functional model of Na, K-ATPase, which is in agreement with our own ideas. The model is based on the assumption that each one of the two subunits of the sodium pump contains a channel which may be filled with three Na^+ ions or with two K^+ ions. The inlets to the channel on the cell side and the extracellular space become successively closed to one type of ions and open to another. The accessibility of the channel to water changes correspondingly. The functioning of the subunits is conjugated but is not synchronous, so that the three Na^+ ions and the two K^+ ions can be transported in different directions at the same time.

The functional activities of membrane proteins seem to be accompanied by changes in their effective Stokes' radius, as was shown by Beece *et al.* (1981). The pulsed spectroscopic technique was used to detect the influence of viscosity on the kinetic parameters of the successive reactions involved in the photochemical cycle of bacterial rhodopsin. The viscosity was varied by varying the glycerol concentration between 20 and 80%, between 240 and 315°K. As the viscosity increased, the rate of two out of four stages in the photochemical cycle becomes slower, possibly as a result of the decrease in the rate of structural transitions of bacterial rhodopsin.

Bürkli and Cherry (1981) were the first to use flash photolysis to obtain data on the structural flexibility of triplet-tagged Ca, Mg-ATPase in the membranes of sarcoplasmatic reticulum. This flexibility is manifested at shorter relaxation times, which characterize the rotation of the protein in the membrane as a whole ($10^{-4} - 10^{-3}$ sec). It seems to result from the tendency to execute independent thermal motion which is displayed by the subunits or subglobules of this protein.

The subject of this book is the dynamic properties of proteins. It must be borne in mind, however, that the laws governing the behavior of macromolecules may be just as valid in supra-molecular systems: membranes, ribosomes, etc. Thus, for instance, a functional model of ribosome, based on the relative displacements of its two sub-units, was proposed by A.S. Spiryn a long time ago. The principle of this model may be formulated as follows:

"The periodic opening and closing of ribosome sub-particles is a drive mechanism, which ensures the realization of all tridimensional displacements of tRNA and mRNA during the process of translation"
(Spiryn and Gavrilova, 1971)

The assumptions underlying the dynamic model of behavior of proteins in water, as discussed in this chapter, will be applied in the discussion and interpretation of the material contained in the following chapters.

Chapter 3

Conformational Mobility of Proteins

Even the most general conception of the protein molecule – which is imagined as a homopolymer with a linear memory – necessarily involves its capability of fluctuation (Lifshits, 1968; Lifshits and Grosberg, 1973). Statistical analysis indicates that the homopolymeric coil executes continuous macroscopic pulsations, and that its density is not a thermodynamically reliable magnitude. If $T < T_{kr}$, a kind of coil-to-globule phase transition takes place under the effect of a sufficiently powerful compressing field.

Let us consider the conditions under which a globule is stable. If γ is the energy of chain interactions per link, the necessary condition for the formation of an ordered structure is $T < \gamma$. Even in such a case, however, local disturbances of the ordered structure may occur in the system; these are accompanied by an entropy change which is given by k in C, where C is the local defect concentration. The generation, disappearance and migration of the defects may be considered as manifestations of the dynamic properties of the globule. Inside the globule the minimum concentration of the defects is proportional to $-N^{-1}$, while on its surface it is proportional to $-N^{-2/3}$, where N is the number of links in the chain.

The free energy of the globule will decrease by $\widetilde{G}_1 = kT \ln N$ and $\widetilde{G}_2 = 2/3\, kT \ln N$. Accordingly, the maximum number of links in the chain at which the globule formed is still stable is

$$N_{kr} \approx \exp\left(\widetilde{G}/kT\right). \tag{3.1}$$

If $N \gg N_{kr}$, such a chain will be capable of forming more than one globule. This may be the explanation for the existence of chains with more than one globule domain, e.g., in the case of immunoglobulins. According to the theory of Lifshits (1968), a rigid linear memory and interaction in the bulk are responsible for the appearance of discrete free energy levels of the globule. If these levels are fairly close together, and are separated from one another by a low activation barrier only, then, according to Blumenfeld (1977), we may expect the appearance of conformational relaxation vibrations which affect individual zones or even entire macromolecules. This worker considers that the rigidity of the memory on the various organizational levels of the biopolymer is responsible for the existence of a small number of discrete degrees of freedom, so that the shift of its segments resulting in conformational transitions must be considered as mainly mechanical.

In view of the bulk effects, the cooperative properties of the polymer chain may be described by the mathematical apparatus of Markov chain theory, i.e., as a sequence of dependent random events. In a simple Markov chain the probability of occurrence of a given event depends on the realization of one preceding event, while in a complex Markov chain it depends on the realization of several preceding events. As a result, the state of a given link in a polymeric chain depends on the state of its neighboring links. Since the locations of the neighboring links are interrelated, their transformations must be interrelated as well. As a result, events taking place in any one segment of the chain produce conformational changes in the chain as a whole. The cooperative polypeptide chain is capable of serving as a channel for the transmission of information about chemical events taking place in one of its remote links. This is accompanied by electronic conformation interactions (Volkenstein, 1975), with structural changes in the biopolymers.

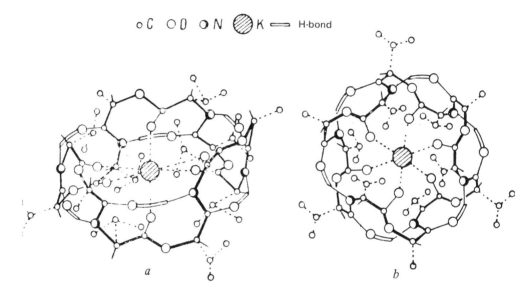

Figure 7. Conformation of the K^+-complex of valinomycin. After Ovchinnikov *et al.* (1974). *a* — side view; *b* — view along the symmetry axis.

A comparison of the microscopic and the macroscopic thermodynamic protein fluctuations leads to the conclusion that the concepts of a protein molecule as a rigid, densely packed structure on one hand, and as a flexible, strongly fluctuating structure on the other, are mutually contradictory only in appearance (Cooper, 1976). This seems to apply to any large systems with the properties of a solid body, but constituted by particles with a relatively small number of atoms.

Even lower-molecular biologically active peptides such as ion carriers — complexones — display structural mobility (Ovchinnikov *et al.,* 1974). A combination of different methods — dispersion of optical rotation, circular dichroism, IR, UV and NMR spectroscopy — made it possible to establish that three forms of valinomycin, A, B and C, forming a 36-membered ring, exist in solution at equilibrium.

The existence of a conformational equilibrium in valinomycin, which is shifted as a result of changed conditions, determines to a large extent the specific behavior of this protein in forming complexes with ions, and during interaction with membranes. In polar media valinomycin rapidly reacts with cations to form an intermediate compound, which then reacts more slowly to form the 'normal' complex (Fig. 7). According to Ovchinnikov *et al.,* (1974), there is an analogy between the reacting (ion — complexone) and the reaction (substrate — active site).

3.1 Studying Hydrated Biopolymers by IR Spectroscopy

This method was used to study protein films, prepared by drying water solutions of lysozyme and bovine serum albumin. When the protein films had reached equilibrium on being saturated with water vapor at room temperature, there appeared an intense absorption band at around 3300 cm^{-1}, which was attributed to the vibrations of the NH-group, and a shoulder at 3450 cm^{-1} due to the vibrations of the hydroxyl groups of water (Buentempo *et al.,* 1972).

The intensity of the second band decreases strongly on drying; this is accompanied by some decrease in the NH-band. It would appear that the amount of the sorbed water affects the condition and the mobility of the peptide groups of globulins. This finding was confirmed by Chirgadze and Ovsepyan (1972a, b), who concluded that water, which plays the role of an active substituent of peptide groups in hydrogen bonds, catalyzes conformational mobility in proteins. In addition, the displacement of water molecules in peptide structures results in the migration of a small free volume, which also activates the mobility.

The similarity between the IR spectra of water in hydrated protein films and those of pure water proves that water-water interactions predominate down to very low water concentrations (0.2 gm H_2O per 1 gm protein), and that the nature of the hydrogen bond between the exposed protein groups and the water does not much differ from that of hydrogen bonds interlinking one water molecule with another. According to IR spectroscopic data, cooling protein films to below $0°C$ is not accompanied by ice formation if the water content is below 0.3 - 0.4 gm water per 1 gram of the protein (Kuntz and Brassfield, 1971).

It is permissible to conclude from the linear variation of the shift in the absorption band of the IR spectrum of water, and from the experimental data, that the interaction of proteins with water resembles the interaction of water with water in the gas phase, i.e., is not as intense as in the liquid phase. Such a conclusion is unexpected in view of the results of studies of water in model systems (sec. 1.2), which indicate that the water molecule becomes stabilized by the effect of various compounds. We believe that the observed effects may result from the decrease in the average strength of hydrogen bonds between the water molecules forming the environment of proteins, produced by exchange reactions between the different water fractions and by the fluctuations of the protein matrix.

IR spectroscopic studies also showed (Falk *et al.,* 1970) that DNA films give a water spectrum which is identical with that of the free water, except that there is a slight shift towards lower frequencies, and that ice is not formed if the water content is below 0.6 gm of water per 1 gm DNA. This method failed to reveal the presence of ice-like water in samples at above-zero temperatures. The changes in IR spectra produced by saturation of

DNA with water have led to the conclusion that the charge groups are solvated first, while the remaining groups are solvated at a subsequent stage.

The absence of typical signs indicating the presence of ice from the IR spectra of proteins is not in contradiction with our postulated existence of a quasi-crystalline phase in biopolymer cavities with a lifetime about 10^{-8} sec. In ice, each molecule undergoes only 10^5 reorientations at $0°C$, and a somewhat larger number of translational displacements (Eizenberg and Kauzmann, 1975). At temperatures above $0°C$ the quasi-crystalline water in biopolymer cavities may be expected to differ from ordinary frozen water by its larger amplitudes of vibration around its equilibrium positions, and by higher reorientation rates, i.e., by a lower average strength of hydrogen bonds. This is why typical features of true ice are absent in the IR spectra of hydrated proteins. Nevertheless, the averaged diffusional distribution and orientation of water molecules in biopolymer cavities may well correspond more nearly to an ice structure than to the structure of ordinary water at the same temperature.

The first data indicating the existence of structural fluctuations of proteins were obtained by the method of deuterium exchange (Linderström-Lang and Schellman, 1959). Conformational mobility of macromolecular segments is the most probable cause of the observed rapid exchange of hydrogen atoms from the interior of the globule for deuterium.

3.2 Fundamental Principles of Hydrogen Exchange Method

The function of reporter groups in the hydrogen exchange method is assumed by the NH-groups of the protein, which are usually distributed in the internal zones of the protein. The H-D exchange reactions in these groups proceed by way of dynamic conformational rearrangements in the protein, as a result of which the NH-groups become accessible to the solvent molecules. These rearrangements are believed (Abaturova and Molchanova, 1977) to parallel the generation of equilibrium defects at isolated points in the crystalline Frenkel-type lattices: in the case of solid inorganic crystals, these are internodal atoms or 'holes' at the nodal points. The probability of occurrence of a local defect will then depend on the energies of non-valent bonds which must be broken, i.e., in the final count, on the packing density of the protein in that part of the molecule. Thus, the variations in the packing density of the protein inside the molecule results in different probabilities of local transconformations in its various parts by a factor of 10^6 - 10^8.

The reaction mechanism of hydrogen exchange in native proteins is usually described by the following equation:

$$N_i \underset{k_{-i}}{\overset{k_{+i}}{\rightleftarrows}} I_i \overset{k_{0i}}{\longrightarrow} \text{exchange} \tag{3.2}$$

where N_i is one of the fundamental states of the proteins, in which the i-th peptide group is inaccessible to the solvent; I_i is one of the transient conformational states in which the i-th peptide group is accessible to the solvent, and the $N_H \rightarrow N_D$ reaction can therefore take place; k_{+1} and k_{-1} are the rate constants of the respective conformational transitions, and k_{0i} describes the rate of the exchange itself.

If $k_{-1} \gg k_{0i}$, and $k_{-1} \gg 10\,k_{+1}$, the experimental rate constant of the observed reaction of proton exchange by the i-th peptide group of the protein is given by the relationship:

$$\beta_i = (k_{+i}/k_{-i})\,k_{0i}, \quad \text{or} \quad \beta_i = \rho_i k_{0i}. \tag{3.3}$$

For this reason the time dependence of the exchange of X-peptidic hydrogen atoms in the protein may be represented by the equation:

$$X_t = n^{-1} \sum_{i=1}^{n} \exp\left(-\beta_i t\right) = n^{-1} \sum_{i=1}^{n} \exp\left(-\rho_i k_{0i} t\right), \tag{3.4}$$

where X_t is the fraction of the peptidic H-atoms which have not undergone exchange after a time t from the beginning of the exchange reaction has elapsed, and n is the total number of peptidic H-atoms. The conformational stability of the protein is determined by the selection of the $N \rightleftharpoons I$ equilibrium constants for all peptide protein groups: $\rho = k_{+1}/k_{-1}$. The magnitude $\rho = \exp(-\Delta G/kT)$ can be readily used to calculate the free energy change ΔG of $N \rightleftharpoons I$ transitions.

The only way of obtaining information about the conformational stage of the exchange is by using the experimentally observed value of the overall exchange rate to calculate the contribution of the chemical stage of the reaction. It is usually assumed that k_{0i} has the same value for all the peptide groups in the protein, and is equal to the rate constant of the exchange in the peptide groups of poly-D,L-alanine. However, it was shown by Molday *et al.* (1972) that the values of k_0 are determined by the effect of the contiguous amino acid residues, and are therefore different for different peptide groups in the protein. These workers proposed a method of predicting kinetic curves of hydrogen exchange of fully denatured proteins from their known original structure. Studies of protein stability carried out by hydrogen exchange method may be supplemented by the method of proteolytic degradation of proteins.

3.3 Results Obtained by Hydrogen Exchange and Proteolytic Studies

It has been shown, in the case of hemoglobin and its subunits, ligand-bonded in different manners, that there is a satisfactory correlation between the rate of proteolysis and the rate of deuterium exchange. In certain cases the rate of proteolytic degradation of the protein is the more sensitive parameter of the two. Thus, for instance, the presence of inorganic phosphates in the solution (0.1 M) of methemoglobin at pH 6.3 largely suppresses the rate of its proteolysis, but is without effect on the rate of deuterium exchange. Moreover, at pH 8.5 the method of proteolytic degradation readily reveals the higher stability of cyanomethemoglobin as compared with aquomethemoglobin, while the rates of hydrogen exchange for these compounds are practically identical (Abaturov and Molchanova, 1977).

A change in the rate of proteolysis also seems to be indicative of a shift in the $N \rightleftharpoons I$ equilibrium, which is accompanied by a change in the accessibility of segments with low packing densities to proteases. The lifetime of the I-state may be expected to be of the order of the time of one revolution of the enzyme — i.e., about 10^{-5} sec.

Abaturov and Varshavskii (1978) reviewed the experimental data obtained by the method of hydrogen exchange from the viewpoint of the equilibrium dynamics of the tri-dimensional protein structure and came to the conclusion that, under quasi-physiological conditions, the changes in the protein structure recorded by this method are mainly local conformational transitions. A complete uncoiling of the protein molecule is not very probable. The estimated rate of torsional, weakly immobilized movements of the side chains on the surface, which mostly consists of rotations around the C-C bonds, yields correlation times of the order of 10^{-9} to 10^{-13} sec.

Abaturov (1976), who studied the deuterium exchange rate as a function of the pH for a number of proteins, came to the conclusion that spontaneous conformational transitions proceeding at fast rates and participated in by the major part of the molecule, are a feature which is common to all globular proteins, the frequencies of such fluctuations varying between 10^2 and 10^8 sec^{-1}.

Abaturov and Varshavskii (1978) distinguish between two types of spontaneous transitions which may take place in proteins: those which are strongly and those which are weakly temperature-dependent. These workers noted that certain structural, temperature-dependent transitions, which affect the accessibility of 20% of the hydrogen atoms in the peptide groups, and which are characterized by the thermodynamic parameters ΔG = 4.2-9.6 KJ/mole, ΔH = 42-84 KJ/mole and ΔS = 42-84 J/mole·°K, represent partial uncoiling processes. In terms of our theory, they may correspond to the fluctuations of protein cavities between 'open' and 'closed' states.

Local transitions, which do not much vary with the temperature, are also possible; for such transitions ΔG = 42-50 KJ/mole, ΔS values are negative, and ΔH values are close to zero. It seems that temperature-independent transitions are not accompanied by changes in the number of hydrophobic contacts. It is known that dispersion effects, which largely stabilize the nonpolar protein zones, do not vary with the temperature. These workers consider, accordingly, that temperature-independent transitions are caused by the change in the state of the groups in the hydrophobic nucleus and by optimization of van der Waals contacts. The perturbation of such an assembly by a ligand or by a spontaneous collapse of internal voids may produce changes in the state of stress of the torsional and valency angles of the polypeptide skeleton, with resulting exposure of internal peptide groups.

The conformational transitions which accompany reactions involving protein + ligand or reactions of enzymatic catalysis are usually multi-stage, temperature-dependent processes. It is not very likely, accordingly, that temperature-independent, conformationally excited states play a determining role in trans-globular conformational transitions or that they determine the biological activity of proteins. In our view, such states may merely favor such transitions if their accumulated number is sufficiently high. The final result may be, for example, a deformation in the geometry of the protein cavity in the 'open' state, a change in the free energy of this state, and a corresponding change of the dynamics of behavior of the cavity, which is temperature-dependent.

According to the hydrogen-exchange data (Abaturov *et al.,* 1977), the intramolecular mobility of methemoglobin sub-units decreases considerably if the H_2O ligand is exchanged for CN. These workers made an important observation, viz., that a change of state affecting only a very small part of the globule – its active site – results in a cooperative change in the stability and intramolecular mobility of the different proteins throughout the globule. This finding is in agreement with the small, but very extensive perturbations in the structure of hemoglobin sub-units, which were recorded by Deatherage *et al.* (1976) by differential Fourier analysis. In our view, this phenomenon is responsible for perturbational interactions between protein cavities; in the case of hemoglobin, this is the reaction between its heme-containing cavities and its central cavity.

Markus *et al.* (1967) were the first to interpret the stabilizing effect of ligands on proteins in terms of our own theory. A study of the interaction between methyl orange and human serum albumin established that bonding two methyl orange molecules with human serum albumin enhances the stability of the protein to hydrolytic attack by five different proteinases. These workers suggested that the ligand reduces the accessibility of the proteases by reducing the frequency of the conformational states in which the peptide bonds are accessible to proteolysis, while the conformational changes are the result of relative shifts in the domains of human serum albumin. The authors concluded that the rate of proteolysis is determined not only by the extent of exposure of the hydrolyzable bonds, but also by the mobility of the corresponding peptide segments. The slower rate of deuterium exchange is in agreement with such an interpretation.

3.4 Fundamental Principles and Applications of the NMR Method

The NMR method makes it possible to record the signals of the nuclei such as ^1H, ^2H, ^{13}C, ^{14}N, ^{31}P, etc. forming part of macromolecules, and to study the mobility of specific chemical groupings in this manner. Such studies are based on the dependence of the time of spin-lattice relaxation T_1 and spin-spin relaxation T_2 of magnetic moments of the nuclei on the characteristic correlation time τ of their motion. The principles of the method will now be clarified.

Consider a system of nuclei with their magnetic moments in equilibrium with an applied magnetic field. The prevailing orientation of the spins, produced by the field, is manifested as a macroscopic magnetic moment, directed along the axis of the magnetic field. If this equilibrium is disturbed in some manner, e.g., by an RF field or by a sudden change in the applied magnetic field, the magnetization of the sample will undergo an exponential relaxation towards a new value, with a characteristic rate constant T_1. This relaxation process is accompanied by an exchange of energy between the nuclei and their environment – 'lattice', and T_1 is known as the spin-lattice relaxation time for this reason. The second relaxation process is the change of state of a system of spins as a result of a sudden application (or removal) of a magnetic tension in the direction normal to the applied magnetic field. The rate of this process is known as the spin-spin relaxation time T_2. If the sample is fully isotropic, T_2 will be unaffected by spin-lattice effects. T_1 and T_2 can be readily and conveniently determined with the aid of NMR pulse spectrometers, or even with ordinary spectrometers, by measuring the effect signal amplitude saturation at a sufficiently high RF field strength, which is determined by the value of T_1. In such a case T_2 is determined from the width of the unsaturated absorption signal. In the simple case of two interacting protons, at a distance r from each other, and rotating istropically, T_1 and T_2 are related to the correlation times τ of this proton pair, rotation as follows:

$$1/T_1 = 3\hbar\gamma^4/10r^6 \left[\tau/(1 + \omega_0^2\tau^2) + \right.$$
$$\left. + 4\tau/(1 + 4\omega_0^2\tau^2)\right], \tag{3.5}$$

$$1/T_2 = 3\hbar^2\gamma^4/20r^6 \left[3\tau + 5\tau/(1 + \omega_0^2\tau^2) + \right.$$
$$\left. + 2\tau/(1 + 4\omega_0^2\tau^2)\right], \tag{3.6}$$

where γ is the hydromagnetic ratio for the protons. The resonance frequency is $\omega_0 = \gamma H_0$, where H_0 is the strength of the constant magnetic field.

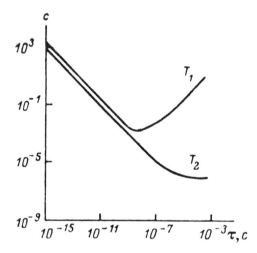

Figure 8. Theoretical variations of T_1 and T_2 with the correlation time τ, calculated at the resonance frequency of 29 MHz. After Carrington and McLaughlin (1970).

The variation of T_1 and T_2 with the correlation time τ of the rotary motion, as calculated by formulas (3.5) and (3.6), is graphically represented in Fig. 8. The rates of the relaxation processes $1/T_1$ and $1/T_2$ are proportional to the square of the value of the interacting magnetic moments and are inversely proportional to the sixth power of the distance between them. If the frequency of translational or rotational motion is high $(1/\tau > \omega_0 = \gamma H_0)$, the values of $1/T_1$ and $1/T_2$ of the nuclei are equal, and are directly proportional to τ.

The mobility of chemical groups in proteins may also be studied with the aid of the Overhauser effect (Emsley *et al.*, 1968), which is the result of suppressing the spin-spin interaction between the carbon nuclei ^{13}C and the protons of the macromolecule owing to the saturation of the proton resonance lines by a strong RF field. This saturation alters the difference in the populations of the levels of magnetic moments of ^{13}C which react with the protons in the magnetic field, and hence intensifies the ^{13}C resonance lines. The magnitude of the observed effect depends on the frequencies of the interacting proton and carbon nuclei, and is therefore a measure of the motions executed by macromolecules (Led *et al.*, 1975).

Aksenov (1977, 1979) studied the NMR data obtained on different proteins, and came to the conclusion that the mobility of aromatic amino acid residues in proteins is much more constrained than that of all other residues. A postulate of our own model is that aromatic and other nonpolar residues participate in the formation of protein cavities. Accordingly, a limitation on their degrees of freedom may ensure the stability of the geometry of these cavities in the 'open' state for a period of time of at least 10^{-8} sec, which is required to organize the water into a quasi-crystalline structure.

A unique contribution to the study of intramolecular mobility of proteins was made by pulse NMR — the spin-echo method — which, while being practically insensitive to the magnitude of the chemical shift, may be used to determine the width of the resonance lines and their relaxation time T_2.

The experimental determination of T_2 from the descent curve of NMR spin echo, and the representation of this curve as a sum of individual components with different T_2 values, is based on the assumption that this descent is exponential for each component. This assumption is in fact realized if the resonance curve is of the Lorentz form.

Aksenov and Kharchuk (1974) and Aksenov and Filatov (1978) studied the decay curves of the spin echo T_2 and the reduction curves of seven globulins: cytochrome C, RNA-ase, lysozyme, DNA-ase, hemoglobin, serum albumin and immunoglobulin G, as well as of the protons of tobacco mosaic virus in D_2O. The decay curves of the spin echo given by the protons of all the proteins thus studied were all constituted by three components: a fast-decaying component, an intermediate component and a slow-decaying component, characterized by three different values of the spin-spin relaxation time T_2. The fast-decaying component of the spin echo is attributed by these workers to the rigid-structures zones in the protein globules, and is due to the rotation of the molecule as a whole. The intermediate component of the decay of spin echo is attributed to those zones in the macromolecule in which an internal motion takes place, caused by the changed distance between nuclear magnetic moments interacting at frequencies of 10^5 to 10^6 sec^{-1}. When the temperature of the solution of lysozyme in D_2O was raised from 4 to 63°C, the intermediate component fraction strongly increased. At room temperatures the proportion of this component is also considerable, and varies between 35% for DNA-ase to 74% for the lysozyme. When the viscosity of the solution increased owing to the increase in serum albumin concentration from 2.5 to 30%, the fraction of the intermediate component decreased from 55 to 34%. The slowly decaying component corresponds to the most mobile, external groupings of the protein, which are probably peripheral, and which constitute 5-10% of the overall protein volume, with correlation times between 10^{-9} and 10^{-10}

Thus, the observed two types of internal motion are in agreement with the idea of segmental mobility of the polypeptide chains in protein macromolecules, which was also confirmed by Linderström-Lang and Schellmann (1959) by the deuterium exchange technique, and with a fast rotation of the surface molecular groups. In terms of our own theory, the intermediate component may also reflect the mobility of protein segments which is due to the fluctuations of its cavities.

It was concluded from the temperature dependence of spin-lattice relaxation time of protein protons, and from the variation of T_1 with the viscosity of the solution and with the molecular mass of the protein, that T_1 is mainly determined by the intensity of the Brownian motion of the macromolecule. Nevertheless, T_1 does not grow without limit as the rotation of the protein molecule slows down, but attains a limiting value of about 150 msec. Hence, T_1 is also determined by the intramolecular mobility of the protein (Filatov, 1978).

The plot of T_1 of the protons of serum albumin in D_2O as a function of the temperature contains two extremal points – at $-100°C$ and at $0°C$. The former minimum is attributed to the rotation of the methyl groups with a correlation time of $2.2 \cdot 10^{-10}$ sec, which was obtained on extrapolating to room temperature. For hydrated serum albumin and RNA-ase the minimum at about $0°C$ corresponds to the motion of protons, with a correlation time of 10^{-8} sec. Dried protein powders do not give this minimum, while their former minimum is deeper and much wider. It was suggested by Filatov (1978) that the minimum around $0°C$ is related to the oscillations of the main chains and side chains of the protein through angles of at least $30°$. The variation of T_1 with the frequency revealed a considerable dispersion of the former in the case of serum albumin and RNA-ase protons. The analysis of the curves obtained also led to the conclusion that the distribution of the correlation times of protein groupings in fact includes the time of about 10^{-8} sec.

Aksenov (1979) studied the processes taking place during the spin-lattice relaxation of protein protons. He distinguished between at least three types of motion in the interior of proteins: rotation of methyl groups ($\tau \sim 3 \cdot 10^{-12}$ sec), oscillation of polypeptide chains ($\tau \sim 10^{-8}$ sec), and the rotational shifts of chain segments or side groups, whose frequency is lower (τ of the order of $10^{-5} - 10^{-6}$ sec). The last-mentioned type of motion is noted only in the presence of water. All the enzymes studied (lysozyme, RNA-ase, α-chymotrypsin and peroxidase) showed a slower decay of the spin echo than that displayed by non-enzymic proteins. This is an important result, which indicates that the internal motion in proteins which act as enzymes is more intensive.

It is to be noted that the value of T_2 may be affected only by motions which alter the average distance between the interacting protons. Accordingly, relative translation of large macromolecular domains may be lost against the background of proton fluctuations in each domain separately. This is shown by the fact that the correlation times obtained for native and for denatured proteins are approximately equal. It may be hoped, however, that this drawback will be eliminated as the method is further developed and rendered more sensitive in future.

NMR studies on staphylococcal nuclease revealed that conformational transitions between the two states in the HisH2 zone are relatively slow. The lifetime of each state at pH 5.33 is estimated as $\tau \geqslant 40$ msec (Markby *et al.*, 1970). The equilibrium constant between the individual states varies with the pH. The same method was also used to show that the frequency of the R ⇌ T fluctuations of modified deoxy-hemoglobin at pH 7 is 10^4 sec^{-1}, while at pH 9 the equilibrium is shifted towards the R-form (Ogawa *et al.*, 1974). These fluctuations are inevitably accompanied by changes in the geometry of the internal cavity of hemoglobin, and by corresponding changes in the properties and the amount of water contained in the cavity.

The ^{13}C NMR method was used in a study of intramolecular mobility of muscle parvalbumin (molecular mass 12,000) and its modified, Ca^{2+}-bonded form (Nelson *et al.*, 1976). Judging by the results of X-ray diffraction analysis, the protein contained several sharply defined domains, and comprised six α-helical parts and two different calcium-bonding centers in the cavities between these segments.

Nelson and his co-workers based themselves on Blumbergen's well-known relaxation theory, which interprets molecular motion in terms of the relaxation parameters T_1 and T_2, and on the Overhauser effect, and obtained the values of intramolecular correlation time for several carbon atoms in parvalbumin. Thus, for instance, the correlation time of rotational motion of the β-carbons of lysine and valine is 10^{-8} sec. Their general conclusion was that the mobility of the peripheral amino acid residues is higher than that of the residues in the interior. If one of the two calcium atoms is removed by EDTA, the result is an enhanced mobility of the side groups of lysine, and certain changes in the secondary structure. The effect of calcium removal is most pronounced in the NMR spectrum of Tyr2, which is locally remote from the Ca bonding center. Spectroscopic measurements indicate that in the presence of two calcium atoms the tyrosine ring is unable to rotate around the C-C bond, but becomes able to do so if one of the calcium atoms is removed.

A study (Snyder *et al.*, 1975) of bovine pancreatic inhibitor by high-resolution NMR showed that the aromatic ring of Tyr35 rotates at about 160 sec^{-1} at pH 5 - 9 and 25°C. A similar study, performed on the same protein (Gelin and Karplus, 1975), showed that the internal tyrosines or tryptophans rotate at 10^5 - 10^7 sec^{-1}. An analysis of the tridimensional protein structure led to the conclusion that this kind of mobility is possible in the presence of small (≃ 0.1 nm) shifts of the groups around these residues, with consequent decrease in steric hindrances to their rotation.

3.5 Gamma-Resonance Spectroscopy and Its Applications

Important information about protein dynamics can be obtained by gamma-resonance (GR) spectroscopy. This applies, in particular, to studies of the dynamics of the active sites of heme-containing proteins, in which the heme iron has been replaced by its isotope ^{57}Fe — a treatment which does not affect the native properties of these proteins. A combination of GR spectroscopy with the spin-label method facilitates the interpretation of the data obtained by each method separately (Likhtenshtein *et al.*, 1975). Frolov *et al.* (1973, 1977) utilized these methods in their studies of dynamic properties of α-chymotrypsin, human serum albumin, lysozyme, collagen, MbO_2, hemoglobin, ferridoxin, etc.

GR spectroscopy is based on the following principle. The probability f' of a recoilless absorption of γ-quanta, and the relative broadening of the GR line ($\Delta\Gamma/\Gamma$), depend on the dynamics of the protein environment of the nucleus, such as ^{57}Fe. The anisotropic motion of ^{57}Fe, whose correlation time is of the same order as its lifetime in the excited state, broadens the resonance absorption line as a result of the Doppler effect. We have

$$\Delta\Gamma/\Gamma = 4K^2D\tau, \tag{3.7}$$

where $\Delta\Gamma$ is the broadening of the GR line, K is the wave number of the γ-quantum, τ is the lifetime in the excited state (for ^{56}Fe $\tau = 1.4\cdot10^{-7}$ sec) and D is the coefficient of diffusion, which may obey the Stokes-Einstein Law.

A necessary condition for the Mössbauer effect to be observable is that the extent of the motion of the γ-quantum-absorbing nucleus be limited; for this reason the experimental work was conducted on frozen solutions and on dry and moistened preparations. The increased vibration amplitude of the nucleus results in the decrease of

$$f' = \exp\left(-\bar{x}^2/\lambda^2\right), \tag{3.8}$$

where $\bar{x}^2$ is the mean square shift of the nucleus towards the γ-quantum and λ is the wavelength of the γ-quantum. This expression is valid if the vibration frequency of the nucleus $\omega \gg 1/\tau \simeq 7\cdot10^8$ sec^{-1}. The absorption spectra of ^{57}Fe-modified hemoglobin and methemoglobin are identical with those of the native substances, and each one contains one ^{57}Fe atom per heme. Modified human serum albumin, lysozyme, α-chymotrypsin and collagen contained, respectively, 3, 2, 3 and 5 labels per protein molecule.

The relative f'-value was determined by measuring the area under the resonance curve, that area at the temperature of liquid nitrogen being taken as unity. The different degrees of hydration were realized by varying the atmospheric humidity of evacuated boxes ($P = 10^{-1}$ mm Hg or 13.33 Pa), in which the proteins had been kept for 48 hours at 30° over aqueous solutions of sulfuric acid of various concentrations (Roslyakov and Khurgin, 1972). The specimen-containing cell was then tightly closed.

Figure 9 shows f' as a function of the temperature for proteins with differing humidity contents. It is seen that f' decreases sharply at -60°C. In addition, a broadening of GR lines is noted above -20°C for human serum albumin and for lysozyme. These temperature intervals are practically the same as those noted for the shifts of fluorescence spectra (sec. 3.7). It is to be noted that there is no correlation between the temperatures corresponding to the incipient decrease in f' or between the temperatures at which f' has decreased to one-half of its original value and the molecular mass of the protein. The authors believe that the effects observed at temperatures above -10°C are due to the mobility of macromolecular domains of lysozyme and albumin.

The values of the relative humidities of the proteins $P/P_s f' = 0.5$ at 25°C, corresponding to a 50% decrease in the probability of the f'-effect, are listed below:

Lysozyme	0.53	HbO$_2$	0.78
Human serum albumin	0.71	MbO$_2$	0.66
α-Chymotrypsin	0.86	Chromatophores	0.73
Collagen	0.98	Ferridoxin	0.75

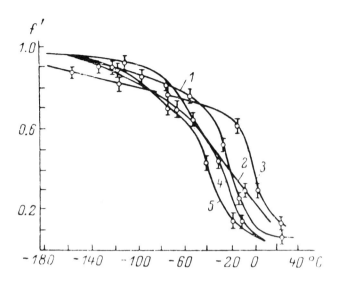

Figure 9. f' as a function of the temperature for proteins hydrated to differing degrees, gm H_2O per 1 gm protein. After Frolov *et al.* (1977).
1 — model iron-sulfur protein (0.2); 2 — α-chymotrypsin (0.3); 3 — collagen (0.48); 4 — human serum albumin (0.32); 5 — lysozyme (0.16).

As in the case of the $f'/(T)$ curves, there is no direct relationship between P/P_S values for which $f' = 0.5$ and the molecular mass of the proteins.

The results obtained by GR spectroscopy which are of the greatest importance to us may be expressed as follows (Frolov *et al.*, 1973, 1977). At 25°C the GR labels in the proteins execute vibrations, which have a high frequency ($v \gg 10^7$ sec^{-1}), a small amplitude (< 0.01 nm) and a small variation of f' with P/P_S, up to $P/P_S = 0.4$; as the hydration is further increased, there appear lower frequency vibrations with a large amplitude (0.01–0.03 nm increase), and f' varies rapidly with P/P_S; in the case of hemoglobin and methamoglobin, the function $f'(P/P_S)$ at $P/P_S > 0.7$ at 25°C is distinguished by low-frequency shifts of nuclear vibrations ($v \cong 10^7$ sec^{-1}), which is manifested as a broadening of GR lines.

The results obtained by this method, like those obtained by several others, indicate that biopolymers must contain water for their dynamic properties to be manifested. The change in the nature of GR vibrations of the protein label, which takes place when $P/P_S > 0.4$, is probably due to the appearance of fluctuations of protein cavities since, from this degree of hydration onwards, they may become filled with water molecules to an extent sufficient to stimulate their mobilities. In hemoglobin and methemoglobin the GR label is located in cavities with an active site; accordingly, at $P/P_S > 0.7$ the observed low-frequency shift of nuclear vibrations more probably reflects the disturbances in the environment of ^{57}Fe, which are introduced by the fluctuations of these cavities.

In the case of strongly hydrated human serum albumin and lysozyme these effects become manifested at a temperature as low as –10°C when — according to the data obtained on model systems (sec. 1.2) and in agreement with our own conceptions – the water near the nonpolar segments of the macromolecules thaws out and the cavities of the macromolecules begin to fluctuate.

Similar results were also obtained by Prusakov *et al.* (1979). These workers studied the systems human serum albumin – water, in which the bonding of the Mössbauer ^{57}Fe label to the protein matrix was of different types (both in the interior and on the surface). The results thus obtained were compared with the results of mobility studies of spin labels.

In the case of solutions of human serum albumin, in which the ^{57}Fe atoms are found in the surface layer of the biomolecule, the curve describing the temperature dependence of the probability f' of the Mössbauer effect showed two distinct intervals. The first interval was in the temperature range between -60°C and -7°C, in which the decrease of f' was not accompanied by a broadening of the Mössbauer line; the decrease of f' in this case could be due to two reasons: incipient mobility of a few links of the polypeptide chain, and the increased mobility of the ^{57}Fe label atoms themselves, if the water molecules near the surface, which begin to melt at -60°C have entered the coordination sphere of the label directly. In the second interval f' decreases more rapidly above -7°C, and there is a broadening of the doublet lines. In the view of Prusakov *et al.*, this effect is caused by the appearance of a new type of motion – vibrations of large molecular fragments, with relatively large amplitudes and periods larger than 10^{-7} sec – as the surface layers of the water begin to melt in depth.

3.6 The Method of Spin Label and its Application

The EPR spectra of all nitroxyl radicals – spin labels and spin probes – are triplets. The observed superfine resolution is due to the fact that the spin of ^{14}N nucleus is unity, while the projections of its magnetic moment onto the quantum axis, given by the external magnetic field H, are 1, 0 and -1. Thus, the local field H_l, which acts on the magnetic moment of the lone electron of the N–O group, may assume three values. Since the resonance condition may be written as

$$h\nu = g\beta \mid H \pm H_l \mid . \tag{3.9}$$

three absorption lines will correspond to the three H_l-values. The width ratio of these lines, and their locations on the magnetic field axis will depend on the averaged anisotropy of the g-factor, and of the hyperfine interaction between the magnetic moments of the lone electron and the nitrogen nucleus.

. According to certain theories, it is possible to use the observed parameters of the EPR spectrum to find the correlation times of the N–O group of the label ($\tau \simeq 1/\nu$), which are a measure of their rotation rates. Having determined the temperature and viscosity relationships of rotational diffusion parameters of protein, conjugated with labels of different lengths and flexibilities, it is possible to distinguish between three surface layers of the water-protein matrix (Likhtenshtein, 1974): the external water layer, with properties much like those of pure water, distinguished by frequencies $\upsilon \simeq 10^{10} - 10^{11}$ sec^{-1}; a layer comprising the protein side groups and the water which is directly bound by these groups ($\nu \simeq 10^{10} - 10^8$ sec^{-1}); and a deeper layer, including the segments of the poly-peptide chain, the side groups 'compressed' by the matrix and the firmly bound water ($\nu \simeq 10^8 - 10^7$ sec^{-1}).

These temperature relationships of the rotational diffusion parameters of the spin label

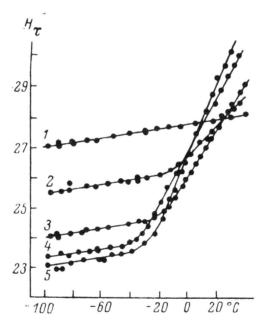

which are attached to Met192 of α-chymotrypsin at different degrees of hydration (P/P_S-values between 0.03 and 0.96) show (Fig. 10) that even at $0.03 < P/P_S \leqslant 0.34$ the label becomes mobile, beginning at -30°C or -20°C, and the mobility increases with increasing moisture content.

Figure 10. The parameter H_T of EPR spectra as a function of the temperature for powdered spin-labeled α-chymotrypsin with various degrees of hydration P/P_S. After Frolov *et al.* (1977). $1 - 0.03$; $2 - 0.34$; $3 - 0.61$; $4 - 0.78$; $5 - 0.95$.

Similar results were obtained for spin-labeled albumin. The rotational diffusion label parameters in dry spin-labeled preparations do not show any major variations with the temperature (Frolov *et al.*, 1977).

Studies of GR and spin-label mobilities at $25°C$ as a function of the extent of hydration, carried out on albumin, hemoglobin and ferridoxin, established that at $P/P_S < 0.6$ the GR mobility of the label increases, while the motion of the spin label is still strongly inhibited ($\nu < 3 \cdot 10^7$ sec^{-1}). At $P/P_S > 0.6$, the rotation rate of the spin label increases to 10^8 sec^{-1} and the amplitude of GR label vibrations increases. In the case of α-chymotrypsin and collagen the GR mobility and the spin-label mobility increase under almost identical conditions ($P/P_S > 0.7$).

The mobility of spin labels begins to be manifested in sufficiently hydrated samples beginning from $-20°C$, at which temperature water may already begin to 'melt' in nonpolar protein zones, and the zones begin to fluctuate. At $25°C$ spin labels respond to the increased mobility of the protein matrix only after the saturation of all surface hydration centers, when the protein cavities continue to fill with water and, as required by the dynamic model, begin to fluctuate.

It would seem, accordingly, that EPR spectra of spin-labeled moistened proteins reproduce their mobility, related to the fluctuations of their cavities. Similar results were obtained by Belonogova *et al.* (1979), who studied the effect of hydration on the temperature dependence of structural mobilities of α-chymotrypsin and serum albumin by the method of spin labels, spin probes and gamma-resonance labels.

Unique information about the fluctuations of protein structure in solution can be obtained by the spin label method, when the EPR spectrum results from the superposition of two spectra representing two different micro-environments. In the case of β-93 spin-labeled hemoglobin it was conclusively proved that the EPR spectrum of the label represents the ability of the C-end of the β-chains, and hence also of their tertiary structure, to fluctuate between two different states (Moffat, 1971). In one of these states the β-145 tyrosyl 'covers' a small cavity in this sub-unit, and forms a hydrogen bond between the imidazole of β-146 histidine and β-94 aspartate, while in the second state the tyrosyl 'opens' the cavity, and the C-end executes free rotations (Moffat *et al.,* 1971).

Under certain conditions the nitroxyl groups contained in the active site of the protein can apparently exist in two states — a less compact and a more compact one. This was the conclusion arrived at by Khodakovskaya and Yakovlev (1975), who studied the EPR spectrum of spin-labeled organo-phosphoric inhibitor of choline esterase, following its localization in the active site of the enzyme. The EPR of the spectrum may reflect the capacity of the active site to fluctuate between 'open' and 'closed' states.

It was noted by Hsia and Piette (1969), who studied the cross-reactivity of antibodies homologous with dinitrophenyl, that EPR spectra of spin-labeled cross-haptens, complexed with the active site of the antibodies, also comprise two types of components, corresponding to a faster and a slower rotation of the N–O group. The most convenient explanation of this finding is a thermodynamic equilibrium between the two states of the active site.

A series of studies, conducted by Käiväräinen *et al.* (1973b) and by Käiväräinen and Nezlin (1976a, b) on immunoglobulins G (IgG), spin-labeled outside their active site, and on dimers of Ig light chains, revealed that the EPR spectra also have A- and B-components, corresponding to the two states of the label. It was suggested that these components indicate the capacity of Fab subunits and of the dimers of Ig light chains to exist in two conformations (cf. sec. 6.3).

Very important and highly interesting results were obtained by the theoretical computation of the nature of the relative mobilities of two protein domains interconnected by a flexible segment (Zientara, 1980). The calculations were based on the model of Brownian diffusion as given by the equation of Smoluchowski. The effect of the orientation phenomena, inter-domainal forces of various types and of the hydration on the process of the domains nearing each other was investigated.

The time of coalescence (association) of the domains significantly depends on the screening of the charged domains by ions and on the extent of hydration. The presence of the hydrate shell constitutes an energetic barrier which prevents the domains from drawing near to each other. Thus, the theory confirms the existence of two (A and B) discrete relative positionings of the bound domains in solution.

A substantial amount of data on the structural flexibility of proteins was obtained by separate determinations of the correlation times of spin-labeled macromolecule and the label which is bound to it by covalent bonds. This method, although not yet generally known, is simple in handling and is therefore certain to find extensive use. For this reason we shall discuss it in more detail and compare it with similar methods already in use.

Separate Determination of the Correlation Times of Spin-Labeled Proteins and their Labels

If the label is rigidly bound to the protein, i.e., if its N-O group has no freedom of rotation with respect to the protein, the correlation time τ_{R+M}, as determined from the EPR spectrum, is equal to the correlation time of the macromolecule τ_M (Shimshick and McConnell, 1972; McCalley *et al.*, 1972). In accordance with the Stokes-Einstein Law:

$$\tau_M = \frac{4\pi}{3k} a_M^3 \eta/T,$$

(3.10)

where a_M is the effective Stokes radius of the macromolecule, η is the viscosity of the solution, T is the absolute temperature and k is the Boltzmann constant.

If the experimental value of τ_M or of a_M is smaller than the calculated value — i.e., than the values based on the assumption that the protein is rigid — this indicates that the protein structure is flexible, i.e., that its domains or subunits are capable of executing thermal shifts with respect to one another.

In the general case the N-O group of the label, which is bound to the protein by a covalent bond, has its own (proper) mobility with respect to the protein surface. The resulting correlation time τ_{R+M}, as determined from the EPR spectrum, will then depend on two types of motion: the motion of the N-O group of the label with respect to the macromolecule, and the rotation of the macromolecule itself. Käiväräinen (1975) proposed a method involving a separate determination of the correlation time of the label with respect to the protein τ_R and of the correlation time τ_M of the macromolecule, based on the variation of the resultant correlation time τ_{R+M} with the ratio of the viscosity of the solvent to the absolute temperature η/T.

If the resultant motion of the N-O group is expressed in terms of its components in the form of averaged spherical functions, it can be shown that

$$\frac{1}{\tau_{R+M}} = \frac{1}{\tau_R} + \frac{1}{\tau_M} \tag{3.11}$$

This equation will be valid if the reorientations of the N-O group with respect to an external system of coordinates during the motion of the label with respect to the macro-molecule on one hand, and during the rotation of the macromolecule itself, on the other, can be considered as two independent processes. Such a situation can be easily imagined, since the 'stem' of the label is not comparable to a rigid rod, but consists of several segments interconnected by 'hinge' elements. This notwithstanding, each such 'stem', just like any other side group, may revolve relative to the protein surface, describing a well-defined cone. Equation (3.11) can be easily rewritten in a form more convenient in practical work:

$$\left(\frac{1}{\tau_{R+M}}\right)_T = \frac{1}{\tau_R} + \frac{(\eta/T)_{st}}{\tau_M} \left(\frac{T}{\eta}\right)_T \tag{3.12}$$

In this equation the correlation time τ_M of the macromolecule is taken under standard conditions (water, $25°C$; $(\eta/T)^{st} = 3 \cdot 10^{-5}$ p/K). The viscosity of the solution was varied by varying the concentration of the added saccharose. The resulting correlation time τ_{R+M} may be calculated from the shift (+1) of low-pole (or high-pole (-1) A-component of the EPR spectra $(H_{+1}(\tau))$ of spin-labeled proteins (Fig. 45) with respect to its maximum shift as $\tau \rightarrow \infty$:

$$\begin{aligned}
\Delta H_{+1}(\tau) &= H_{+1}(\tau \rightarrow \infty) - H_{+1}(\tau) \\
\Delta H_{-1}(\tau) &= H_{-1}(\tau \rightarrow \infty) - H_{-1}(\tau)
\end{aligned} \tag{3.13}$$

with the aid of calibrational relationships (Kusnetsov and Ebert, 1974 or Kuznetsov *et al.*, 1971). However, τ_{R+M} is more conveniently found from the equation of Goldman *et al.* (1972):

$$\tau_{R+M} = a(1-S)^b \frac{32}{A_z} \tag{3.14}$$

where the experimental parameter S is

$$S = \frac{A_z^1}{A_z} = \frac{H_{-1}(\tau) - H_{+1}(\tau)}{H_{-1}(\tau \rightarrow \infty) - H_{+1}(\tau \rightarrow \infty)} \tag{3.15}$$

and a and b are constant coefficients which, for the jumpwise diffusional model, are $2.55 \cdot 10^{-9}$ and -0.615 respectively. These values were calculated for the line width of the EPR spectrum of ~ 3 gauss (Goldman *et al.*, 1972). These conditions are fully realized for immobolized A-components of the EPR spectra.

If the effective Stokes radius of the part of the label which interacts with the solvent is much smaller than the effective Stokes radius of the macromolecule (a), and if the viscosity of the solvent at a constant temperature is significantly different from the effective micro-viscosity of the protein matrix (b), the variation of τ_R with η/T may be neglected. In such a case $1/\tau_{R+M}$ varies linearly with T/η, and the tangent of the slope angle of the line may be used to find the effective Stokes radius a_M and the correlation time of the macromolecule reduced to standard conditions τ_M. If the value of η/T is sufficiently large, and the condition (b) is no longer met, the variation of $1/\tau_{R+M}$ with T/η is no longer linear. At the intersection point of the extrapolated linear segment of the experimental function with the ordinate axis $1/\tau_{R+M} = 1/\tau_R$. τ_R is a more objective criterion than τ_{R+M} which describes the interaction between the label and the macromolecule.

It is important to note that the τ_M-values obtained by this method are very close to the results obtained by independent, generally accepted methods – hydrodynamic methods, fluorescence depolarization method and other methods. Very often it is the variations of τ_R and τ_M rather than their absolute values which are of interest.

The enthalpy and entropy of activation of label rotation may serve as supplementary parameters characterizing the interaction of the label with its micro-environment. These values are found with the aid of the Eyring-Polanyi equation:

$$\tau_R = \frac{h}{kT} \exp(-\frac{S}{R}) \exp(\frac{H}{RT}) \tag{3.16}$$

The similarity of experimental dates presentations as the functions: $1/\tau_{R+M} = f_1(T/\eta)$ or $(2A_z - 2A'_z)^{-b} = f_2(T/\eta)$ or $(\Delta H_{\pm_1})^{-b} = f_3(T/\eta)$

The experimental parameters such as $2A'_z(T/\eta)$ and $\Delta H_{\pm_1}(T/\eta)$ may be determined directly from EPR spectra of spin-labeled protein at different values of T/η and $2A_z(T/\eta = 0)$ at frozen solutions of protein.

We can express the equation (3.14) in such forms:

$$\frac{1}{\tau_{R+M}} = \frac{A_z}{32a}(1 - \frac{A'_z}{A_z})^{-b} = \frac{1}{\tau_R} + \frac{(\eta/T)_{st}}{\tau_M}(\frac{T}{\eta})_T \tag{3.17}$$

or:

$$a_0(2A_z)^b(2A_z - 2A'_z)^{-b} = \frac{1}{\tau_R} + \frac{(\eta/T)_{st}}{\tau_M}(\frac{T}{\eta})_T \tag{3.18}$$

where:

$$a_0 \equiv A_z/32a$$

Taking into account that: $2A_z - 2A'_z = \Delta H_{+1} + \Delta H_{-1}$ and using the condition of jump-way diffusion of N-O group (Kuznetsov, Ebert, 1974): $\Delta H_{+1} \approx \Delta H_{-1}$, eq. (3.18) may be rewritten as:

$$a_0 A_z^b(\Delta H_{\pm_1})^{-b} = \frac{1}{\tau_R} + \frac{(\eta/T)_{st}}{\tau_M}(\frac{T}{\eta})_T \tag{3.19}$$

In the region of moderate viscosity we obtain three expressions to $tg\varphi$ of experimental lines from eq. (3.12); (3.18) and (3.19):

$$tg\varphi_1 = \frac{3 \times 10^{-5}}{\tau_M} = \frac{\Delta(1/\tau_{R+M})}{\Delta(T/\eta)}$$

$$tg\varphi_2 = \frac{3 \times 10^{-5}}{a_0(2A_z)^b \tau_M} = \frac{\Delta(2A_z - 2A_z')^{-b}}{\Delta(T/\eta)} \qquad (3.20)$$

$$tg\varphi_3 = \frac{3 \times 10^{-5}}{a_0 A_z^b \tau_M} = \frac{\Delta(\Delta H_{\pm 1})^{-b}}{\Delta(T/\eta)}$$

The value of τ_M for standard condition (water, 25°, $(\eta/T)_{st} = 3 \times 10^{-5}\ P/K$) may be easily calculated from one of the eq. (3.20).

When spin-label is rigidly bound to protein, its N-O group continuously diffuse with the same correlation time ($\tau_{R+M} = \tau_M$). For such case the values: $a = 5.4 \times 10^{-10}$ G/s; $(-b) = 1.36 \approx 3/2$. If the spin-label has its own mobility respectively to the macromolecule and the character of such mobility correspond to the model of jump-way diffusion, then the values: $a = 25.5 \times 10^{-10}$ G/sec, $(-b) = 0.615$ (Goldman et al., 1972).

For all investigated spin-labeled proteins HSA-SL, Hb-SL, lysozyme -SL et al., the values of τ_M, calculated on basis of the model of "jumps" were much close to the date of independent methods, as compared with values obtained from the model of continuous diffusion of N-O group.

The value of τ_R for all this cases was about or less than 10^{-8} s.

Other Methods of Calculation

Wallach (1967) proposed a method for the determination of τ_{R+M} from the known correlation time τ_M and from the flexibility parameters of the label, which in turn depend on its stereochemistry. Thus, for instance, if the rotation takes place around one internal axis of the label only, this worker showed that

$$\tau_{R+M} = \tau_M \cdot \frac{1}{4} (3 \cos^2\theta - 1)^2 \qquad (3.21)$$

where θ is the angle between the rotation axis of the label and the orientation axis of the N-O group of the label.

Thus, if τ_{R+M} and the angles between the label axes around which rotation takes place are known, τ_M can be determined. However, this approach is only valid on the assumption that the internal rotations are either much faster or slower than the rotation of the molecule itself. In reality this is not always the case. The magnitude of the angles θ may depend on the effect of the micro-environment which is usually unknown.

Dudich et al. (1977) modified the method of Hubbel and McConnell (1971), devel-

oped for a fast anisotropic motion of spin probes of protein membranes. Such a modifi-
cation, in conjunction with the variation of the parameters $A_{||}$ and $A_\perp$ (eqs. (3.22) and
(3.23)) with viscosity, makes it possible to determine the correlation times of spin-labeled
macromolecules if the N-O group of the label takes part in two processes at the same
time: the fast anisotropic rotation around the protein, with $\tau_R < 10^{-9}$ sec, describing a
cone with the apex angle of 2α, and the slow, isotropic rotation together with the protein, with $\tau_M \geqslant 10^{-8}$ sec. The condition $\tau_R \gg \tau_M$ must also be met.

Calculations based on this model do not yield the actual value of τ_r or its variation.

If these three conditions are met, the EPR spectrum of spin-labeled protein at $\alpha \leqslant 50°$
will be a two-component spectrum. However, the fact that the EPR spectrum is a two-
component spectrum is not in itself proof of a fast anisotropic label rotation, since it may
also be the result of the label being in two states with different micro-viscosities with respect to the protein (Fig. 46). Such a tendency was repeatedly confirmed by X-ray diffraction analysis.

Experimental Criteria of Applicability of Various Models

Before any given model can be used, its applicability criteria must be experimentally confirmed. Such criteria include:

1. The specific temperature dependence of the parameters constituting the anisotropic
'gyrostat' model:

$$\overline{A}_{||} = A_0 + \frac{B}{3}(\cos^2\alpha + \cos\alpha) \qquad (3.22)$$

$$\overline{A}_\perp = A_0 - \frac{B}{6}(\cos^2\alpha + \cos\alpha) \qquad (3.23)$$

where A_0 is the isotropic hyperfine interaction constant, and B is a constant the value of
which depends on the properties of the radical.

2. The value of the parameter:

$$R = \Delta H_{+1} / \Delta H_{-1} \qquad (3.24)$$

where the significance of $\Delta H_{\pm 1}$ is the same as in equation (3.13).

3. The value of η/T at which the variation of ΔH_{+1} or A_{max} with $(T/\eta)^{2/3}$ begins to
deviate from linearity.

It follows from Hubbel and McConnell's (1971) theory of anisotropic gyrostat that if
the parameter $\overline{A}_{||}$ decreases with increasing temperature (owing to the increase in the
rotation cone angle 2α), the parameter $\overline{A}_\perp$ should increase. In the case of spin-labeled
hemoglobin, human serum albumin and IgG, whose spectra have the two-component
shape, it was shown that this condition is not met. This is the first argument against the
gyrostat model.

If the value of R (eq. (3.24)) is close to unity, the model of jumpwise diffusion of the

N-O group (Kuznetsov and Ebert, 1974) is 'operative'. According to this model, the correlation time of the N-O group is the averaged time between its random angle reorientations. The randomness of this angle is also an indication of the isotropic nature of the mobility of N-O group around the protein. This condition was met in all the objects investigated by ourselves – i.e., equation (3.12) was applicable. This is the second argument against the gyrostat model.

It was experimentally shown that $1/\tau_{R+M} = f(T/\eta)$ or $A_z' = f(T/\eta)^{2/3}$ begin to deviate from linearity only at very high viscosities ($\eta/T \simeq 5 \cdot 10^{-4}$ p/K) (corresponding to 47% saccharose at 20°C) of protein solutions spin-labeled with 2,2,6,6-tetramethyl-N-1-hydroxylpiperidine-4-amino-(N-dichlorotriazine). It is easily shown that this η/T-value represents 10% of $(\eta/T)_{eff}$ of the water-protein matrix in the label-bonding segment. According to the data yielded by isotropic diffusion method, the τ_R-value at 20°C of a spin label with an effective Stokes radius of $a_R = 0.4$ mm is about 10^{-8} sec. Substituting these values of a_R and τ_R in eq. (3.10), we find that $(\eta/T)_{eff} \simeq 5 \cdot 10^{-3}$ p/K.

It follows that the deviation of $1/\tau_{R+M} = f(T/\eta)$ from linearity violates condition (b) above, but is in full agreement with the model of isotropic motion of the N-O group of the protein-bound label.

On the other hand, if the mobility of the label were $\tau_R \simeq 10^{-9} - 10^{-10}$ sec, as postulated by the anisotropic gyrostat model, the nonlinearity would begin at a much earlier stage (at saccharose concentrations below 8% at 20°C). However, this is never actually noted, which is the third objection to the 'gyrostat' model.

It can be readily shown from the NMR data that yet another condition of applicability of the 'gyrostat' model, viz., $\tau_R \ll \tau_M$, fails to be met (cf. para. 3.4).

The above analysis shows that the fast anisotropic gyrostat model, unlike in the case of membranes, is not applicable to spin-labeled proteins (Dudich *et al.,* 1977). The data obtained by using equation (3.12), based on the model of jumpwise isotropic diffusion of the N-O group of the label, on the contrary, are shown by all available evidence to be reliable.

The results thus obtained reveal that the subunits of oxidized methemoglobin, immunoglobulins (Käiväräinen, 1975a; Käiväräinen and Nezlin, 1976a) and the domains of serum albumin are fairly mobile. These studies will be discussed in the following chapters.

An intramolecular mobility of a similar type was demonstrated with the aid of a fluorescent label inserted between the domains of aspartate-transaminase subunits (Polyanovskii *et al.,* 1970), pepsin 'nuclei' (Glotov *et al.,* 1976a, b), and between subunits of human luteinizing hormone (Bishop and Ryan, 1975).

3.7 Application of Luminescence to the Study of Protein Dynamics

Luminescence methods also furnish important information on the intramolecular mobility of proteins. The generation of fluorescence and phosphorescence may be explained by means of Fig. 11. After the chromophore has passed into one of the excited singlet states S_2, S_3 ... within $10^{-14} - 10^{-15}$ second, the nuclear configuration becomes equilibrated, and changes occur in the dispersive, electrostatic and electrodynamic interactions of the chromophore with its nearest surroundings within a time of the order of 10^{-12} sec. This may be explained by non-radiative transition of the molecule to the lower excited level S_1. After $10^{-8} - 10^{-10}$ sec residence in this state, the molecule may return to the singlet ground level S_0, with or without the emission of radiation, the respective probabilities being q_1 and P. Another possibility is the non-radiation transition from the S_1-level to the metastable level T_1. Such a transition reverses the direction of the spin and is known as intercombination conversion.

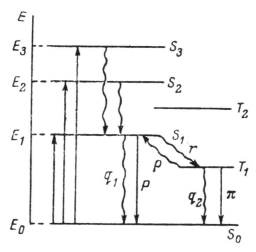

Figure 11. Diagram of the electronic levels of the molecule and inter-level transitions.

The probability of the $T_1 \rightarrow S_1$ transition, which is also accompanied by spin inversion, is small, since the lifetime of the electron in the excited state is rather long — 10^{-3} to 10^{-1} sec. The radiative transition $S_1 \rightarrow S_0$ corresponds to fluorescence, while the radiative transitions $T_1 \rightarrow S_0$ give rise to phosphorescence.

The properties of a fluorophore may be described in terms of energetic characteristics, expressed as the location and the shape of the radiation spectrum. The location of the spectrum is described as the location of a peak within some wavelength or frequency range. If the transition to the excited state takes $10^{-14} - 10^{-15}$ sec, and involves a major change in the electric dipole moment of the chromophore, the dipole field of the surroundings will not be at equilibrium, and will undergo relaxation to a new equilibrium state. The relaxation time of this process is determined by the dynamic properties of the medium. This is accompanied by a change in the locations of the excited and the ground states, with a shift in the fluorescence spectrum relative to the absorption spectrum. The size of this shift will depend on the dipole moment of the transition $\Delta\mu$. Fluorescence spectra of tryptophane are best suited to the study of the structural mobility of this protein, since the moment of indole transition in tryptophane ($|\Delta\mu| \simeq 4D$) is much larger than that of the benzenic chromophore of phenylalanine or the phenolic chromophore of tyrosine.

For a given number of dipoles in the vicinity of the chromophore the location v_C of the peak of the fluorescence spectrum is given by the ratio between the lifetime of the excited chromophore τ_f and the period of dipole relaxation τ_ρ (Mazurenko, 1973):

$$v_c = v_\infty + (v_0 - v_\infty) \tau_f / (\tau_f + \tau_\rho), \qquad (3.25)$$

where v_∞ is the value of v_c if $\tau_\rho \ll \tau_f$, and v_0 is the value of τ_c if $\tau_\rho \gg \tau_f$.

The magnitude of the shift of the fluorescence spectrum will thus depend on whether the lifetime of the chromophore in the excited state is long enough for the dipoles of the medium to have become reoriented in the direction of the new field produced by the dipole of the chromophore. As a result, when protein solutions are frozen and the mobility of chromophore environment decreases (τ_ρ increases), the fluorescence spectrum is shifted towards lower frequencies. It was shown by Permyakov and Burshtein (1975) and by Permyakov (1977) that each protein has its own characteristic curve of location of the fluorescence spectrum as a function of the temperature (Fig. 12). As a rule, two spectral shift zones are noted: the high-temperature zone (between -20°C and 0°C) and a low-temperature zone (between -20°C and -90°C). When H_2O had been exchanged for D_2O, the shifts were affected in both these zones: the shifts in D_2O in both zones took place at temperatures higher by 5 - 10°C than in H_2O. This indicates that both types of shifts are related to the freezing of water, which interacts with the protein in two different ways. According to our own theory, the high-temperature shift is caused by the freezing of the water in contact with nonpolar segments, including the polar zones of the protein. The low-temperature shift corresponds to the freezing of the water interacting with the charged polar groups on the surface of the protein, its cavities, and to the stabilization of their fluctuations. This leads to immobilization of the internal dipoles due to the cooperative properties of biopolymers.

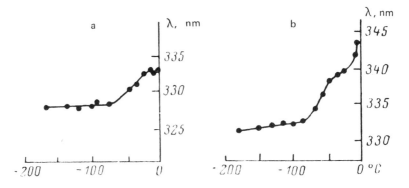

Figure 12. Location of the fluorescence spectrum of the aqueous solution of cobra as a function β-lactoglobulin (a) and neurotoxin II/of the temperature at pH 6.5. After Permyakov (1977).

Spectrum shifts were absent in all protein powders dried over P_2O_5 for a long period of time between $0°C$ and $-196°C$. As the proteins were gradually moistened, temperature shifts began to appear, at first in the low-temperature zone, and attained their peak value when the water content had attained about 0.3 gm per 1 gm protein. This was followed by the appearance of the shifts in the high-temperature zone (between $-20°C$ and $0°C$), which kept increasing until the water content had attained 0.6 gm H_2O per 1 gm protein (Permyakov, 1975).

If we assume that the thermodynamically favored sequence is for the water to become absorbed on the polar surface groups of the macromolecule first, and on the nonpolar protein cavities only afterwards, the observations are seen to agree with the interpretation of the shift sequence given above.

An important fluorescence parameter is its quantum yield, i.e., the ratio between the number of quanta radiated by the system to the number of the excitation quanta absorbed by the system in unit time. Experiments carried out on model compounds showed that the excited singlet states of indolic and phenolic chromophores may become deactivated by three different mechanisms (Burshtein, 1977 a, b): by radiation of a light quantum (rate constant k_f); by intercombination conversion to the triplet state (rate constant k'_{ST}); and by reversible photochemical or pseudo-photochemical reactions between the excited chromophore and side molecules and the quenching groups (rate constant k'_i). The quantum yield will then be

$$q = k_f/(k_f + k'_{ST} + \Sigma k'_i) \quad \text{or} \quad 1/q = 1 + k_{ST} + \Sigma k_i, \qquad (3.26)$$

where $k_{ST} = k'_{ST}/k_f$ and $k_i = k'_i/k_f$.

Since k_{ST} is independent of the temperature, equation (3.96) may be written as follows:

$$1/q(T) = 1 + k_{ST} + f(T) \sum_i k_i = a + bf(T), \qquad (3.27)$$

where a and b are constants if $f(T)$ are equal for all $k_i = k_i f(T)$.

It has been experimentally shown that the function $f(T)$ corresponds to T/η, where T is the absolute temperature and η is the viscosity of the solution. It was found that the function T/η is a satisfactory description of the fluorescence yield of the tryptophane and tyrosine residues as a function of the temperature in a number of proteins (Fig. 13). This shows that the temperature dependence of non-radiative transitions in the protein is limited not so much by the specific features of the chromophores and their nearest environment as by certain dynamic laws which govern the protein structure. The reciprocal of the lifetime of the excited chromophore $1/\tau$ also varies linearly with T/η.

According to Burshtein, the quenching collisions of the excited chromophore with the neighboring groups take place during the structural fluctuations of the globule, which are related to the translational motion of its sufficiently large parts with respect to one another. A consequence of our own model is the variation of the fluctuation frequency of protein cavities with the viscosity of the solvent, since these are accompanied by changes in the effective volume and geometry of the macromolecule.

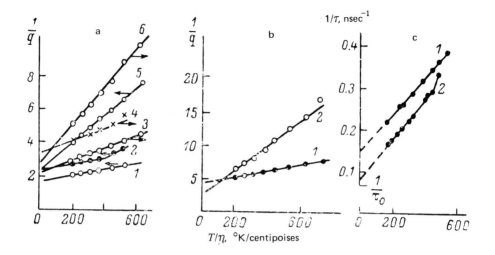

Figure 13. Linear variation of the reciprocal quantum yields (a, b) and of fluorescence lifetime (c) in proteins with T/η of the solvent. After Burshtein (1977b).
a): 1 — RNA-ase, pH 6.0; 2 — human serum albumin, pH 6.1; 3 — L-asparaginase, pH 4.0; 4 — azurin, pH 4.25; 5 — neurotoxin, pH 7.4; 6 — neurotoxin, pH 6.3. b): 1 — cytotoxin, pH 6.0; 2 — carp parvalbumin, pH 6.4. c): ribosomal preteins: 1 — S-4; 2 — S-7.

It is important to note that the characteristics of temperature quenching of denatured proteins are qualitatively different from those observed for native proteins. In denatured proteins, the dependence of the quenching processes on T/η is, as a rule, no longer linear. The reason for it may be the disruption of globular domains whose relative motion determines the fluctuations of protein clefts between the 'open' and the 'closed' states.

In addition to being caused by the interaction between protein chromophores and the neighboring groups, fluorescence quenching may also be produced by collisions with low-molecular compounds present in solution. Iodide, nitrate, cesium and thiocyanate ions, molecular oxygen, acrylamide and other substances are often used as quenchers of protein fluorescence (Burshtein, 1977 a, b).

Judging by the magnitude of the quenching effect, the accessibility of internal chromophores constitutes 2-5% of the total accessibility of the surface, i.e., free low-molecular chromophores. The results of such experiments are interpreted with the aid of the Stern-Volmer equation:

$$Q = q/q - 1 = K_{SV}C = \tau_0 k_q C = \tau_0/\tau - 1, \qquad (3.28)$$

where Q is a characteristic of the quenching effectiveness; q_0 and q are quantum yields of the fluorescence in the absence and in the presence of the quenching agent at concentration C, respectively; K_{SV} is the proportionality constant of the Stern-Volmer equation; τ_0 and τ are the fluorescence lifetimes in the absence and in the presence of the quenching agent, respectively; and k_q is the bimolecular rate constant of the reaction between the quenching agent and the excited chromophore, which is a measure of the accessibility of the chromophore to the dissolved quenchers.

Saviotti and Galley (1974), who studied the nonlinear variation of the magnitude $Q = \tau_0/\tau - 1$ of the phosphorescence with the concentration of the oxygen quencher, came to the conclusion that the internal chromophore becomes accessible to O_2 as a result of 'fragmentations' and 'collapses' of the protein structure. An alternative interpretation of this effect is that the quencher is capable of diffusing inside the protein, as though the latter were a viscous liquid (Eftink and Chiron, 1976).

The frequency of occurrence of structural disturbances, which result in quenching collisions and molecular diffusion through protein vacancies, should be of the same order as the reciprocal of the excitation time $10^9 \geqslant v \geqslant 10^7 \text{ sec}^{-1}$. The dimensions of the vacant sites formed will be a few Angstrøm units, while the rotation of the groups ensuring dipole relaxation is probably of the order of a few degrees (Burshtein, 1976b).

3.8 Results of X-Ray Diffraction Studies of Protein Mobility

X-ray diffraction studies of proteins yield indirect information about their dynamics. Illegible diffraction patterns of certain parts of crystallized proteins, e.g., the C-ends of oxyhemoglobin (Perutz, 1970) and Tyr248 in carboxypeptidase (Lipscomb, 1974) are the result of their mobility, i.e., of the delocalization of their electron density.

Morgan and Peticolas (1975) studied electron diffraction as a function of the radiation frequency, and came to the conclusion that the cavity of a macromolecule resembling lysozyme opens and closes at the rate of 10^4 sec^{-1}. Vibrations of large molecular segments at rates of the order of 10^7 sec^{-1} were detected by Raman spectroscopy in chymotrypsin and pepsin by Frushour and Koenig (1974) and in lysozyme by Genzel *et al.* (1976). For review, see Shimanouchi (1976).

The stabilization of one of the conformers by crystallization may be the reason for the asymmetry of Bence-Jones protein revealed by X-ray diffraction, even though this protein consists of two chains with identical primary structures (Schiffer *et al.*, 1973). The dimeric allosteric enzyme — hexokinase B — in the crystalline form is also asymmetric (Steitz *et al.*, 1976). The two subunits are turned with respect to each other by an angle of 156° around the molecular symmetry axis, and are displaced by 1.36 nm along this axis. This means, firstly, that the environments of the two sub-units are not identical and, secondly, that such a conformational state may exist in solution.

Moult *et al.* (1976) compared the structures of egg lysozyme in triclinic and in tetragonal crystals, at 0.25 nm resolution, and determined the location of the water which is firmly bound to these proteins. A juxtaposition of the electron density charts of the two types of crystals revealed that the differences in the internal protein zones are small, but that there are significant conformational differences in the principal chains and in the chains on the surface. These differences result from the different intermolecular effects in triclinic and in tetragonal crystals: in the former the lysozyme molecules are more dense-ly packed, and the contact points are more numerous.

In several cases the conformations of the long flexible side chains show major differ-ences in the two protein structures, e.g., Arg14, Lys33, Phen38, Arg61, Arg114 and Arg128. There are also certain differences between the conformations of the principal chains, especially so in the β-loop area between the residues 44 and 50. Except for 101, all residues constituting the active site have remained unchanged.

About one-third of the total amount of solvent molecules in the two crystalline forms were identified on the differential charts. About 140 water molecules were found in the tetragonal, and about 110 in the triclinic form. Of these, 60 molecules occupied the same locations, including the water molecules in the active site and three internal molecules (Imoto *et al.*, 1972). Almost all the molecules observed had three ligands, two of them almost invariably forming part of the protein. Except for a small area around Lys13 and Lys33, no second layer of solvent molecules was observed anywhere. It should be borne in mind, however, that the X-ray method responds only to the states of water molecules which are sufficiently fixed. Almost all the accessible polar and charged groups interact with water. In both structures a number of very large peaks on electron density charts around the positively charged side chains may be assigned to the bound nitrate or chlor-ide anions.

In the view of Moult *et al.* (1976), the observed conformational lability of lysozyme may have a functional significance. Similar results were obtained in a comparative X-ray study of the iso-forms of subtilysine (Drenth *et al.*, 1971) and chymotrypsin (Tulinsky *et al.*, 1973). In the latter case conformational variations occurred as a result of changes in pH.

It must be borne in mind, accordingly, that the observed X-ray structure depends on the conditions of crystallization. Comparative structural analyses of a number of pro-teins in solution and of protein crystals by the X-ray scatter method (Fedorov *et al.*, 1978) indicates that domain mobilities become stabilized during the crystallization of proteins. The stabilization of the protein structure in the crystal was also repeatedly con-firmed by other workers (Rapli, 1973).

Theoretical methods for the calculation of small-amplitude motions in proteins by modelling the molecular structure dynamics are now used to an increasing extent (McCammon *et al.*, 1977). It has been shown in this way (McCammon *et al.*, 1979) that the orientation fluctuations of the tyrosine rings in the interior of bovine pancreatic in-hibitor of trypsin are still as large as $\pm30°$ in picosecond time intervals ($\sim10^{-12}$ sec). In this particular case, numerical integration over the time interval of $(0\text{-}7)\cdot10^{-12}$ sec was performed on the equations of motions of the 454 atoms of bovine pancreatic inhibitor (exclusive of hydrogen atoms), which contained empirical free-energy functions. When the integration interval was extended to 100×10^{-12} sec (Karplus and McCammon, 1979), a strong dynamic interaction was noted between the N- and the C-ends of the bovine pan-creatic inhibitor. The mean square fluctuations of the atoms of this inhibitor, calculated for the temperature of $306°K$, is about $7.5\cdot10^{-3}$ nm^2. However, the authors left out of account the possibility of large protein fluctuations, produced by the relative motions of large parts of the molecule, as well as the possible interaction between the protein and the solvent.

Frauenfelder *et al.* (1979) obtained information on the micro-mobility of the constituent atoms of crystallized protein by studying the electron density distribution in their diffraction patterns. This method was applied to the study of mean square thermal shifts of metmyoglobin atoms in the crystal $<\chi^2>$ at several temperatures between 220 and 300°K. The values of $<\chi^2>$ did not exceed $3 \cdot 10^{-3}$ nm^2 and increased from the center of the protein towards its periphery. As a rule, the charged and the polar groups are more mobile than the nonpolar. The residues in contact with heme have an $<\chi^2>$ of $1.2 \cdot 10^{-3}$ nm^2. A similar analysis of the diffraction pattern of a lysozyme crystal revealed that the largest mean square atom shifts take place in the active site of the enzyme (Artymiuk *et al.* (1979).

The analysis of the diffraction patterns of hemoglobin between –20°C and +25°C made it possible to determine the mean square displacements of 1261 non-hydrogen atom (Douzou, 1980). It was found that the heme zone is more rigid, while the protein periphery is less so. The heme-containing cavities in hemoglobin typically display strong atomic vibrations.

A comparison between the structures of horse carboxyhemoglobin and methemoglobin at 0.28 nm resolution (Heidner *et al.,* 1976) revealed that the SH-group of the F9(93)β-cystein in methemoglobin is in a state of equilibrium between two extremal states, in one of which it is exposed to the solvent, while in the other it is located inside the tyrosine 'pocket' between the spiral-shaped F and H segments.

Esipova (1973) concluded from the X-ray diffraction data obtained for proteins that it is the groups located at the points of flexure of the chains which display the highest local mobilities. In fact, the delocalization of electron density in lysozyme was recorded near Lys21, located on a chain bend; similar results were also obtained for chymotrypsin and trypsin (Singler *et al.,* 1968). A special analysis, performed on a number of proteins, showed that certain groups of the active site are located at the bends, which indicates that these cavities have conformational mobility (Esipova and Tumanyan, 1972). Glycine seems to play a special role in the manifestation of "flexible" properties of protein.

The first results of X-ray diffraction analysis now being published indicate that two domains which usually form the active centers become reoriented and draw close to each other while reacting with ligands. The active center cavities pass from the 'open' to the 'closed' state, and water is displaced from the cavity as a result. This has been demonstrated for lysozyme (Phillips and Levitt, 1974), hexokinase (Bennett and Steitz, 1980) and phosphoglycerate kinase (Pickover *et al.,* 1979).

There are grounds for hoping that the number of such studies will increase in the near future.

The results of studies on protein dynamics may be summarized as follows. Three main types of mobility may be distinguished in protein structures: shifts of large parts of a macromolecule (domains) relative to each other, at frequencies between 10^4 and 10^7 sec^{-1}, which are responsible for the fluctuations of protein clefts including the active sites; motions of surface groups, which are usually polar and are charged, at frequencies of 10^9 - 10^{11} sec^{-1}; migration of local disturbances, i.e., of defects in the protein domains.

There is no doubt that water plays what may be a determining role in the manifestation of dynamic properties by biopolymers. By interacting with proteins, water saturates all types of hydration centers simultaneously (external groups, both polar and non-polar, as well as clefts). The probability of arrangement of water molecules over the sorption sites is determined by the arrangement of the sorption energy on these sites. The data reported above indicate that the sorption of water takes place on the external (polar, charged) groups first; however, model studies of water sorption show that the sorption of water by the cavities becomes thermodynamically more favored as these become filled and cooperative effects are intensified (sec. 1.2). If the amount of the available water is sufficient, the protein cavity may well successfully compete for water with the external polar groups in the course of subsequent hydration.

Certain specific features of the dynamics of interaction between proteins and water are manifested only in native proteins, and disappear with their denaturation. The biopolymer and its hydrate shell form one integral cooperative system. The mobilities of any protein segments and those of the water molecules with which they interact are interrelated, and may be expected to have similar correlation times. There is probably a correspondence between the distribution of the correlation times specific to the protein structure in solution, and a corresponding distribution of correlation times of individual fractions of water molecules, which interact with biopolymer segments of differing mobilities.

Chapter 4

Interaction between Proteins and Water

If the dynamic model of behavior of proteins in water is accepted, the system: protein + bound water must be considered as one single whole. The changes in the state of the water and macromolecular segments solvated by it are clearly interrelated and interdependent. The properties of water interacting with the outer surfaces of the domains may be quite different from those of the water contained in the cavities. In our view, the geometry and dimensions of these cavities must be adapted to the structural parameters of the water.

It may be expected that changes in the hydrations of proteins resulting from their transformations will largely depend on the shift of the equilibrium between the 'open' and 'closed' states of the clefts in either direction. We shall begin by considering certain general problems concerning the hydration of biopolymers.

4.1 Hydration of Biopolymers and Methods of its Study

'Bound' water, or 'hydrate shell', is understood to mean the water in the vicinity of the surface of the macromolecule, whose properties differ from those of the 'bulk' water in solution. A simple method of studying the thermodynamic properties of water in any system is the determination of its activity a. If, under the experimental conditions employed, the water vapor behaves more or less as an ideal gas, we have $a = P/P_0$, where P_0 is the vapor pressure of water under standard conditions and P is the pressure of the water vapor above the surface of the solution or the powder being studied.

The interaction between proteins and water may be studied with the aid of the so-called absorption isotherm, which shows the amount of the water sorbed by the protein (in gm H_2O per 1 gm of protein) as a function of the relative water vapor pressure P/P_0. The desorption isotherm, which is obtained when P/P_0 is progressively decreased, is usually located above the absorption isotherm. This hysteretic effect may be caused by a strong decrease in the conformational mobility of the protein in the dry state, as a result of which the true equilibrium between the biopolymer and water only becomes established when P/P_0 is increased. A typical sorption curve is shown (Fig. 14) for collagen.

Khurgin *et al.* (1972) conducted a gravimetric study of the interaction between α-chymotrypsin and lysozyme and water vapor. The hysteresis between the water sorption and desorption curves obtained by these workers is attributed to the conformational changes taking place in the protein molecule as a result of dehydration. The polar groups on the surface act as sites of the stronger sorption. As the sorption sites become progres-

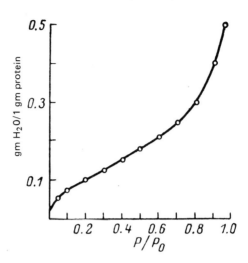

Figure 14. Isotherm of water sorption by collagen at 25°C as a function of the relative pressure of water vapor. After Bull (1944).

sively filled, a network of hydrogen bonds is formed by the water molecules. This may be accompanied by the formation of the native structure of the protein. According to the authors, the hydrate hull thus formed is highly cooperative.

Studies (Mrevlishvili *et al.,* 1975) conducted with the aid of a low-temperature adiabatic calorimeter revealed that in highly concentrated collagen solutions the melting point of water decreases with decreasing water content. The interval and shape of the heat absorption curve obtained during the fusion of bio-polymer solutions depend on their identity and condition. Thus, the curves obtained for proteins differ from those obtained for nucleic acids.

According to Syrnikov (1974), the degeneration of the phase transition of the first kind in concentrate solutions is caused by the interaction of water with the biopolymer. The water in a concentrated solution of a biopolymer may be considered as a system with a very large surface area — and it is known from the thermodynamics of small-sized systems that sharp phase transitions of the first order disappear if the contribution of the surface energy to the total energy of a system is sufficiently large.

Microcalorimetry is a method for determining the extent of biopolymer hydration by two techniques, both of which are based on the fact that, as distinct from free water, phase transitions never occur in hydrate shells. In working with moist powders it is assumed that the amount of the bound water is equal to the maximum amount of the water (in gm H_2O per 1 gm protein) at which the peak heat absorption, corresponding to the phase transition which occurs in free water only, just fails to be noted. The first technique is based on this principle.

The second technique is based on the comparison between the experimentally determined amount of heat required to heat up the frozen solution from any sufficiently low temperature to just above zero with the calculated amount of heat which would be needed to heat pure water in the amount equal to that contained in the biopolymer solution. The difference between these two values makes it possible to estimate the amount of non-freezing water, i.e., of the water which has interacted with the biopolymer.

The NMR method has also proved to be a useful way of studying the hydration of biopolymers. The chemical shift of the protons in biopolymer solutions, as recorded by the NMR method, is an indication of the time-averaged state of the hydrogen bonds in the system. In protein crystals and in other macroscopically oriented systems a number of discrete water signals are noted, which is an indication that hydrogen bonds are anisotropic. Thus, for instance, the NMR spectrum of the water in collagen fibers is a doublet (Dehl and Hoeve, 1969). If anti-freeze agents are employed, the hydration can be studied at temperatures as low as $-100°C$ (Fratiello and Schuster, 1967).

The NMR signal given by the non-freezing hydrate shell of the macromolecule was isolated by freezing the water in protein solutions (Kuntz *et al.*, 1969). The method is based on the fact that, during freezing, the width of the line given by free water increases to 10^5 sec^{-1}, while the width of the signal given by the bound water remains at about 10^3 sec^{-1} at $-35°C$.

The amount of the water which has failed to freeze is a linear function of the protein concentration within a wide range of values. The integral intensity of the absorption signal given by water protons can be used for immediate calculation of the degree of hydration of the protein. The data which have been obtained for a number of proteins are in satisfactory agreement with the calorimetric results (Table 4). If it is assumed that each polar group in the protein is capable of interacting with one water molecule, the amount of the water calculated in this way amounts only to about one-fourth of the hydrate shell of the protein as recorded by NMR or by calorimetric methods.

We shall describe below a modification of the NMR method, involving the use of several concentrations of D_2O, by which separate determinations of the water fractions sorbed by polar and nonpolar parts of the protein can be made.

TABLE 4. Hydration values of proteins (gm H_2O per 1 gm protein) obtained by NMR and by calorimetric methods.

Protein	NMR ($-35°$)	Calorimetric
Egg albumin	0.33	0.32
Bovine serum albumin	0.40	0.32, 0.49
Hemoglobin	0.42	0.32
Lysozyme	0.34	0.3
DNA	0.59	0.61

When the degree of hydration is determined by NMR below zero temperatures, it must be borne in mind that the appearance of the discrete narrow resonance line given by the hydrate water which has not frozen will depend not only on the fairly high mobility of its molecules, but also on whether their lifetime in this state is sufficiently long — i.e., longer than the reciprocal width of the resonance absorption band, expressed as frequency (Aksenov, 1979).

A PMR Method for the Study of Hydration of Proteins at Below-Zero Temperatures Involving the Use of Low Concentrations of D_2O

It is known that the association constant of D_2O with protein is higher than that of H_2O (Lobyshev and Kalinchenko, 1978). Accordingly, if D_2O is added to a solution of a protein, even in small amounts, it may be expected that the HOH in its hydrate shell will be partly replaced by DOH and DOD. There is evidence to show that the substitution first affects that part of the water which solvates nonpolar segments and cavities. This is confirmed by the following considerations: model experiments show that the solvation of the charged (Swain and Bader, 1960) and of the strongly polar (Nemethy and Scheraga, 1964) groups by D_2O is weaker than that of nonpolar groups, while the solvation of the nonpolar segments is more intense. If the H_2O molecules in the cavities form cooperative systems, they should be easily displaced by D_2O, since D_2O has a stronger tendency to do so (Lobyshev and Kalinchenko, 1978).

Since it is this part of the hydrate shell which plays a special part in the dynamic model, its quantitative evaluation is highly important.

With this purpose in view we modified the method of Kuntz *et al.* (1969) for the determination of the degree of hydration of proteins. Our approach involved a comparative analysis of the integral intensities of water absorption bands in ordinary protein solutions containing various (0 – 30%) concentrations of D_2O, between -5°C and -40°C. Full allowance was made for the fact that the fraction of H_2O in the hydrate shell which became displaced by D_2O can no longer be detected by the PMR method. Its amount can be estimated from the weakening of the PMR signal of bound water. Absolute hydration values (H_2O) may be calculated from the PMR signal of an ethalon solution (4.5 M LiCl + 0.01 M Mn^{++}). At -35°C its integral intensity is 50 M of the non-frozen water.

The difference between the peak areas of protein solutions containing 3 - 5% D_2O and those containing more D_2O (up to 30%) usually disappears as soon as the temperature is decreased to between -45 and -50°C, i.e., when the water in the vicinity of polar and of charged segments has frozen. Judging by model experiments on porous zeolites, such water, if contained inside cavities, only freezes at between -80 and -100°C (para. 1.3). It is at exactly these temperatures that the difference between the signals of samples not containing D_2O and those containing 3-5% D_2O disappears. This confirms that isotopic substitution takes place, in the first instance, in the cavities. The number of protons ΔN^{H_2O} displaced from the protein on the addition of D_2O comprises the protons of the water displaced from the cavities Δn_c, on nonpolar surface zones Δn_n, and on polar Δn_{pg} and charged Δn_{cg} groups:

$$\Delta N^{H_2O} = \Delta n_c + \Delta n_n + (\Delta n_{pg} + \Delta n_{cg}) \qquad (4.1)$$

If the affinity to D_2O of the cavities is much stronger than that of the other sorption centers, we have, on adding small (up to 5%) amounts of D_2O:

$$\Delta N^{H_2O} \approx \Delta n_c \qquad (4.2)$$

Accordingly, the difference between the integral intensities of the frozen solutions of D_2O-free proteins and proteins containing low concentrations of D_2O is proportional to the amount of water contained in protein cavities. As the D_2O concentration increases, D_2O will replace H_2O on the surface of low-polar and nonpolar parts of the macromolecule as well. Results of studied carried out on model compounds quoted above seem to indicate that the water bound to the charged protein groups will be next to be displaced by DOH and DOD, as the concentration of dD_2O increases.

The experimental method just described may prove very useful in the studies of the distribution of water in the internal (nonpolar) and in the external (polar) areas of biological membranes.

We shall now discuss yet another new method for the study of protein hydration, this time at above-zero temperatures.

*A Method for the Study of Protein Hydration by Combined NME and Spin-Label Methods** *

We proposed a method for studying protein hydration based on the determination of the difference between the spin-spin relaxation rates of protons in protein solutions $(1/T_2^{S+M})$ and of the pure solvent $(1/T_2^S)$ if the correlation time of the protein τ_M and the viscosities of the solutions η/T are known.

The correlation time of the protein τ_M may be obtained by separate determinations of the correlation times of spin-labeled proteins and the labels to which they are bound (p. 54).

The method is based on the assumption that the half-width of the PMR line of the protein solution $(\Delta v = 1/\pi \cdot T_2)$ is determined solely by the mobility of two fractions of water in solution — the free water and the water bound to the protein — and by their respective fractions $(1 - q)$ and q. It is known (Abetsedarskaya *et al.*, 1977) that the interchange between these two fractions may proceed at a rate higher than 10^4 sec^{-1}. The rapidly exchanged protons of the protein may also make some small contribution to the value of q. This model of the two states of water is clearly a simplification. Owing to the thermal motion of the amino acid residues of the protein, no water molecules (n) are in fact rigidly affixed to the protein surface. For this reason n should be regarded as an effective parameter, which characterizes the hydration of the protein.

However, this two-state model is convenient in calculations. It is frequently employed for the determination of n and of the correlation times of the protein-bound water molecules by the method of pulsed NMR.

It is probable that n is actually the truest description of the water in protein cavities with a limited translational mobility.

According to the two-state model, the resultant value of T_2^{S+M} for protons in protein solution is expressed in terms of T_2^S of the pure water and T_2^M of water molecules rigidly bound to the protein, as follows:

$$\frac{1}{T_2^{S+M}} = \frac{1-q}{T_2^S} + \frac{q}{T_2^M} + \frac{4\pi^2 q(1-q)^2 \Delta^2}{K_{S \rightleftharpoons M}^{H_2O}} \tag{4.3}$$

---

* This study was carried out by the author in collaboration with A.S. Goryunov and G.A. Sukhanova.

where $K_{S=M}^{H_2O}$ is the rate constant of the exchange between the free and the bound water, and Δ is the difference between the chemical shifts of the free and the protein-bound water.

Since the exchange between the protons of the hydrate shell and the free water takes place rapidly $(K_{S=M}^{H_2O} > 10^4 \text{ sec}^{-1} \ll \Delta)$ (Abetsedarskaya et al., 1967), the last term in equation (4.3) may be neglected.

If the PMR instrument employed is operating under stationary conditions, the frequency of the Larmor precession $\omega_0 = 2\pi \cdot \nu_0 = 3.8 \cdot 10^8$ rad/sec. If the correlation time of rotational diffusion of the macromolecule under study is $\tau_M \geqslant 3 \cdot 10^{-8}$ sec, the following equation

$$\omega_0^2 \tau_M^2 \gg 1 \tag{4.4}$$

will be valid for water molecules rigidly bound to the protein.

If condition (4.4) is met for water which is rigidly bound to the protein, we have (Carrington and McLaughlin, 1970):

$$\frac{1}{T_2^M} = \frac{18}{40} \frac{q_N^4 \beta_N^4}{\hbar r^6} \cdot \tau_M = c\tau_M^{H_2O} \tag{4.5}$$

where g_n is the g-factor of the proton (= 5.585); β_N is the nuclear magneton, $\hbar = h/2\pi$ is Planck's constant, and r is the distance between the protons in a water molecule (1.58 Å). The numerical value of the dipole-dipole interaction constant is $c = 2.37 \cdot 10^{10}$ sec^{-1}.

In view of (4.5), equation (4.3) may be written as follows:

$$qc\tau_M^{H_2O} = \frac{1}{T_2^{S+M}} - \frac{1-q}{T_2^S} \tag{4.6}$$

If the protein concentration in solution is less than 5%, q will be much smaller than unity.

The usual practice is to employ correlation times reduced to standard conditions (water, 25°C, $\eta/T = 3 \cdot 10^{-5}$ p/K).

The value of $\tau_M^{H_2O}$, determined under nonstandard η/T conditions, is interconnected with the corresponding value $\tau_M^{H_2O\,(st)}$ under standard conditions by the relationship

$$\tau_M^{H_2O} = \tau_M^{H_2O\,(st)} \frac{\eta/T}{3 \times 10^{-5}} \tag{4.7}$$

In view of (4.7), and since $q \ll 1$, we may write down equation (4.6) as follows:

$$q\tau_M^{H_2O\,(st)} = 1.27 \times 10^{-15} \left(\frac{1}{T_2^{S+M}} - \frac{1}{T_2^S} \right) \cdot \frac{T}{\eta} \tag{4.8}$$

using the numerical value of C. In this equation $1/T_2^{S+M}$ is found from the half-width of the water band in the protein solution, while $1/T_2^S$ is found from the half-width of the PMR signal of the pure solvent at the same T/η value. T/η may be varied by varying the temperature or by varying the viscosity through the addition of glycerol or saccharose.

If, for these T/η values, the values of $\tau_M^{H_2O(st)}$ (which, by definition, are equal to the correlation times τ_M of the macromolecule) are known, q may be evaluated; conversely, if information is available on the number of H_2O molecules rigidly bound to the protein (such an information may be obtained, for example, by NMR relaxation method), it is possible to calculate τ_M.

If the fraction q of the water molecules bound to the protein, the molar concentration of water (= 55.5 M) and of the protein [C] are known, the number of H_2O-molecules rigidly sorbed on one protein molecule can be easily calculated: $n = q \cdot 55.5/[C]$.

The method just presented is simple and may be used in laboratories with an NMR spectrometer working under stationary conditions at their disposal. The fluorescence polarization or the spin-label methods are also indispensable in finding $\tau_M^{H_2O}$.

The potentialities of this method of studying the hydration of biopolymers and its variations will be exemplified in para. 5.2 for non-denaturation thermally induced transitions taking place in human serum albumin.

The chemical shift of the non-freezing water is about one-millionth part larger than that of bulk water, whereas the difference between the chemical shifts of the water vapor and water at room temperature is about six-millionths. Accordingly, the effect given by the hydrogen bonds which link water with the protein surface will be weaker than that given by hydrogen bonds in free water. This conclusion is in agreement with results obtained by IR spectroscopy. As in the case described in para. 3.1, the effects actually obtained may depend on the fluctuations of the protein surface and on the rapid exchange processes between the water molecules belonging to the different fractions of the hydrate hull, which diminish the average strength of the hydrogen bonds in water.

The amount, the mobility and the activation parameters of water as recorded by the NMR method proved to be highly sensitive to the conformational state of the molecules. It was noted in the case of a number of proteins that the amount of the "non-freezing" water increases by about 10% as a result of denaturation. This is accompanied by a 3-fold to 5-fold broadening at $-35°C$ (Kuntz and Brassfield, 1971). The NMR method may be used to calculate the thermodynamic parameters of water in any one of its states from the temperature dependence of its line width (Table 5). It is noteworthy that the enthalpy and entropy of mobility activation are negative for moist powders, while being positive in solutions. ΔG remains almost unchanged.

It was noted during a study of specific heat capacities of lysozyme, chymotrypsin and egg albumin, both in solution and dried, that the specific heat capacity of the water associated with proteins is higher by 30% than that of the free water. It would appear that this difference is due to the interaction of the water with the nonpolar groups of the protein molecules (Suurkuusk, 1974). X-ray diffraction analyses of a number of proteins showed that hydrophobic groups are encountered on the protein surface much more frequently than predicted by the Fisher model (Klotz, 1970). There is experimental evidence to show that, out of 13 different proteins chosen at random, at least 9 contained accessible nonpolar segments, which are capable of forming hydrophobic bonds with *n*-hexylamine- or phenylalanine-substituted agarose in the presence of 3.3 M NaCl (Hofstee, 1975).

TABLE 5. Activation parameters of "bound" water, free water and ice.
After Kunz and Kauzmann (1974).

System	τ, sec	ΔH, KJ/mole	ΔS, J/mole·°K	ΔG, KJ/mole
Lysozyme powder, gm H_2O/gm:				
0.34	$9.5 \cdot 10^{-10}$	-12.15	-113.1	21.8
0.54	$6.5 \cdot 10^{-10}$	-1.68	-75.4	21.0
Solution:				
bovine serum albumin frozen				
at $-25°C$	$3.5 \cdot 10^{-8}$	21.00	-16.8	26.0
Hemoglobin	$1.4 \cdot 10^{-10}$	30.60	41.9	18.0
Liquid water	$8.4 \cdot 10^{-12}$	16.80	22.6	10.0
Ice, $0°C$	$2 \cdot 10^{-5}$	53.20	40.2	42.3

It is seen, accordingly, that the interaction of water with nonpolar parts of biopolymer molecules is a natural phenomenon. Our task will be to gain an understanding of its mechanism and its functional significance. In this context, the reactions between water and protein clefts are of special importance.

4.2 Protein Cavities (Clefts) and their Interaction with Water

All proteins which have been examined by X-ray diffraction techniques showed at least one deep nonpolar cleft — the active site. In addition, many proteins were found to contain 'cracks' not carrying direct functional charges, and which may be filled with water. These include inter-domain cavities, and also the spaces between the sub-units of oligomeric proteins.

Studies of proteins by the X-ray diffraction method have the structure of the active site as their principal object, while the properties and dimensions of their other cavities are studied to a lesser extent or are disregarded altogether. We must therefore limit our discussion to the general structural features of the active sites. When the cavity has passed into the 'closed' hydrophobic state, it is shielded from exposure to water by reorientation of one or more amino acid residues, and also as a result of rotation of its constituent domains.

During the crystallization process, the equilibrium between the 'open' and 'closed' states of the cavity may undergo a shift towards the latter state, and this may be accompanied by displacement of the ordered water. Esipova *et al.* (1978), who studied the parameters of thermal transitions in crystalline pepsin as compared to pepsin solutions, found that protein denaturation in solution is accompanied by a heat effect, which is much larger than that observed in the crystal. However, the fusion temperature of the protein structure is the same in both cases. In the opinion of these workers, the heat effect of protein fusion is smaller in the crystal than in solution, because of the disruption in the structure of some of the ordered water, which interacts with the macromolecule in solution during crystallization. Takizava and Hayashi (1976) arrived at a similar result on studying the enthalpy of isothermal crystallization of lysozyme. It is not surprising, accordingly, that the quasi-crystalline water in protein cavities cannot be revealed by X-ray diffraction techniques in all cases.

The upper part of the cavity often contains polar amino acid residues. This is in fact the structure of the conical-shaped active site of carbonic anhydrase B, which is 1.2 nm deep, and also of its iso-enzyme — carbonic anhydrase C (Kannan *et al.*, 1975). A similar structure is displayed by the active site of thermolysin (Matthews *et al.*, 1974) and by carboxypeptidase (Hartsuck and Lipscomb, 1971; Lipscomb, 1974). The entry to the nonpolar active site of MOPC315 immunoglobulin, which is saturated with aromatic residues, also contains polar groups — aspartate, lysine, arginine and histidine residues (Dower *et al.*, 1977).

The cavity of the substrate-bonding domain of each of the four sub-units of D-glyceraldehyde-3-phosphate dehydrogenase is also not exclusively nonpolar. The part of the cavity near the protein surface is ringed by hydrophilic residues, except for Leu303 and Pro121 (Buehner *et al.*, 1974). The bottom of the active site of chymotrypsin is less polar than its mouth (Nedev *et al.*, 1974). Thus, the presence of a relatively polar 'belt' at the mouth of the active site is a fairly general structural feature of such sites.

The function of this polar belt may be to produce a kind of phase separation interface between the quasi-crystalline water in the nonpolar part of the 'open' cavity containing the active site and the free water of the solution. Moreover, polar groups may act as a special kind of crystallization centers, bringing about an orientation of the water molecules located on the fringe of the cavity, which is best adapted to its geometry.

Any amino acid exchanges taking place at the active site, unless equivalent in polarity and in volume, will affect the activity of the protein — not only on account of the disturbed stereo-specificity with respect to the ligand but, in our opinion, also owing to the distortion in the geometry of the cavity, with resulting disturbance of the quasi-crystalline packing of the water molecules contained in it. Since the cavities are mainly constituted by nonpolar amino acid residues, their exchange frequency must be much lower than that of the polar residues, since they are more likely to change the protein properties. The substitution of a polar by a non-polar amino acid and *vice versa* may be expected to occur more rarely than substituting a polar or a nonpolar group by a group of like polarity. This has been demonstrated by Volkenstain (1975).

These results have been interpreted as an indication of the importance of hydrophobic interactions in maintaining the protein structure in its native state. It may be expected, moreover, that the nonpolar residues forming the active site and other clefts exert a preserving effect, in addition to the effect of the residues forming the 'hydrophobic nucleus'.

It was shown by Lim and Ptitsyn (1970) that one nonpolar amino acid in myoglobin or in hemoglobin may be exchanged for another if their volumes are more or less equal. The volume of the hydrophobic nucleus as a whole, and the volume elements comprising three residues are more constant than the volumes of individual residues. As a result, mutually compensating amino acid exchanges take place even in small elements of the nonpolar matrix. This type of compensation may also ensure the preservation of the overall geometry of nonpolar protein cavities.

If these ideas are accepted, the dimensions and the volume of a cavity ought to correlate with the dimensions and the volume of an elementary ice lattice cell. Now the size of
an active site may vary between 0.6 - 0.8 nm and 2.0 nm or even more (Table 6), while
the dimensions of unit latice cell of crystalline ice, containing 4 H_2O molecules, are
$a = 0.448$ nm and $c = 0.731$ nm. Hence, the size of protein cavities ought to be large enough
to accommodate a sufficient number of water molecules, which then form a cooperative
system of hydrogen bonds with segments of cavities and with each other (Table 6). The
frequently occurring linear dimension of the cavities (1.5 nm) is a multiple of the largest
dimension of the elementary lattice cell of ice (0.731 nm). The bottom of the cavity may
contain a metal ion, whose charge and location during the functional activity of the
protein may determine the stability of the ordered water in the cavity; it may be bound
by OH^- or by H_2O as ligands.

TABLE 6. Dimensions of active site cavities and central protein cavities as determined by X-ray
diffraction method.

Protein	Size of cavity, nm	N^{H_2O}	Literature references
Active site of			
Fab (MOPC 603)	1.5 x 1.2 x 2.0	112	Davies *et al.*, 1975
Fab (NEW)	1.5 x 0.6 x 0.6	17	Amzel *et al.*, 1974
Cytochrome C	2.1 x 0.8 x 0.6	31	Dickerson and Timkovich, 1975
Carbonic anhydrase (depth)			
C	1.5	>20	Kannan *et al.*, 1971
B	1.2	20	Kannan *et al.*, 1975
Cavity of concavalin A	1.5 x 0.9 x 0.6	25	Hartman and Ainsworth, 1973
Central cavity of			
α-chymotrypsin dimer at pH 4 (diameter)	0.8 - 0.9	–	Tulinsky *et al.*, 1973
Hemoglobin	(0.8 - 1.0) x 2.0 x 5.0	250-312	Perutz *et al.*, 1960; Perutz, 1965, 1968

N^{H_2O} – the amount of H_2O in cavities (calculated by the author).

In a computerized X-ray diffraction study of the surface layer of α-chymotrypsin, a
tridimensional model of the enzyme, constituted by 0.3 nm cubic cells, was constructed
and analyzed; as a result, the popular view, according to which proteins consist of a
hydrophobic nucleus shielded from the solvent by a mobile hydrophilic shell was shown
to be erroneous (Nedev and Khurgin, 1975). The average atomic density of the nucleus of
α-chyotrypsin is practically the same as the density of the surface layer. This indicates
that the protein structure is exposed, and the protein thus has sufficient conformational
mobility to display enzymatic activity. A considerable part of the surface which is accessible to water is nonpolar. About 50 water molecules are able to penetrate into the inner
regions of the protein (Nedev *et al.*, 1974). The active site cavity, with a surface of 3
nm^2, contains only two polar side groups. In order to enable the formation of the specific
complex, water is displaced from this cavity by the substrate (Blow *et al.*, 1974).

Thus, Fisher's (1964) model, according to which all hydrophobic residues are situated inside the globule and are out of contact with water, is merely a gross approximation, which fails to fully reflect the nature of the interactions between the proteins and the hydrate shell.

Multiglobular allosteric proteins contain not only cracks in each subunit, but also a larger central cavity, common to all subunits (Blundell and Cutfield, 1973). Thus, for instance, the hemoglobin tetramer contains a 5.0 nm long void, passing right through the molecule. As the result of the bonding of the oxygen in hemoglobin, the geometry of the central cavity changes, as a result of which the volume of the tetramer is reduced by 8% (Perutz, 1970; Perutz and Ten Eyck, 1971).

The subunits of D-glyceraldehyde-3-phosphate dehydrogenase comprise two domains: the coenzyme NAD^+ (nicotinamide adenine dinucleotide) is localised on the first domain, in a special 'pocket'; the second domain, which fulfils catalytic and regulatory functions, contains an even larger nonpolar cavity (Buehner *et al.*, 1974). The reaction with NAD reduces the volume of the tetramer by 7.08% (Durchslag *et al.*, 1969) and the degree of hydration by 15.7% (Sloan and Velick, 1973). The reason for the latter decrease is, very probably, the displacement of some of the water from the central cavity into the external medium, as a result of the equilibrium shift towards the 'closed' state.

Large-sized voids were observed between the variable and the constant domains of the Fab subunits of immunoglobulins (Poljak, 1975), and between the C_H2 constant domains forming the Fc subunit, and between Fab and Fc-subunits (Huber, 1976). The presence of nonpolar cavities in proteins was also noted by a number of workers studying the solubilization of hydrocarbons by proteins. Volynskaya *et al.* (173) used this method to reveal four types of nonpolar zones in bovine serum albumin; the volume of one of these zones was 3 nm^3. The overall volume of nonpolar cavities in bovine serum albumin is 7 nm^3 or about 6% of the volume of the globule (Pchelin *et al.*, 1958). A study (Pchelin *et al.*, 1962, 1969) of the solubilization of benzene by pepsin revealed a sharp drop in catalytic activity, accompanied by a decrease in the specific optical rotation of the protein. Obviously, benzene is sorbed on the nonpolar active site. The solubilization of benzene by egg albumin enhances the resistance of the protein to thermal and acidic denaturation (Izmailova *et al.*, 1963), most probably owing to the constraints on the mobility of the cavities when these are filled with benzene.

It would appear that only native proteins contain nonpolar cavities. In solutions of 8M urea and 50% dimethylformamide benzene was not bound by serum albumin (Izmailova *et al.*, 1966). The capacity for sorbing considerable overall volumes of nonpolar molecules is also displayed by immunoglobulins, lysozyme (Volynskaya *et al.*, 1973) and gelatin (Pchelin *et al.*, 1958).

A computerized calculation of the densities of the internal zones of nine protein molecules from the atomic coordinates determined by X-ray diffraction method showed that the densities of the internal zones, which account for 20-50% of the total mass of the molecules, were 10-25% lower than their experimental values obtained during a study of aqueous protein solutions (Kauzmann *et al.*, 1974). The calculations were conducted as follows. All the atoms in the molecule were identified, and the location of the center of gravity of each protein was calculated with allowance for the mass of hydrogen. The masses of all atoms located in arbitrarily designated shells at a distance of 0-0.2, 0.2-0.4, 0.4-0.6 nm and farther away from the mass center were added together, and the sum was divided by the volume of the corresponding zone, thus yielding its average density. When the size of a shell had increased to a point where it began to include the molecules of the solvent, the density dropped sharply. The external radius of the shell, past which this decrease was noted, was denoted as R_{max}. The computed average density of the sphere ρ_{calc} with the radius R_{max} was then compared with its experimental density ρ_H. It was found, for all the proteins studied — lamprey myoglobin and hemoglobin, carboxy-peptidase, chymotrypsin, chymotrypsinogen, lysozyme, RNA, staphylococcal nuclease, subtilysin — that ρ_{calc} was lower than ρ_H by 0.10-0.36 gm/cm^3. Kaumann *et al.* considered three possible explanations for this discrepancy.

1. The true density of the protein is in fact the calculated value, but the protein is enveloped by a very dense 'water jacket'. However, it was calculated that a change in $\Delta\rho$ by electrostriction during the interaction between water and charged surface groups accounts for only a small part of the difference $\rho_H - \rho_{calc}$ between the densities.

2. The external zones of the proteins are denser than their internal zones, since the hydrophobic groups are preferentially localized on the inside, while the polar groups are preferentially localized on the outside of the molecule. The distribution of these atoms was analyzed, when it was found that the presence of the somewhat larger number of hydrophobic groups in the interior of the protein would not account for the discrepancy in the density values.

3. The internal regions of the protein tend to hold a large number of water molecules. Were the difference $\rho_H - \rho_{calc}$ to be explained by this effect, the quantity of H_2O in protein cavities would have to be quite large — between 96 molecules for RNA-ase and 304 molecules for chymotrypsinogen. If the contribution of the other two factors is allowed for, this amount becomes smaller, but would still remain the principal factor.

It is difficult to reveal the presence and tridimensional network of water molecules in fluctuating protein cavities by the X-ray diffraction method, since the diffraction pattern represents the averaged image of the conformational states of the protein. If this pattern is sharp, we may conclude that the equilibrium between the conformers has undergone a considerable shift towards one of them.

Numerous data indicate that the mobility of protein molecules in the crystalline state is restricted (Rapli, 1973). If, during crystallization, the cavities have become stabilized in the 'open' state, water may be detected if the resolution is not less than 2 nm. Thus, for instance, the conical-shaped active site of carbonic anhydrase, 1.5 nm deep, was found to contain eight water molecules, some of which were not directly bound to the functional groups in the cavity. The bonding of the competing inhibitor by this enzyme is accompanied by the displacement of water (Kannan *et al.*, 1971).

An ice-like water structure was noted in the active site of carboxypeptidase A. As in the preceding case, the bonding of the substrate results in the water being displaced from the cavity (Lipscomb, 1974). Water molecules were identified in the active site cleft of lysozyme (Phillips and Levitt, 1974). When a polysaccharide substrate is bound, at least some of them are displaced. The mobility of the water out of direct contact with the solvent, and constituting a structural element in the protein, is relatively low.

It would appear that H_2O molecules contained in myoglobin globules are of this type (Takano, 1977). During a study of the F_V-fragment of Bence-Jones protein, formed by the variable domains, three water molecules were detected in the small nonpolar cavity between the domains (Epp *et al.*, 1975). Thirteen internal water molecules were found in chymotrypsin (Birktoft and Blow, 1972), while ten such molecules were found in chymotrypsinogen. In the last-named case, it was shown that the internal water undergoes free exchange with the bulk water (Weber *et al.*, 1974).

The distribution of the ten water molecules in the internal cavity of carboxypeptidase A was studied in the greatest detail. Of these, two water molecules were found to be bound to the neighboring four molecules by hydrogen bonds (as in ice I), seven molecules formed three hydrogen bonds, while one was present as H_3O^+ and neutralized the negative charge on the Glu108 residue (Hartsuck and Lipscomb, 1971).

The experimentally determined functional dependence of the time of spin-spin relaxation, activation energy of water molecules, and electrical conductivity on the extent of hydration of gelatin, starch and cellulose indicated that the sorption of water on these biopolymers initially takes place in cavities with cluster formation; as the cavities fill up completely, the associated water molecules become stabilized and their mobility decreases (Gamayunov *et al.*, 1975). If, for instance, the volume of a protein cavity is $0.6 \times 1.0 \times 1.0$ nm = 0.6 nm^3, the cavity may contain 20 water molecules, since the volume of one H_2O molecule is about 0.03 nm^3.

The Monte Carlo method was used to calculate the equilibrium distribution of the water molecules in the active site of carbonic anhydrase B, the atomic coordinates of which had been determined by X-ray diffraction analysis (Clementi *et al.*, 1979). The locations of the cavity-forming residues were assumed to be fixed. Twenty water molecules were placed in the cavity, and their displacements, which decrease the free energy of the system, were modelled at 300°K. The radial distribution probability function of the Zn^{++}-water molecules (between 0.15 and 0.45 nm) contained four peaks, the first three corresponding to one H_2O molecule each, while the fourth peak corresponded to four water molecules. The calculations were performed for Glu105 in the ionized state (at high pH-values). The results indicated that, at a distance of a fraction of one nm away from Zn^{++}, the structure of the water in the cavity was highly ordered. It was calculated that the average energy of interaction between water and water in the cavity is about 10.5 KJ/mole, the corresponding value for the free water being about 35.6 KJ/mole. This situation may well correspond to the 'closed' state of the active site of crystallized protein, in which, in accordance with our theories, the cavity-water interaction is thermodynamically disfavored.

In a molecule of aspartate carbamoyltransferase (mol. mass 307,000) — which consists of six catalytic and six regulatory monomers — the width of the water-filled central cavity is about 2.5 nm, while its length is 5.0 nm. It is considered that the active sites of the enzyme are accessible to substrate only from the central cavity, but not from the external solution (Warren *et al.*, 1973). This cavity is connected with the external medium by means of six channels, about 1.5 nm in diameter. The bonding of the allosteric inhibitor of cytosine phosphate with the regulatory subunits results in conformational changes of the enzyme, which are such that the stationary concentration of the substrate in the cen-

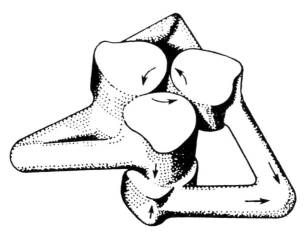

Figure 15. A model of the quaternary structure of aspartate carbamoyl transferase. After Cohlberg *et al.* (1972).

tral cavity decreases. This may be caused by a decrease in the rate of diffusion of the substrate through the channels or by the deformation (contraction) of a large internal cavity (Fig. 15). Each catalytic monomer is shown as a plane structural unit. The regulatory monomers are cylindrical-shaped, and associate in pairs at a sharp angle to form dimers.

* * *

The results outlined above justify the following conclusions:

1. Besides the active site cleft, protein molecules also contain other cavities which may be filled with water.

2. The structures of the active clefts in many proteins have certain features in common: the enhanced nonpolarity of the 'bottom' of the cleft, and a much more highly polar 'belt' at its 'mouth'. The charged groups inside the cavities may be neutralized by H_3O^+ or by OH^-. The water content of such clefts is usually not less than 20 water molecules.

3. Ligands, substrates and competing inhibitors are capable of displacing water from the active cleft. Internal protein cavities containing essentially unchanged water are also present.

4.3 The Dynamics of Water-Protein Interaction

In view of the types of intra-protein mobilities described in Chapter 3, and of the main postulates of the dynamic model of behavior of proteins in water, it is reasonable to expect that the mobility of the water molecules caused by biopolymers will be mainly determined by four processes: (i) changes of state taking place in water molecules in the course of their displacement and sorption in protein cavities, when the latter fluctuate at the rates of $10^4 - 10^7$ sec^{-1}; (ii) reorientations of water molecules caused by the rapidly ($10^9 - 10^{11}$ sec^{-1}) rotating polar side chains of the proteins; (iii) exchange between the water fractions participating in the fluctuations of the cavity and solvating the side groups, and (iv) the rotation of the macromolecule as a whole, which generates a component of motion in the occluded and environmental water molecules, whose correlation time depends on its effective Stokes radius and the viscosity of the medium.

According to X-ray diffraction data, water is contained in the interior of a large number of proteins. Thus, for instance, 10 water molecules were found in carboxypeptidase A, 13 molecules in α-chymotrypsin (Birktoft and Blow, 1972) and 10 molecules in chymotrypsinogen. This water is located in distinctly nonpolar environment and, judging by crystallographic results, forms a fairly stable system.

Water of this type (11-12 molecules) was probably found in lysozyme by the method of advanced drying *in vacuo* (Rao and Bryan, 1975). Papain also contains 31 specifically bound water molecules, most of which form internal hydrogen bonds (Berendsen, 1972).

The question which immediately arises is whether this water constitutes an element of the rigid molecular matrix, similarly to the non-exchangeable peptidic hydrogen or is capable of exchanging with the water in the protein environment. In the latter case it might be expected that the internal water would enhance the conformational lability of the protein in solution.

Weber *et al.* (1974) employed an original method for the determination of exchange capacity between the internal and external water in a solution of chymotrypsinogen. The lyophilic protein was dissolved in $H_2^{18}O$, was fully denatured and then fully restored to its native state, after which it could be expected that 10 molecules of $H_2^{18}O$ would be found in the interior of chymotrypsinogen. The solution was then passed through a G-25 Sephadex column, which had been equilibrated by a buffer prepared with ordinary water. The eluted protein was immediately frozen, lyophilized and its $H_2^{18}O$ content determined. It was found that during its passage through the column (not more than 20 minutes), the internal $H_2^{18}O$ had undergone a complete exchange with the ordinary external water. The authors noted, however, that this effect need not necessarily occur in crystals as well, since the conformational mobility of the proteins is restricted in such a case. It is probable that the water contained in such intra-protein cavities, which is out of contact with the outside environment, will be less mobile than the water contained in cavities or 'clefts'.

A promising technique for studying the nature of protein-water interaction is the so-called 'molecular dynamics' method. Information obtained by X-ray diffraction analysis about the distribution pattern of the atoms in the protein is fed into a computer, random coordinates are assigned to the freely moving water molecules, and the energy of their interaction with the protein is calculated for every moment. Finally, the computer generates a pattern corresponding to the minimum free energy of this interaction and to the most probable distribution of the water molecules. The study of the pattern thus obtained showed that some molecules sorbed on the protein are fairly stable, while the others may execute a limited amount of motion, and the remainder behave as fully free water in the protein crystal.

Mrevlishvili *et al.* (1975) used the NMR method in combination with the calorimetric method in his study of the state of water in collagen fibers. As a result, he proposed the following model for the hydration of collagen. The hydration hull comprises two fractions: an internal layer of highly ordered water, which is oriented in the form of H_2O chains along the fiber axis (the correlation time of this fraction was 10^{-8} sec, while its amount was 30% of the dry mass of the protein), and a disordered external hydrate layer (20% of the mass of dry protein). The NMR data indicated that a rapid exchange took place between these fractions.

These authors subsequently noted (Mrevlishvili and Sharimanov, 1978) that a model which postulates the presence of only two components in collagen solutions gives rise to contradictions. Thus, according to NMR data, the decrease in the extent of hydration, resulting from the conversion of the macromolecules to the coiled form, was from 0.13 to 0.04 gm H_2O per 1 gram of protein, while according to the calorimetric data it increased from 0.48 to 0.61 gm H_2O per 1 gram of protein.

Analysis of the experimental data obtained by different methods indicates that the water in collagen solutions contains three different components, with different correlation times: $10^{-1} - 10^{-11}$, $10^{-9} - 10^{-8}$ and $10^{-7} - 10^{-6}$ sec (Cooke and Kuntz, 1974), which have been attributed to the free water, to the bound water q_1 and to the specifically bound water q_2 respectively. If there is a rapid exchange between these components, the observed relaxation rates may be expressed as follows:

$$1/T_1 = (1 - q_1 - q_2)/T_1^{(0)} + q_1/T_1^{(1)} + q_2/T_1^{(2)}, \qquad (4.9)$$

$$1/T_2 = (1 - q_1 - q_2)/T_2^{(0)} + q_1/T_2^{(1)} + q_2/T_2^{(2)}, \qquad (4.10)$$

where $(1 - q_1 - q_2), q_1$ and q_2 are the respective fractions of the three water components, and $T_1^{(0)}$, $T_2^{(0)}$, $T_1^{(1)}$, $T_2^{(1)}$ and $T_1^{(2)}$, $T_2^{(2)}$ are the respective relaxation times.

Using these equations together with the experimental data for the helix-to-coil transition, Mrevlishvili and Sharimanov (1978) concluded that this transition is accompanied by a simultaneous decrease in the amounts of the water in the outer hydrate hull of the triple helix and of the specifically bound water. However, these workers overlooked a possible relationship between the observed water dynamics and the dynamics of the central cavities in the collagen molecule.

In the case of DNA, in which water also seems to be an important structural component, it has been shown (Malenkov et al., 1975) that the constant of the equilibrium between the more compact A-form and the less compact B-form varies with the extent of hydration or with the relative vapor pressure. X-ray diffraction data indicate that at relative humidities below 75-80% the A-form predominates, while at higher humidity values it is the B-form which becomes stabilized. The $A \rightleftharpoons B$ equilibrium is also affected by various agents, such as alcohols and various ions, which produce a reversible shift of this equilibrium to the left (for a possible mechanism of such effects see sec. 5.1). Under physiological conditions, the $A \rightleftharpoons B$ equilibrium in DNA shows a strong shift to the right (Ivanov et al., 1975). Similar considerations probably apply also to collagen.

By taking advantage of the fact that the relaxation rate of water protons is enhanced by paramagnetic metal ions, Fabry et al. (1970) showed that coordinated H_2O or OH^- are present in the first coordination sphere of the active site of carbonic anhydrase. Since the activity of this enzyme is preserved when the Zn(II) in its active site is replaced by

Co(II), it follows that Co(II) derivatives of this enzyme are suitable for such investigations.

Co(II) carbonic anhydrase considerably accelerates the relaxation rate T_1^{-1} of the water protons. On the other hand, this rate is considerably reduced when water-displacing ligands such as azide or ethoxy-ol-amide are added. This confirms that the effect is due to the interaction between water and the paramagnetic center. The calculated distance between proton and cobalt is 0.22 - 0.25 nm, while the lifetime of the proton in this state is 10^{-5} sec. The shape of the curves showing the specific relaxation rate of bovine and human (Co(II) carbonic anhydrase B (Fig. 16) as a function of the pH seems to reflect the transition of the cavities from one state to another.

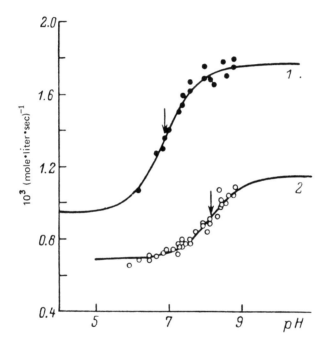

Figure 16. Specific relaxation rates of bovine (1) and human (2) Co(II)-carbonic anhydrase B as a function of the pH. After Fabry *et al.* (1970). Arrows indicate the pK-values. 0.1 M tris-HCl, 0.1 M sulfate. 25°C.

Changes in the absorption spectra between 450 and 700 nm and in circular dichroism spectra (Linskog and Nyman, 1964; Coleman, 1968) in this pH range also indicate that a rearrangement of the coordination sphere does in fact take place. Not only conformational rearrangements, which may represent an equilibrium shift between the 'open' and 'closed' states of the active site in either direction, but also the dissociation of the coordinated water molecule, may be the explanation for the observed relationships. Ward (1969, 1970), who used ^{35}Cl as the NMR probe, showed that at high pH values Cl⁻ competes with OH⁻ for a place in the coordination sphere of the metal.

EPR spectra of a spin-labeled competing inhibitor of carbonic anhydrase — a sulfona-mide derivative — sometimes reflect the ability of the active site to exist in two discrete states (Fig. 17). In one of these states there are larger steric hindrances to the rotation of the label, while in the other such hindrances are weaker (Wee *et al.*, 1976). These may possibly represent the 'closed' and the 'open' states of the cavity containing the active site.

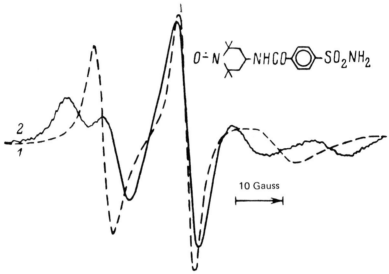

Figure 17. EPR spectra of spin-labeled sulfonamide inhibitor (20 μM) in the active site of human carbonic anhydrase (262 μM) I(1) and II(2). After Wee *et al.* (1976). 0.1 M phosphate buffer, pH 7.4.

If this interpretation is correct, it follows from the EPR spectra that the equilibrium between the 'open' (b) and the 'closed' (a) states of the active site of the highly active human carbonic anhydrase II, containing a spin-labeled inhibitor, is strongly shifted towards the 'closed' state, while in the case of the low-active carbonic anhydrase I it is shifted towards the 'open' state. It may be recalled in this connection that according to X-ray diffraction data the active site of human carbonic anhydrase C contains a highly structurized group of water molecules (Liljas *et al.*, 1972). It seems that in the case of low-active enzymes, the active site becomes stabilized in the state which is 'open' to water. Another explanation is the rapid (Va = bz 10^8 sec^{-1}) exchange between two states of its active site.

A study of protein solutions by the method of dielectric dispersion revealed two absorption zones of high-frequency EM signals: $\simeq 10^8$ and 10^{11} sec^{-1}. The lower frequency is attributed to the contribution of the 'bound' water, while the higher one is attributed to the 'bulk' water. The analysis of these data is a very complex task, since it is difficult to allow for the mobility of the side chains in the proteins. Lysozyme powders with different degrees of hydration were studied between 10^7 and 10^{11} sec^{-1}, when it was found that for water contents below 0.3 gm per 1 gm of protein the value of the dispersion was $\simeq 1.7 \cdot 10^8$ sec^{-1}; after the water content had increased to 0.5 gm/gm, the frequency increased somewhat (to $2.5 \cdot 10^8$ sec^{-1}) and a new dispersion ($\simeq 10^{10}$ sec^{-1}) appeared (Harvey and Hoekstra, 1972). The dielectric constant showed a linear increase as the water content was raised to 0.3 gm/gm, after which it decreased rapidly as the humidity content of the sample increased beyond this value. A simple model was proposed to explain these results. According to the model, the protein contained two layers of hydrate water: in the first layer ($\simeq 0.3$ gm/gm) a typical rotation rate of water molecules was $2 \cdot 10^8$ - $5 \cdot 10^8$ sec^{-1}, while in the second layer the mobility of the water was lower by about two orders than in ordinary water ($\simeq 10^{10}$ sec^{-1}). The thickness of the hydrate hulls is probably variable, since the protein contains various types of sorption centers, with varying water-bonding energies.

Randall *et al.* (1978) obtained interesting results in their study of the dynamics of the hydrate hull of a fully deuterated protein of C-phycocyanin (molecular mass 190,000) by the method of quasi-elastic neutron scattering. The analysis of the data obtained for a film-like preparation with a relative humidity of about 90% at 24°C revealed that water molecules near the protein surface execute jump-like shifts. The time between successive jumps was $\tau_0 = 5$ - 10 nsec, while the magnitude of a jump was $a = 0.55$ - 0.60 nm. A second protein preparation, with a lower hydration value, which had been prepared by drying the solution of the protein in water, yielded the values $\tau_0 = 15 \cdot 30$ nsec, and $a = 0.7$–0.9 nm. These results were explained as being due to the jumps of the water molecules between the primary sorption centers on the surface of the molecule. However, the structural mobility of the protein, including the fluctuations of its cavities, may also have contributed. The interpretation of these results could be clarified to some extent by quantitative evaluations of the fraction of the molecules executing the jumps at different hydration values.

The most detailed information on the properties of the solvate shells of proteins was obtained by the NMR method, especially in the pulse mode, in which the spin-echo could be determined. The study of water mobility is based on certain relationships which exist between the experimental values of relaxation parameters and the molecular motion of protons (cf. sec. 3.4). Most NMR studies of hydration of proteins were carried out in a constant magnetic field, i.e., at a single value of the resonance frequency ω_0. A more accurate method for the determination of τ is based on the experimental determination of T_1 and T_2 as a function of ω_0. It was thus found that 13 - 20 water molecules are relatively firmly bound to apotransferase (Koenig *et al.*, 1975). In another series of studies, these workers compared the relaxation rates of ^{1}H, ^{2}H and ^{17}O in aqueous solutions of lysozyme and hemocyanin. They noted that the dispersion in the relaxation rates of all these solution components, normalized with respect to their relaxation rates in pure water, is practically identical. The calculations were based on a simplified model of two states of water in solution — with the water molecules firmly bound to the proteins, and in the free state.

The number n of the bound water molecules was found to be 3 for lysozyme and 850 for hemocyanin. If it is assumed that these water molecules are not bound to the protein in a perfectly rigid manner, the n-values will be somewhat larger. The average lifetime τ_M of water molecules on lysozyme at 22°C is

$$7.8 \cdot 10^{-7} \sec \gtrsim \tau_{M} \geqslant 4.2 \cdot 10^{-8} \sec. \left.\rule{0pt}{14pt}\right\} \qquad (4.11a)$$

and on hemocyanin at 25°C

$$1.5 \cdot 10^{-6} \sec \gtrsim \tau_{M} \geqslant 1.6 \cdot 10^{-7} \sec. \qquad (4.11b)$$

Judging from the data available on the dielectric relaxation of protein solutions, the remaining water molecules which react with the charged and the polar groups on the protein surface (150 for lysozyme and $6.5 \cdot 10^{4}$ for hemocynain), display dynamic properties much like those of free water. However, the correlation time of their rotary motion ($10^{-9} - 10^{-10}$ sec) are somewhat longer than those of free water. These results are readily accounted for by the dynamic model; the hemocyanin molecule is several times larger than the lysozyme molecule, and contains many cavities which keep fluctuating between the states which are 'open' and 'closed' to water. An 'open' cavity, filled with ordered water, rotates together with the protein as one whole. The approximate number of such water molecules may be characterised by n.

The application of the two-state model of water to NMR studies of a number of protein solutions yielded the following values for the correlation times of the bound water molecules: egg albumin $1.03 \cdot 10^{-8}$; egg protein $2.0 \cdot 10^{-8}$; RNA-ase and bovine serum albumin 10^{-8} to $3 \cdot 10^{-8}$ (Abetsedarskaya *et al.*, 1967). These values are close to the specific correlation times of the respective proteins, but the authors obtained similar τ_{M}-values in solutions of proteins with strongly differing molecular masses. It may be assumed, therefore, that the mobility of the protein-bound water is not solely due to the rotation of the molecule as a whole, but also possibly to the fluctuations of its cavities.

Aksenov (1979) employed the spin-echo method in his study of the variation of $1/T_2$ of the water protons in protein solutions with the viscosity, pH and ionic strength. The recorded scatter of activation energy values for the variations of T_1 and T_2 of the water protons was between 0 and 25 KJ/mole for the various samples. The data obtained for the water protons in protein solutions were compared with the dynamic parameters found for the proteins themselves. A correlation was found between the values of T_2 of the protons in water molecules in contact with the protein surface, and those of the protons in macromolecules. It was concluded that bound water in biopolymer solutions may exist in several different states.

The spin-echo method was also used by Linström and Koenig (1974) and by Thompson *et al.* (1975) in their studies of the spin-lattice and spin-spin relaxation rates of protons in erythrocytes which contained hemoglobin in its oxy- and deoxy-forms. They found that the bulk (98%) of the erythrocyte water (Type I) had a correlation time of about $3 \cdot 10^{-11}$ sec, corresponding to the correlation time of free water; 1.3 - 1.5% of the water (Type 2) had a correlation time of $2 \cdot 10^{-9} - 4 \cdot 10^{-9}$, and probably represented the hydrate hull on the surface of hemoglobin molecules. Type 3 water, which accounted for only 0.2% of the total amount, had a correlation time of 10^{-7} sec or longer. When the hemoglobin passes from the oxygenated to the deoxygenated state, the correlation time of these protons becomes longer, probably owing to the stabilization of the central cavity in hemoglobin molecules in the 'open' state, and to the increase in the lifetime of the water cluster contained in it.

Hsi and Bryant (1975b) and Hsi *et al.* (1976) studied the NMR relaxation of the protons in monoclinic lysozyme crystals at several temperatures; these workers also noted three types of water — one free and two bound.

At temperatures above the melting point of water in the crystal, the fading curve of the spin echo had two relaxation times only. Some of the protons, corresponding to 0.17 gm H_2O per gram of protein, i.e., to 132 water molecules to 1 protein molecule, represented the less mobile fraction, with a relaxation time of $T_1 = 28$ msec, while the remaining water protons showed a mobility similar to those in free water ($T_1 = 240$ msec) (fraction III).

At temperatures below 273K for the bound unfrozen water also two relaxation times were observed. At 230K the more mobile fraction I constitutes 0.20 gm H_2O per 1 gm protein (161 H_2O molecules per one protein molecule) and has a T_1 of 68 msec, while the slow fraction II (0.03 gm H_2O per gram of protein or 24 H_2O molecules per one protein molecule) has a T_1 of 2 msec. Heat-denatured or urea-denatured samples display only one protein-relaxation process under similar conditions. It is evident that the observed relaxation times of water are directly related to the mobility of the protein matrix and side groups and not to the rotation of the macromolecule as a whole, since such rotations are restricted in protein crystals.

An analysis of the $T_1 : T_2$ ratios, performed for all three water fractions, showed that there is a rapid exchange between fraction II and fraction III, while the bulk of the solvate hull (fraction I) is incapable of participating in such exchanges.

Especially noteworthy is the comparison of the variations of T_1 with the temperature of fractions I and II in frozen specimens. It is important to note that the T_1 of fraction I, which constitutes 83% of the measured amount of the water, does not alter significantly as the temperature is decreased from 0 to -60°C, whereas the T_1 of fraction II (17%) decreases up to -40°C, and remains constant thereafter. The decrease in T_1 indicates an increase in the correlation time of water molecules.

The different course of temperature variations of T_1(I) and T_1(II) is caused by the different nature of the water sorption sites of these fractions. When applying the dynamic protein model, it is natural to identify the mobile fraction I with the water bound to the polar and the charged protein groups, while the low-mobile fraction II is identified as the water contained in protein cavities, which probably becomes stabilized in the 'open' state when the water environment of the protein has frozen. The correlation time of water molecules in fraction I at 230°K is $3.3 \cdot 10^{-9}$ sec. Water II, contained in protein cavities, also preserves some of its mobility under these conditions.

The existence of three types of water protons was confirmed by studies of powdered proteins and frozen protein solutions (Hsi and Bryant, 1975a; Hsi *et al.*, 1975). Thus, for instance, a study of NMR relaxation of frozen lysozyme solutions yielded results very similar to those quoted above for the crystals of this protein. One of the two protein-bound water fractions (the faster fraction I) corresponds to 0.28 gm H_2O per 1 gm protein, i.e., to 223 H_2O molecules per 1 protein molecule, while the slow fraction II corresponds to 0.06 gm H_2O per 1 gm protein or to 52 H_2O molecules per 1 protein molecule.

Experiments conducted on model compounds (Chapter I) showed that at any temperature below 0°C the water bound to the polar protein groups is more mobile than the water bound to the nonpolar protein groups.

It is seen that in view of the distribution of the correlation times of water molecules in protein solutions, frozen solutions, crystals and powders, we may distinguish between two fractions of water molecules which are directly bound to the protein: one of them (80-90% of the total amount) has a correlation time of 10^{-9} - 10^{-10} sec when extrapolated to above-zero temperatures, while the other fraction (10-20%) has a correlation time of 10^{-6} - 10^{-8} sec. The former fraction seems to be directly related to the mobility of the side groups on the surface of the proteins, while the latter, less mobile fraction may be identified as the water in the protein cavities.

It has been shown, with the aid of thermodynamics of small systems, that if the properties of the solvent and the conformational state of the macromolecules are interdependent, fluctuations of major volume fractions of the solution may result (Syrnikov, 1974). The solution, which is a large system, is considered as an assembly of small systems which are constituted by macromolecules immersed in the solvent. A two-state model of the protein and of the solvent is used.

The thermodynamic potential of a large system is given by the equation:

$$\Phi_6 = N_s \left[c\mu_1 + (1-c)\,\mu_2 \right] + wN \left[b\mu_A + (1-b)\,\mu_B \right] + wkT \left[b \ln b + \right.$$
$$\left. + (1-b) \ln (1-b) \right] + N_s kT \left[c \ln c + (1-c) \ln (1-c) \right] + wkT \ln x, \quad (4.12)$$

where N_s is the total number of solvent molecules, c is the concentration of the solvent in its first state, μ_1 and μ_2 are the chemical potentials of the first and second states of the solvent respectively, w is the total number of the macromolecules, N is the number of links in the macromolecular chain, b is the relative concentration of the macromolecules in state A, μ_A and μ_B are the chemical potentials per link of the macromolecules in states A and B respectively, and $x = w/N$ is the concentration of the macromolecules in solution.

The magnitude b may be regarded as the internal parameter of a large system. The mean square deviation for such parameters is given (Myunster, 1962) by the equation:

$$\overline{(b - \bar{b})^2} = kT \left(\partial^2 \Phi_6 / \partial b^2 \right)^{-1}. \quad (4.13)$$

Since it is considered that the A- and B-states of the macromolecule affect the properties of water in different manners, i.e., that the chemical potentials of the solvent are functions of parameter b, we obtain

$$\overline{(b - \bar{b})^2} = \frac{kT}{N_s \left[c \dfrac{\partial^2 \mu_1}{\partial b^2} + (1-c) \dfrac{\partial^2 \mu_2}{\partial b^2} \right] + wkT \dfrac{1}{b\,(1-b)}}. \quad (4.14)$$

after the second derivative of Φ_b has been substituted in (4.5).

The second term in the first part of the denominator is much larger than unity. Even if the first term is positive, the changes in the concentrations of the macromolecules in states A and B will be small. However, if at certain biopolymer concentrations $(x_{cr} = w/N_s)$ the conditions become such that

$$c\, \frac{\partial^2 \mu_1}{\partial b^2} + (1-c)\, \frac{\partial^2 \mu_2'}{\partial b^2} = -x_{cr} kT\, \frac{1}{b\,(1-b)}\,, \qquad (4.15)$$

the denominator of equation (4.6) becomes zero and exceedingly large fluctuations in the concentrations of A- and B-states may be expected.

From the point of view of the dynamic model, which is propounded in this book, this result may be interpreted as a possible fluctuation of the constant $K_{A \rightleftharpoons B}$ between the A- and B-conformers of the protein, where the A-conformer represents a protein with 'closed' nonpolar cavities, while the cavities in the B-conformer are 'open'. Since the entropy, enthalpy and heat capacity, and hence also the chemical potential of water undergo changes during its sorption or desorption by the fluctuating protein cavities, all the pre-conditions are met.

We may conclude, accordingly, that the information now available on the mobility and the distribution of the various water fractions in protein solutions, and on the structure and conformational mobility of the macromolecules themselves, does not contradict the dynamic model of behavior of proteins in water. We shall show in the chapters which will follow that the changes in the properties of proteins produced by specific and non-specific external agents may also be interpreted in terms of this model.

Chapter 5

Effect of Perturbing Agents on the Dynamic Properties of Proteins

We shall now consider, from the aspect of the dynamic model of behavior of proteins in water, the effect on proteins of agents which, when present in small doses, perturb the system protein+water as one whole without, however, impairing its native properties. We shall assume, for the sake of simplicity, that all the protein clefts execute the A ⇌ B transitions simultaneously, since the corresponding changes in protein conformation may then be regarded as fluctuations between the A- and B-conformers. This assumption does not affect the validity of the final results.

According to Käiväräinen (1977, 1978b), the total free energies of A- and B-conformers may be represented as:

$$G_A = G_0 + G_A^{B+H_2O},$$
(5.1)

$$G_B = G_0 + G_B^{B+H_2O} + P(x)\left[G_B^{H_2O} - G_B^{H_2O.(cl)}\right],$$
(5.2)

where G_0 is that part of the free energy of the protein and of its hydrate shell which is independent of the state of its cavities; $G_A^{B+H_2O}$ and $G_B^{B+H_2O}$ are the parts of the free energies of the protein with its hydrate shell in states A and B respectively, which are complementary to G_0; $G_B^{H_2O} - G_B^{H_2O(cl)} \equiv G^{cl}$ is the energy of clusterophilic interactions, which represents the difference between the free energies of the liquid water in protein cavities $G_B^{H_2O}$ and of the water in quasi-crystalline state $G_B^{H_2O(cl)}$; and $P(x)$ is the probability of impairment of the ordered water structure in the cavities, which is a function of the concentration and activity of the perturbing agent x.

In view of (5.1) and (5.2), the constant of the equilibrium between A- and B-conformers may be expressed as follows:

$$K_{A \rightleftharpoons B} = \exp\left\{-\frac{G_A^{B+H_2O} - \left[G_B^{B+H_2O} + P(x)G^{cl}\right]}{RT}\right\}.$$
(5.3)

Let us consider the physical mechanism responsible for the increase of $K_{A \rightleftharpoons B}$ under the effect of perturbing agents which follows from the dynamic model of the protein. According to this model, the principal difference between the A- and B-conformers is the presence of ordered water in the cavities of the B-conformer. Accordingly, the reason for the shift in the equilibrium A⇌B must be sought in the presence of agents which perturb the water participating in A ⇌ B transitions, and its interaction with the nonpolar cavities of the B-conformer, i.e., its effect on $P(x)$.

We shall assume, to begin with, that interactions between the nonpolar protein clefts and the quasi-crystalline water are thermodynamically favored over the interactions between these cavities and the dissociated cluster. In other words, $G_B^{H_2O} > G_B^{H_2O(cl)}$ and $G^{cl} > 0$, i.e., the clusterophilic interaction contributes to the stabilization of the B-conformer.

The conditions which must be met for the interaction between water and the nonpolar protein zones to be thermodynamically most favored: $- P(x) = 0$; $G_B = min$ — are highly exacting, not only as regards the mutual orientations of water molecules in the clusters, but also as concerns the geometry, stability and structure of nonpolar cavities. Clusters enclosed in nonpolar cavities are highly cooperative systems. Accordingly, the appearance of even a small number of defects in the quasi-crystalline cluster lattice, brought about by the presence of the perturbing agent, is sufficient for the clusters to increase by G^{cl} (Fig. 18). When the effect of the perturbing factor has disappeared, the quasi-crystalline structure of the water in the nonpolar cavities may be restored if the jump from G_B to G^{cl} is not accompanied by a thermal fluctuation of the protein with an energy larger than $G_{B \to A} - G^{cl}$. It is clear that the probability of such a fluctuation is higher than that of a fluctuation with the energy $G_{B \to A}$.

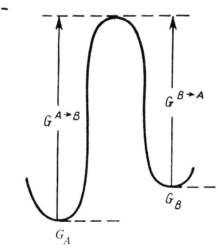

Figure 18. Profile of the free energy of a protein capable of existing as A- and B- conformers. After Käiväräinen, (1978b). G_A and G_B are the free energies of the respective conformers.

Perturbing agents may initiate a phase transition of water in two ways: by acting on the structure of the water cluster directly, and by acting on the conformation of the protein including that of protein cavities. The intensification of the activity of the external factors (increase in ion concentration or in the temperature) increases the probability of occurrence $P(x)$ of perturbances which are large enough to 'melt' the clusters and hence to increase the value of $K_{A \rightleftharpoons B}$ (cf. eq. (5.3)). In our opinion, such an increase of $K_{A \rightleftharpoons B}$ is the principal factor operative in pre-denaturation process. According to Blumenfeld (1977), van't Hoff's equation cannot be used to calculate the enthalpies of structural non-denaturation transitions in proteins. Our own model leads to a similar conclusion. In fact, van't Hoff's equation is usually written in the form:

$$K = \exp (\Delta S / R) \exp (-\Delta H / RT), \tag{5.4}$$

hence

If

$$\frac{dK}{dT} = K \left(\frac{\Delta H}{RT^2} - \frac{1}{RT} \frac{\partial \Delta H}{\partial T} + \frac{1}{R} \frac{\partial \Delta S}{\partial T} \right). \tag{5.5}$$

$$\frac{1}{RT} \frac{\partial \Delta H}{\partial T} = \frac{1}{R} \frac{\partial \Delta S}{\partial T}, \tag{5.6}$$

equation (5.5) assumes the form which is usually employed for the calculation of ΔH:

$$dK/dT = K \Delta H / RT^2, \tag{5.7}$$

or

$$d \ln K/d (1/T) = -\Delta H/R. \tag{5.8}$$

Since

$$\Delta G = \Delta H - T \Delta S, \tag{5.9}$$

the condition (5.6) means that

$$\frac{\partial \Delta G}{\partial T} = \frac{\partial \Delta H}{\partial T} - T \frac{\partial \Delta S}{\partial T} = 0, \tag{5.10}$$

i.e., $\Delta G(T) = $ const.

According to the views advanced in this book (eq. (5.3))

$$\Delta G = G_A^{B+H_2O} - \left[G_B^{B+H_2O} + P (T, x) G^{cl} \right]; \tag{5.11}$$

so that

$$\frac{\partial \Delta G}{\partial T} = \frac{\partial G_A^{B+H_2O}}{\partial T} - \frac{\partial G_B^{B+H_2O}}{\partial T} - P (T, x) \frac{\partial G^{cl}}{\partial T} - G^{cl} \frac{\partial P (T, x)}{\partial T}. \tag{5.12}$$

If we assume that in the pre-denaturation range the values of $G_A^{B+H_2O}$, $G_B^{B+H_2O}$ and G^{cl} are relatively independent of the temperature, we have:

$$\partial \Delta G / \partial T = -G^{cl} [\partial P (T, x) / \partial T]. \tag{5.13}$$

It follows from the model that, in the general case, $\partial P(T,x)/\partial T > 0$, i.e., condition (5.10) is not met and van't Hoff's equation is, strictly speaking, not applicable. However, in certain cases there is agreement between the values of ΔH of the denaturation transition as directly determined by calorimetry, and as found according to van't Hoff (Privalov, 1975; 1979; Privalov, 1979; 1981). This may be explained by postulating that at the beginning of the denaturation the probability of perturbance of the ordered structure of the water in protein cavities attains its maximum value and remains unchanged thereafter, i.e., $\partial P(T,x)/\partial T \cong 0$. It also means that the thermal denaturation of globular proteins may be regarded, to a good approximation, as a transition between two states – the native and the denatured state.

It may be readily shown how the shortening of the lifetime of the B-state under the effect of disturbing agents, responsible for the increase in $P(T,x)$, may affect the time-averaged free energy $\bar{G}$ of the protein, i.e., its stability.

The complete, time-averaged free energy of the protein which is fluctuating between two conformers A and B may be written as:

$$\bar{G} = G_0 + \bar{G}_{AB},\qquad(5.14)$$

where G_0 is that part of the free energy of the protein which is independent of the fluctuations of its nonpolar clefts, while $\bar{G}_{AB}$ is the free energy of the protein which varies with the state of its cavities.

Let the protein spend a fraction f_A of unit time as the A-conformer; it will then spend a fraction $1 - f_A$ of unit time as the B-conformer. We are assuming that the lifetimes of protein cavities in the 'closed' A-state and in the 'open' B-state are much longer than their lifetimes in the intermediate states. We then have

$$\bar{G}_{AB} = f_A G_A + (1 - f_A) G_B = (1 - f_A)(G_B - G_A) + G_A,\qquad(5.15)$$

where G_A and G_B are the free energies of the cavities in the 'closed' and 'open' states respectively. If the free energy of the protein changes under the effect of specific or non-specific agents, we have

$$\Delta \bar{G} = \Delta G_0 + \Delta \bar{G}_{AB}.\qquad(5.16)$$

ΔG_0 is always non-negative, since G_0 is the minimum free energy of the rigid matrix of the molecule. Since $\Delta(G_B - G_A) = -RT \Delta \ln[(1 - f_A)/f_A]$, $\Delta \bar{G}_{AB}$ may be written as follows:

$$\Delta \bar{G}_{AB} = -(G_B - G_A)\Delta f_A + RT\Delta \ln f_A + \Delta G_A;\qquad(5.17)$$

Accordingly, since ΔG_A is non-positive and since, as a result of the perturbance effects, Δf_A is positive, $\Delta \bar{G}_{AB}$ will be negative if

$$|(G_B - G_A)\Delta f_A| > |RT\Delta \ln f_A|.\qquad(5.18)$$

This inequality condition is always met if

$$G_B - G_A' \geqslant RT.\qquad(5.19)$$

If, owing to the action of external agents, the stability of the protein is enhanced, this means that under the given conditions the value of the decrease of $\bar{G}_{AB}$ is larger than any possible increase of G_0:

$$\Delta\bar{G} = \Delta G_0 + \Delta\bar{G}_{AB} < 0. \tag{5.20}$$

However, as the protein-perturbing effect becomes stronger, ΔG_0 may increase, until the sign of this inequality is reversed and denaturation of the protein begins. It may be expected that the plot of the stability of the protein as a function of the concentration or of the strength of the perturbing agent will be bell-shaped.

A very attractive hypothesis of adaptive changes in the conformational flexibility of protein molecules accompanying changes in the environmental temperature was advanced by Aleksandrov (1975). The basic idea of this hypothesis is that the level of conformation flexibility (mobility) varies with the temperature so as to partly compensate for the destabilizing effect of the temperature. In other words, the molecule is capable of remaining in its functional condition even if the external conditions deviate somewhat from the normal. It follows from the dynamic model that such an adaptation may be attained by changes in the equilibrium constants of the fluctuating protein cavities ($K_{A \rightleftharpoons B}$) caused by the temperature variations of hydrophobic and clusterophilic interactions.

5.1 Changes in the Conformation and Stability of Proteins Brought about by Perturbing Agents

Results of experimental studies on the effect of perturbing agents on the conformational properties of proteins are in full agreement with the consequences of the dynamic model of the protein. We selected, in the capacity of nonspecific perturbing agents, a number of substances liable to affect the dynamic properties of the immunoglobulin of light chains dimers and the Fab fragments of immunoglobulin G: sodium dodecyl sulfate, sodium chloride, ammonium sulfate, as well as temperature. The effect of the salts (1.2 M NaCl and 0.75 M $(NH_4)_2SO_4$) on spin-labeled light-chain dimers, and heating their solutions from 25 to 52°C resulted in changes in the EPR spectra similar to those produced by sodium dodecyl sulfate (Fig. 19)

Judging by the unchanged absorption spectrum of protein solutions, it appears that the changes in the conformational properties of the proteins effected by the agents just listed do not result in the denaturation of the molecules. When the concentration of sodium dodecyl sulfate in solution of spin-labeled immunoglobulin G was raised to 0.2 M, and the pH adjusted to 14 by the addition of alkali, the EPR spectra indicated major disturbances of the opposite kind in the structure of immunoglobulin G (Fig. 20).

EPR spectra of native immunoglobulins G and their constituent parts result from the superposition of two spectra corresponding to two different states of the label — the A-state and the B-state. These states very probably originate from the capacity of the light-chain dimer immunoglobulins and Fab fragments to exist as A- and B-conformers (Käiväräinen and Nezlin, 1976a, b); these correspond to the 'closed' and 'open' states of protein cavities, which are formed between V- and C-domains (Fig. 42).

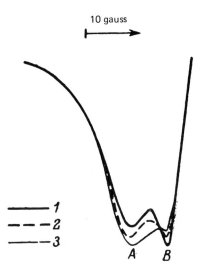

Figure 19. The lower field part of EPR spectra of spin-labeled immunoglobulin G in the absence (1) and in the presence (2, 3) of sodium dodecyl sulfate. After Käiväräinen (1978b). Curves 2 and 3 represent, respectively, concentrations of $5 \cdot 10^{-3}$ and 10^{-2}. 0.1 M phosphate buffer; pH 7.5; 22°C.

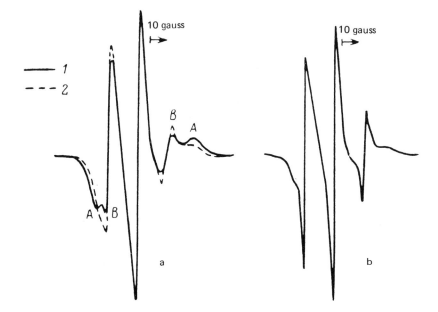

Figure 20. EPR spectra of solutions of spin-labeled native bovine immunoglobulin G at pH 7.5 (a) and at pH 14 + 0.2 M sodium dodecyl sulfate (b). After Käiväräinen (1978b). Curve (a): (——) in the absence of sodium dodecyl sulfate; (------) in the presence of 0.2 M dodecyl sulfate.

The effect of the agents employed by ourselves on the spin-labeled protein may be interpreted as a shift in the equilibrium between the conformers of light-chain dimers and Fab-fragments towards the A-conformer Fig. 19). The variation of the ratio between the areas under the A-components and the total area of EPR spectrum (K') was selected as the quantitative measure of this shift. If it is assumed that $K' = aK_{tr}$, where a is the proportionality coefficient and K_{tr} is the true equilibrium constant, the magnitude $\Delta K/K'$ is an expression of the change in K_{tr} between the A- and B-states of the label (Table 7). The variation of K' during the reaction between Fab-fragment-labeled antibody with the antigen, i.e., during the specific action in IgG, is also shown for comparison (Käiväräinen and Nezlin, 1976a).

The question which was considered of interest was the effect of changes in the equilibrium constant on the stability of the protein. It follows from the plot (Fig. 21) of the turbidity of IgG solution as a function of the temperature that NaCl (1.2 M) and $(NH_4)_2SO_4$ (0.75 M), which cause a leftward shift of the A ⇌ B equilibrium constant, stabilize IgG both against aggregation and against thermal denaturation. However, sufficiently large concentrations of ammonium sulfate (1.5 M or higher) impair the native structure of the protein, as is seen from the absorption by IgG solutions at 280 nm. Thus, for instance, the optical density D^{280} of IgG and IgG + 0.75 M $(NH_4)_2SO_4$ solution at 23°C is 0.200; that of the solution of IgG + 1.5 M ammonium sulfate is 0.400, while that of the solution of IgG + 2.5 M of this salt is 0.720.

The increase in the stability of IgG under the effect of moderate concentrations of perturbing agents indicates that the condition $G_B - G_A \geqslant RT$ is fulfilled. If the concentration of ammonium sulfate is sufficiently high, the protein becomes denatured, as indicated by the increase in D^{280}. Accordingly, the positive increase ΔG_0 is larger than the negative; both $\Delta \overline{G}_{AB}$ and $\Delta \overline{G}$ in equation (5.20) are positive.

That the variation of the protein stability with the salt concentration can in fact be represented as a bell-shaped curve has been demonstrated in detail in the case of albumin. The presence of methyl orange as stabilizer imparts additional stabilization (Käiväräinen and Käiväräinen, 1978).

The study was performed on a preparation of human serum albumin, manufactured by 'Reanal', which had been desalted on a Sephadex G-25 column. Ammonium sulfate $(NH_4)_2SO_4$ was employed as nonspecific perturbing agent, while methyl orange was used as specific perturbing agent, since it forms complexes with human serum albumin in the molar ratio of 2:1, the association constant being about 10^5. The turbidity of the solution at 650 nm was studied in one series of experiments; in another series, the changes in the optical density at 280 nm were investigated (Figs. 22-24).

It is seen that the structure of human serum albumin is stabilized in the presence of 0.5 M ammonium sulfate, but becomes destabilized if this salt is present in concentrations of 1 M. It is seen in Fig. 22 that the plot of the stability of human serum albumin as a function of ammonium sulfate concentration is bell-shaped. The incipient aggregation temperature T_{cr} was taken as the measure of stability. As expected, bound methyl orange enhanced the temperature-stabilizing effect of ammonium sulfate.

The stabilization of the structure of human serum albumin by methyl orange is also manifested as increased stability to salting-out by 2.5 M ammonium sulfate (Fig. 23). It is

TABLE 7. Effect of various agents on the effective equilibrium constant of the label K' and its relative variation $[(K'-K'_k)/K'_k] \cdot 100\%$

Spin-labeled preparations	Light Chain Dimers					Bovine Immunoglobulin G			Antibody, 25°C	
	Controls, 20°C	1.2 M NaCl	(NH₄)₂SO₄ 0.75 M	(NH₄)₂SO₄ 1.5 M	52°	Controls, 22°C	5·10⁻³ M sodium dodecyl sulfate	10⁻² M sodium dodecyl sulfate	Controls	Antibody + hemoglobin
K'	0.193 ± ±0.002	0.228 ± ±0.002	0.232 ± ±0.007	0.282 ± ±0.002	0.240 ± ±0.002	0.190 ± ±0.002	0.200 ± ±0.002	0.218 ± ±0.002	0.210 ± ±0.002	0.248 ± ±0.002
$(K'-K'_k/K'_k)\cdot 100\%$	-	18±3	20±3	46±3	23±3	-	5±3	15±3	-	18±3

Note: In the salt solutions the concentration of the spin-labeled chain dimers of rats was 2·10⁻⁴ M at pH 7.5. At 1.5 M concentration of ammonium sulfate the proteins were salted out and the solution became turbid. The spin-labeled bovine immunoglobulin G was dissolved to produce a concentration of 1.3·10⁻⁴ M in 0.1 M phosphate buffer at pH 7.5. The specific complex of rabbit antibodies with human hemoglobin were formed without a noticeable precipitation. In the control solution the antibody was replaced by nonspecific rabbit immunoglobulin G. Concentrations of the antibody and of nonspecific immunoglobulin G was 5.3·10⁻³; hemoglobin concentration was 1.2·10⁻⁴ M. 0.1 M Medinal buffer; pH 7.5.

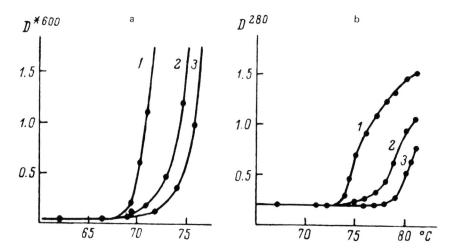

Figure 21. Turbidization D^* at 600 nm a, 15 mg/ml) and absorption D at 280 nm b, 0.15 mg/ml) of IgG solutions as a function of the temperature (Käiväräinen, 1978b).
1 — control solution of IgG; 2 — control solution of 0.75 M ammonium sulfate; 3 — control solution + 1.2 M NaCl. 0.1 M phosphate buffer, pH 7.5.

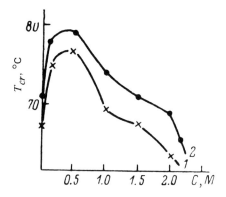

Figure 22. Incipient aggregation temperature T_{cr} as a function of ammonium sulfate concentration (Käiväräinen and Käiväräinen, 1978).
1 — human serum albumin; 2 — 1 molar pt human serum albumin + 2 molar pts methyl orange. The turbidization at 600 nm served as the measure of the aggregation.

interesting to note that, as may be concluded from the absorption by the protein at 280 nm (Fig. 24), processes which accompany macromolecule aggregation in the presence of ammonium sulfate do not significantly affect tyrosine and tryptophan environments. It is only when the salt concentration attains 2.6 M that globules of human serum albumin become unrolled to a significant extent. It would seem, accordingly, that the aggregation produced by the salting-out effect, is mainly connected with the changes in the condition of peripheral zones of the molecule, or changes in the water activity of solvent.

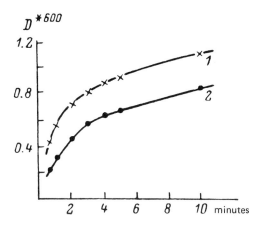

Figure 23. Stabilizing effect of methyl orange on the salting out of human serum albumin by 2.5 M ammonium sulfate at 22°C (Käiväräinen and Käiväräinen, 1978).
1 — human serum albumin; 2 — 1 molar pt human serum albumin + 2 molar pts methyl orange.

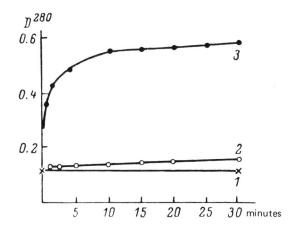

Figure 24. Effect of salting-out concentrations of ammonium sulfate on the absorption of human serum albumin at 280 nm and 22°C (Käiväräinen and Käiväräinen, 1978).
1 — control solution; 2 — 2.5 M; 3 — 2.6 M.

Thus, the effects of specific and nonspecific agents on the stability of albumin are in agreement with the predictions based on the dynamic model of behavior of proteins in water.

According to our interpretation, the perturbing agents which impair the condition $\Delta G^{H_2O} \simeq 0$ for water phase transition of the first order, displace the $A \rightleftharpoons B$ equilibrium of the fluctuations of protein cavities. Such situations may result not only from cavity deformations and destabilization of water clusters, but may also be due to changes in the free energy of water surrounding the protein. One agent thus affecting the environmental water is polyethylene glycol with MM > 6000, whose molecules are coiled and occlude large amounts of water (Polson, 1977), besides tending to fluctuate and thus to perturb the properties of the environment. Two other factors may also be responsible for the $A \rightleftharpoons B$ shift: the effect of excluded volume due to the polymer, and Brownian collisions between the protein molecule and polyethylene glycol, which reduce the stability of 'open' cavities. If the constant of dissociation of the oligomeric protein into dimers or single subunits is larger when its central cavity is 'open' than when it is 'closed', then the shift of the $A \rightleftharpoons B$ equilibrium to the left may be expected to stabilize the oligomeric form. This is probably the reason for the shift to the left of the tetramer $\rightleftharpoons$ dimer equilibrium of phosphorylase B in the presence of polyethylene glycol, which was observed by Kurganov *et al.* (1979). In a sense, the addition of polyethylene glycol is equivalent to an increase in the concentration of the protein itself. The adsorption of protein on the surface may also affect the $A \rightleftharpoons B$ equilibrium and the fluctuation frequency if one of these states becomes stabilized as a result.

Substituting D_2O for H_2O may be expected to reduce the free energy of proteins, since it is known (cf. Chapter 1) that D_2O forms stronger hydrogen bonds. It was in fact observed by Lobyshev and Shnol' (1975) that the denaturation temperature of collagen and RNA-base in D_2O is higher by $2 - 4°C$ than in H_2O.

Lobyshev and Kalinichenko (1978), who studied the factor enhancing the stability of biopolymers in D_2O, came to the following conclusions: the intramolecular hydrogen bonds in proteins and in nucleic acids may become weakened or else remain unchanged; the intensification of the hydrophobic and of salt effects in D_2O is obviously too small to account for the stabilization effect; the major factor responsible for the stabilization is the stronger bonding of D_2O molecules to the protein, resulting in the stabilization of the conformational mobility.

Two mechanisms of protein stabilization by water binding are possible: formation of water bridges between the donors (X) and acceptors (Y) of the hydrogen bond of the type

$$X–H\cdots\underset{\underset{H}{|}}{O}–H\cdots Y \quad \text{and} \quad X–H\cdots\underset{\underset{H}{|}}{O}–H\cdots O\underset{\diagdown H}{\overset{\diagup H\cdots Y}{}}$$

(Lewin, 1974), and formation of ordered water structures during its interaction with the protein. In either case, the isotopic stabilization effect may be regarded as an increase in the bonding constant of hydrate water by the biopolymer.

The effective values of the enthalpy of thermal denaturation transitions of collagen in H_2O and D_2O, calculated according to van't Hoff, are 1060 and 1512 KJ/mole respectively, and their difference is 452±155 KJ/mole. The difference in the enthalpy values of fusion of ice I, constituted by D_2O and H_2O, is 0.29 KJ/mole.

If it is accepted that the change of state of some of the bound water, which accompanies the process of protein denaturation, resembles a phase transition of the first order, this numerical value may be utilized to determine the number of water molecules participating in such a transition (Lobyshev and Shnol', 1975): $nH_2O = 453/0.29 \simeq 1500$, i.e., 0.075 gm H_2O per 1 gm of protein. The total amount of the water bound by collagen, as determined by NMR and colorimetric methods, is 0.35 gm per 1 gm of protein. Thus, about 17% of the bound water is capable of undergoing phase transitions of the first order when the protein is denatured.

The numerical results thus obtained should be treated with caution, since the fusion enthalpy of the structure of the water bound by the protein is known to be smaller than that of the transition ice → water. It should also be borne in mind that the viscosity of D_2O is 20% higher than that of H_2O, which may well reduce the conformational mobility of the protein, thus enhancing its stability.

Lobyshev and Shnol' (1975) also studied the dependence of the denaturation temperature on the D_2O concentration, and showed that the calculated value of this dependence approaches the experimental value if the bonding constant of heavy water is about 10 times as high as the respective constant for ordinary water. This result may be acceptable in view of the fact that the heat of adsorption of heavy water on chymotrypsin is 7 times as high as the corresponding figure for ordinary water. The fact that the experimental curve is nonlinear may be the result of the cooperative nature of D_2O bonding by protein cavities and by nonpolar surface zones. The nonlinear dependence of the temperature at which the mobility of tryptophan residues is 'frozen' — as evidenced by fluorescence spectra — on the concentration of D_2O in human serum albumin also indicates that protein is preferentially solvated by D_2O (Lobyshev and Kalinichenko, 1978).

The increase in the concentration of macromolecules in solution results in an increased frequency of collisions, which may bring about a mutual deformation of the protein structure and of the geometry of the structure-forming cavities. Therefore, in accordance with the mechanism described above, an increase in the concentration of the solution may be expected to shift the $A \rightleftharpoons B$ equilibrium to the left, i.e., reduce the extent of hydration and increase the compactness of the proteins.

In fact, studies of hydration of myoglobin by dielectric methods showed that when the protein concentration was increased from 77 to 161 mg/ml, the extent of hydration decreased from 27±0.06 to 0.23±0.03 gm H_2O per 1 gm of protein (Grant *et al.*, 1974). That the compactness of bovine serum albumin and of hemoglobin increases with increase in their concentrations is indicated by a decrease in their specific volumes (Bernhardt and Pauly, 1975), which can be interpreted as a decrease in the extent of protein hydration with increasing frequency of collisions. It is assumed in this context that the volume of the hydrate water is larger than that of free water on account of the formation of ordered Frank-Evans structures around the nonpolar groups. These results may be compared with the data of calorimetric studies, which indicate an increase in the extent of hydration at lower temperatures. The respective specific volumes of bovine serum albumin and of hemoglobin at 25°C, extrapolated to infinitely dilute solutions, are 0.73604 and 0.75460 cm^3/gm.

Another consequence of our model is the mutually opposite, i.e., the mutually compensating changes in the heat capacity, enthalpy and entropy of the protein and of the water interacting with it in the course of their conformational changes. In fact, if it is assumed that the A-conformer, which is more compact and has a smaller number of degrees of freedom, has lower S-, H- and C_p-values than the B-conformer, it follows that as the A $\rightleftharpoons$ B equilibrium is shifted to the left under the effect of heat or other perturbing agents, the time-averaged heat capacity of the protein will decrease, while that of the water will increase as a result of the displacement of the water from the cavities (Käivär-äinen, 1975b). Resolution of the resulting effect into its component parts is a very difficult experimental task. Alanina *et al.* (1978) attempted to differentiate between the thermal effects of denaturation and of dehydration, i.e., the intramolecular fusion of DNA films. By varying the vapor pressure in the calorimetric cell, these workers reduced the dehydration temperature to below the denaturation temperature of biopolymer molecules. Under these conditions the recorded increase in the heat capacity may be attributed to the dehydration of DNA in its native state. It was found that the heat of dehydration of native DNA specimens is 1257 J/one DNA, while that of the denatured samples was only 838 J, the moisture content being 30% in both cases. This indicates that native DNA contains a large amount of specifically bound water.

The results obtained by Zavyalov *et al.* (1977a, c), who studied the temperature transitions of Bence-Jones protein and its variable and constant parts by differential adiabetic scan microcalorimetry, are an indirect confirmation of the compensation effects taking place between water and protein in the pre-denaturation range. Graphs of the temperature dependence of solutions of intact Bence-Jones protein — which simulate the Fab subunit of IgG and the dimers of the variable and constant domains obtained by its proteolysis — at various pH values, indicate that the plot of the heat capacity of these proteins at pH 7.40 as a function of the temperature is linear between 20 and 40°C, C_p displaying a slight tendency to increase (Fig. 25a). Results obtained on Bence-Jones proteins, their Fv-fragments and isolated Fab-subunits by perturbation spectroscopy indicate at the same time that the cavity forming the active site 'collapses' at pH 7.35 if the temperature is raised from 25 to 35°C. This is accompanied by a considerable decrease in the accessibility of tyrosins to water, ethylene glycol and iodine (Trotskii *et al.*, 1973; Zavyalov *et al.*, 1975, 1977a). There is little doubt that such a change is accompanied by a change in the heat capacity of the protein molecules. The unchanged value of the resultant heat capacity may be explained by the compensating effect of the change of state of the solvent.

The small rise in the resultant value of C_p with increasing temperature in the pre-denaturation range may mean that the increase of C_p due to the fusion of water is somewhat larger than its decrease due to the conformational changes in the protein. However, under certain conditions this water-protein thermodynamic compensation may be impaired. A situation may be imagined in which, as a result of external factors such as pH, the equilibrium between the states of protein cavities may not be shifted to an extent which would permit the change in protein C_p to compensate the perturbance in the structure of water. Under such conditions, if the action of heat in this interval brings about the fusion of the water clusters in protein cavities, it may be expected that the C_p-function will no longer be linear. It would appear that this is what actually happens when the temperature of solutions of Bence-Jones protein and its 'halves' is varied in the pre-denaturation

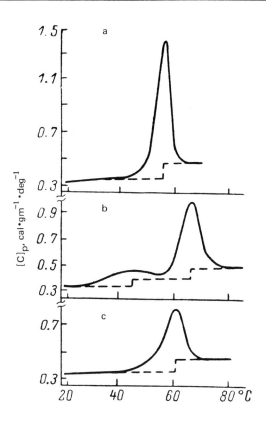

Figure 25. The temperature dependence of partial heat capacity of Bence-Jones protein. pH of solutions: (a) 7.40; (b) 3.23; (c) 2.00. After Zavyalov *et al.* (1977a).

range from 25 to 50°C at pH 3.23 (Fig. 25b). It had been previously shown (Zav'yalov *et al.*, 1975) that at pH ≤ 6.0 the temperature transition of immunoglobulins, with its resulting increase in compactness, was absent in this interval. It is important to note that the resulting enthalpy of denaturation of Bence-Jones protein and its 'halves' at pH 7.4 is equal to the sum of the enthalpies of the pre-denaturation and denaturation processes occurring at pH 3.23. Accordingly, the denaturation of these proteins at pH 7.4 represents a symbatic impairment of protein conformation and of the structure of the water contained in its cavities.

These workers (Zavyalov *et al.*, 1977a) also described a case in which the external environment of the proteins (pH 2.00) is such that, according to our model, they lose their structured water as a result of changes in the geometry of their cavities occurring in the temperature interval studied (Fig. 25c). The denaturation enthalpy of such proteins will then reflect only the changes in the degrees of freedom of their amino acid residues and of polypeptide chain segments. In fact, its actual value (410 KJ/mole) proved lower than the enthalpy of denaturation of these proteins in their native form at pH 7.4

(586 KJ/mole) by about the magnitude of the heat effect which is observed in the pre-denaturation range at pH 3.23 (165 KJ/mole), and which we attributed to the change of state of the water. Thus, all the above experimental results may be interpreted in terms of the dynamic model of behavior of proteins in water, without giving rise to contradictions.

It was noted by Privalov and Khechinashvili (1974), who used the method of differential scan microcalorimetry, that in the case of several globular proteins, the plot of the energy of stabilization of the native structure as a function of the temperature has a maximum around 20-30°C (Fig. 26). This confirms our own conclusion to the effect that if the inequality (5.19) is fulfilled, the stability of the protein as a function of the magnitude of the protein structure-impairing effect has a bell-shaped form.

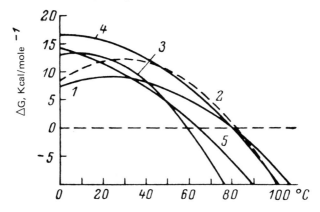

Figure 26. Temperature dependence of the free energy of stabilization of the native structure of globular proteins. After Privalov (1975).

1 — cytochrome; 2 — myoglobin; 3 — chymotrypsin; 4 — lysozyme; 5 — RNA-ase.

This method also gave interesting results when applied to the study of the thermodynamic properties of a fibrous protein - procollagen (Privalov, 1968; Privalov and Tiktopulo, 1970). These workers noted that the cooperative system of water molecules takes part in the stabilization of the molecule. When neutral salts are added, the thermal absorption peak in solution splits up into two, but the overall area remains unchanged (Fig. 27). This effect is similar to that just described for Bence-Jones proteins at pH 3.23 (Fig. 25b), and may be due to the destabilization of the ordered water structure by the neutral salt acting as perturbing agent.

These workers (Privalov and Tiktopulo, 1969; Privalov *et al.,* 1971) showed by special experiments that the first peak is related to a rearrangement of the protein as a result of which it becomes more compact. In terms of the dynamic model this means that the equilibrium A ⇌ B has shifted to the left. As the ionic strength and the pH of procollagen solutions approach the physiological values, the first peak shifts towards higher temperatures and merges with the second, which corresponds to the denaturation of the protein

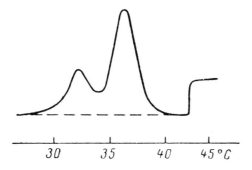

Figure 27. Heat absorption by a salt solution of procollagen.

itself. It would appear that in such cases the structures of the ordered water and of the protein undergo a cooperative fusion. It was shown by Zavyalov *et al.* (1978) that the perturbing effect exerted on the protein structure by inorganic ions resembles that of the temperature.

That denaturation and rearrangement of the water shell of the protein are interrelated processes has been confirmed by NMR data. In the case of collagen fibers, the increased mobility of the water molecules, as manifested by the narrowing of the 60-MHz NMR signal, is symbatic with the protein denaturation as determined calorimetrically by measuring the heat absorption peak (Fig. 28).

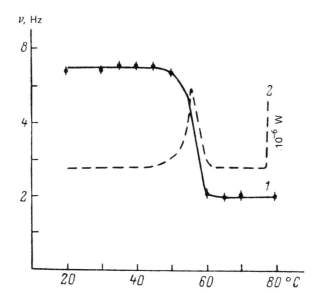

Figure 28. Temperature dependence of the NMR line width (1) and heat absorption (2) for collagen fibers. After Mrevlishvili *et al.* (1975).

In their studies of the thermodynamics of pre-denaturation transition in RNA-ase based on deuterium exchange kinetics, Tiktopulo and Privalov (1975) assumed that deuterium exchange takes place according to the scheme

$$N \underset{k_{-1}}{\overset{k_1}{\rightleftarrows}} D \overset{k_e}{\longrightarrow} D^* \underset{k^{-2}}{\overset{k_2}{\rightleftarrows}} N^*, \qquad (5.21)$$

where N is the compact state, D is the uncoiled state, D^* and N^* are deuterated states and k_i are rate constants, with $k_1 + k_2 \gg k_e$. Then

$$K = (D + D^*)/(N + N^*) = 1/k_e \, [\text{d} \, (D^* + N^*)/\text{d}t]. \qquad (5.22)$$

Since k_e as a function of the pH and temperature is known, K under various conditions may be determined from the rate of hydrogen exchange.

van't Hoff graphs plotted by this method from the $K(t)$ relationship and from the change in the optical density are approximated by two straight lines (Fig. 29). Tiktopulo and Privalov deduced from this fact that the observed deuterium exchange is the result of two qualitatively different processes, one of which (pre-denaturation interval) has an uncoiling enthalpy of 1.5 Kcal/mole (6.28 KJ/mole), while the respective values in the second (denaturation) interval are 86 and 120 Kcal/mole (360.3 and 503 KJ/mole) at pD of 2.6 and 5.1 respectively.

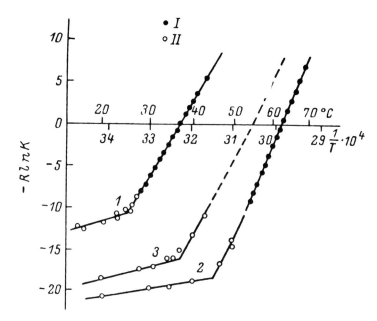

Figure 29. van't Hoff graph plotted from deuterium-exchange data (I) and optical absorption data (II). 1,2 — after Tiktopulo and Privalov (1975); 1 — pD 2.6; 2 — pD 5.1; 3 — after Nakanishi and Tsuboi (1974).

Judging by the magnitude of the enthalpy value, the denaturation is accompanied by the rupture of all hydrogen bonds originally present in the protein. In the pre-denaturation range, deuterium exchange indicates only very small changes in the compact structure of protein caused by non-cooperative rupture of isolated hydrogen bonds, which cannot account for the macroscopic changes in the protein observed in the pre-denaturation zone. The concentration of partly uncoiled molecules in the pre-denaturation temperature range is very low. In this range the number of simultaneous bond ruptures does not increase with increasing temperature, and it is only the rate of the scissions which increases. From the viewpoint of this dynamic model this process may favor a concentration of defects in the structure of nonpolar cavities, impairment of clusterophilic interactions with water and the shift of $A \rightleftharpoons B$ equilibrium to the left.

The same approach was subsequently employed in studies of the correlation between the structural mobility of collagen and the physiological temperature of the animal from which it had been taken (Tiktopulo *et al.*, 1979). The equilibrium constants of micro-opening of structure of collagen in the skin of codfish, grass lake frog and rat, determined at the physiological temperatures of these animals, differed from each other much less than did their values calculated for the same temperature ($25°C$).

The authors concluded that the objective of evolutionary selection of a certain amino acid composition of these collagens is to produce an optimum mobility level of the proteins, which determines their native properties. This conclusion is in agreement with the stipulation of the dynamic model (sec. 2.1), according to which physiological conditions must ensure not only a geometric, but also a dynamic correspondence between the 'open' protein clefts and the water clusters they contain.

Thermo-induced pre-denaturation conformational changes in proteins are a fairly common occurrence. The fluorescence polarization curves between 0 and $40°C$ are S-shaped for most proteins. The corresponding conformation transition is usually noted in a narrow temperature range ($\Delta t = 10\text{-}20°C$, and is accompanied by a decrease in fluorescence polarization, and in a number of cases by a small short-wave shift of the peak of the fluorescence spectrum. Such changes are fully reversible. The experimental results can be interpreted as a result of mutual transitions between two or three structural forms of the proteins under the effect of heat (Mazhul *et al.*, 1970; Chernitskii, 1972).

5.2 Effect of Temperature on Protein Dynamics and on the Interaction between the Protein and the Surrounding Solvent

The NMR relaxation method has yielded results which, in our view, are highly important, explaining as they do the dynamics of the interaction between biopolymers and water (Ratnikova *et al.*, 1975). It is seen (Figs. 30 and 31) that a sharp decrease in the diffusion coefficient D of the water in collagen and egg albumin takes place when the temperature is raised from about -11 to $-3°C$. This temperature range also comprises the maximum of heat absorption by procollagen powder containing 90% water (Fig. 32), which is interpreted as being due to the fusion of the water not bound to the protein (Mrevlishvili *et al.*, 1975). It is now thought that the bound water is not capable of undergoing phase transitions. However, if this effect was related to the fusion of free water (Figs. 30 and 31), D would increase instead of decreasing, and would not be noted for hydrated collagens with 30% water; in the case of egg albumin, T_2 and D undergo similar changes between +15 and +21°C. Ratnikova *et al.* (1975) explained these effects as follows. There are two types of water sorption sites in proteins — I and II — which have different T_2 of water: $T_2(I) \gg T_2(II)$, i.e., the mobility of phase II is much lower than that of phase I. Exchange may take place between the two and the experimental value of T_2 is given by

$$1/T_2 = f_I/T_2\,(I) + (1 - f_I)/T_2\,(II), \qquad (5.23)$$

where f_I is the concentration of the water molecules bound to type I sites. In such a case the jump of T_2 on segment b, on passing from branch c to branch a may be caused by the sudden acceleration in the exchange rate (Figs. 30 and 31).

It was concluded (para. 4.2) from the NMR relaxation data obtained for frozen protein solutions that the water localized in protein clefts (type II) is the less mobile fraction, while the water sorbed by the charged polar surface groups (type I) is more mobile. Accordingly, from the viewpoint of the dynamic model, the observed decrease in D is the result of the intensified exchange between the internal and the surface water fractions due to the increased A $\rightleftharpoons$ B fluctuation rate, and to the thermal activation of the mobility of the solvated surface groups. Owing to the cooperative properties of proteins these processes are interconnected, so that the inclusion of a fast exchange between the fractions may resemble a phase transition.

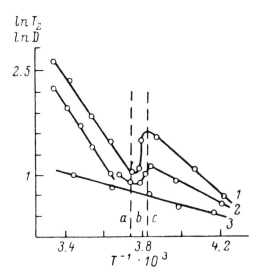

Figure 30. Logarithm of spin relaxation in T_2 of water protons and the logarithm of the diffusion coefficient of water ln D in collagen containing 130% (1), 30% (2) and 10% (3) of water as a function of the temperature. After Ratnikova *et al.* (1975).

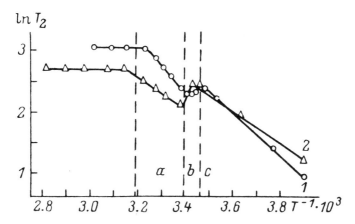

Figure 31. In T_2 of water protons in a commercial (1) and predried (2) samples of egg albumin as a function of the temperature. After Ratnikova *et al.* (1975).

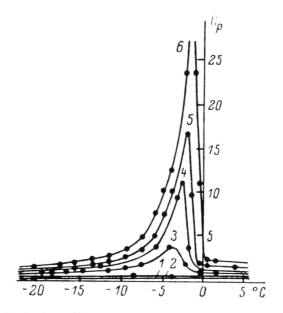

Figure 32. Heat capacity C_p of procollagen containing various water concentrations as a function of the temperature. After Mrevlishvili *et al.* (1975).
1 − 0; 2 − 0.35; 3 − 0.64; 4 − 1.0; 5 − 1.3; 6 − 2.0 gm water per 1 gm protein.

As a result of the exchange there is a tendency to a 'smoothing' of the differences between the thermodynamic properties of the fractions I and II. This means that the entropy of the ordered water in the open nonpolar protein cavities increases and $\Delta S_{B \to A}^{H_2O}$ of the water displacement by the jump $B \to A$ decreases. Such an effect may be noted not only in moistened protein powders, but in solutions as well; however, in the latter case there is an additional effect — possible exchanges between the water bound to the protein and the free water.

Since the probability of occurrence of anisotropic fluctuations in the liquid depends on the number of jump-like reorientations of the individual groups of its molecules (H_2O, for example) (Shakhparonov, 1976), it may be expected that the decrease of $\Delta S_{B \rightleftarrows A}^{H_2O}$ which accompanied the inclusion of the exchange between the two bound water fractions may produce changes in the macroscopic parameters of the protein solution such as the average strength of the hydrogen bonds between solvent molecules.

IR spectroscopic studies of serum albumin solutions (Khaloimov, 1974) are of special interest in this connection. It was shown (Khaloimov *et al.*, 1976a, b) that pre-denaturation conformational rearrangements of proteins are accompanied by changes in the properties of their water environment. Band shifts in the vibration spectrum are an indication of the strengthening or weakening of the hydrogen bonds between the water molecules.

The valent-deformational band of water absorption at 5180 cm^{-1} was used to study the heat-produced transitions of human serum album heavy meromayosin (HMM), actomyosin and frog muscle tissue. Similar results were obtained on studying the overtone band at 6800 cm^{-1} and the deformational-vibration associative band at 2130 cm^{-1}. In all the proteins thus studied a rise in temperature initially resulted in a steady high-frequency shift in the peak of the 5180 cm^{-1} band, owing to the weakening of the hydrogen bonds interconnecting the water molecules by the thermal motion of the latter. Subsequently, within certain temperature ranges (32-36°C for HMM, 37-43°C for albumin, 48-55°C for actomyosin and 22-28°C for frog muscle tissue) the band shift suddenly changes direction and proceeds from higher to lower frequencies. It is seen, accordingly, that these conformational transitions reinforce the H-bonds of the water in solution.

After a corresponding decrease in the frequency of the valent-deformational band within a rather narrow temperature interval, the frequency resumes its steady rise as the temperature is further increased (Fig. 33). The changes in the spectroscopic properties of water corresponding to a strengthening of hydrogen bonds were explained by Khaloimov *et al.* (1976b) as due to the rupture of the hydrophobic bonds in the globule, as a result of which the nonpolar groups come into contact with water and stabilize it. However, such an interpretation is in contradiction with the well-known fact that hydrophobic interactions become stronger and not weaker at higher temperatures. Moreover, it is not at all clear why, in the temperature interval below the observed transition range, the mobility of water molecules in protein solutions is sometimes higher than that of molecules of pure water.

Our own interpretation of the observed effects is different. It is based on the dynamic model of behavior of proteins in water and does not involve these contradictions. We shall assume, to begin with, that the mobility of the solvent molecules may be affected by the exchange between the ordered water and the free water, as a result of the fluctuations of protein clefts. In accordance with the model, a cleft fluctuation from the 'open' B-state to the 'closed' A-state causes water molecules to be suddenly expelled into the environmental medium, while A → B transition is accompanied by the reverse process. The resulting sudden change in the entropy of the water molecules increases the probability of

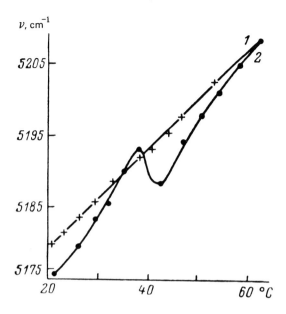

Figure 33. Temperature shift of the valent-deformational band in pure water (1) and in 10% solution of serum albumin. After Khaloimov *et al.* (1976b).

density fluctuations in the surrounding solvent, which may result in an additional perturbance of the system of hydrogen bonds in the water surrounding the protein, and bring about a shift in the vibration spectrum of the water in the protein solution.

As the physiological temperature is approached, $\Delta S_{B \rightleftharpoons A}^{H_2O}$ increases. Under the resulting conditions, the mobility of the water in the protein solution may exceed that of the molecules of pure water (Fig. 33). Our own studies of solutions of antibodies, cytochrome C and hemoglobin (Chapter 6) have shown that this interpretation is the correct one. The inflexion on the curve in Fig. 33 between 38 and 43°C, which enhances the strength of hydrogen bonds, may result from a cooperative inclusion of the exchange between two protein-bound water fractions with differing mobilities, in accordance with the mechanism described above. As a result, $\Delta S_{A \rightleftharpoons B}^{H_2O}$ decreases, which means that the probability of the fluctuations in the solvent, which are responsible for the weakening of the hydrogen bonds, decreases as well. This process is accompanied by a decrease in the lifetime of the B-state, since the clusterophilic interactions are impaired, and by an increase in the lifetime and stability of the A-state as a result of intensified hydrophobic interactions. At the same time there is a considerable shift of the A $\rightleftharpoons$ B equilibrium to the left. The subsequent gradual increase in the mobility of water molecules in solution may be explained by thermal activation of solvated surface residues. The similar nature of the functions described in Figs. 31 and 33 indicates that such effects are of general occurrence.

The perturbing agents – i.e., the various additives to protein solutions – may favor the inclusion of the exchange, and thus enhance the frequency of A ⇌ B fluctuations and labilize the surface groups. It may be expected that the transition temperature will decrease as a result. In fact, the presence of dioxane is known to affect serum albumin solutions in this way (Khaloimov *et al.*, 1976b).

The interpretation just given to the functions presented in Figs. 31 and 33 is confirmed by a number of other studies. Thus, when the temperature of a serum albumin solution was raised from 30 to 60°C, the peak of the fluorescence spectrum shifted towards shorter wavelengths, corresponding to the increase of tryptophanyl hydrophobite environment (Mazhul' *et al.*, 1970). This may be a result of the leftward shift of the A ⇌ B equilibrium.

A study of heat-induced conformational transitions in serum albumin by the method of spin probe (Kuznetsov and Ebert, 1975; Kuznetsov, 1976) revealed a characteristic change in the polarity of the environment of the N–O group of the probe and its mobility in the same temperature interval (Fig. 34a). The parameters of the EPR spectrum indicate that this transition caused the N–O group to pass from an aqueous to a hydrophobic zone which, in terms of our theory, is tantamount to a leftward shift in the A ⇌ B equilibrium. The variation of the correlation time of the probe with the temperature very clearly indicates that this transition is a two-stage process. Since the probe employed was hydrophobic, it very probably becomes localized in one of the nonpolar clefts of the albumin, and reflects the changes in its dynamic behavior.

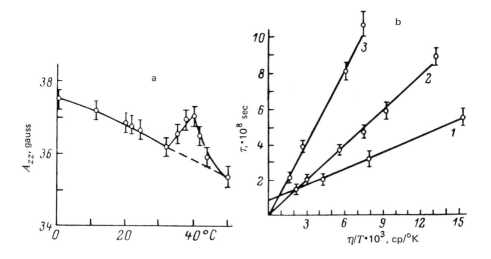

Figure 34. Variation of A_{zz}, which is a measure of the polarity of the environment of the spin probe, adsorbed on serum albumin (a), and its correlation time τ(b) with the temperature and with η/T. After Kuznetsov and Ebert (1975).
1 – 39°C; 2 – 23°C; 3 – 50°C.

In the sharp transition zone, which was attributed to the inclusion of the exchange between the water in nonpolar protein zones and in the non-ordered solvate hull, there is a marked increase in the mobility of the spin probe relative to the macromolecule. The polarity of the environment of the N–O group of the probe increases in parallel; when the temperature is further increased, the polarity decreases strongly, as does the mobility of the probe (Fig. 34b). These results confirm our interpretation, since the destabilization of the 'open' state of the clefts due to the inclusion of the exchange may be expected to reduce the free energy of activation of B → A transition, i.e., to enhance the frequency of cleft fluctuations. As soon as that frequency becomes greater than $1/\tau$, where $\tau = 30$ nsec is the correlation time of the rotary motion of the albumin as a whole, a 'slippage' of the probe with respect to the protein is noted. It would appear that the N–O group then becomes more accessible to water. A further increase in the rate of exchange results in a degeneration of the 'open' state of the clefts, and in the stabilization of the compact A-conformer of the protein by way of enhanced hydrophobic interactions and diminished flexibility; this is manifested as an increase in the effective Stokes radius of the albumin at $50°C$ as compared with its radius at $23°C$ (Fig. 34b).

We used the method of separate determination of the correlation times of spin-labeled proteins and of the labels to which they are bound, as described in para. 3.6, to study the heat-induced transitions taking place in human serum albumin between $5°C$ and $44°C$ in the presence of D_2O and of perturbing agents*. The results indicated a cooperative increase in the flexibility of the albumin at $20°C$ and at $35°C$ (Fig. 35). The radical 2,2,6,6-tetramethyl-N-1-hydroxypiperidine-4-amino-(N-dichlorotriazine) was employed as the label. It is covalently bound to human serum albumin in the ratio of 1:1, probably by the histidine residues.

A possible effect of bonding by heavy metals (present as impurity) on the EPR spectra of labeled albumin was tested for by adding EDTA to the protein solution. The extent and the rate of the transitions in human serum albumin were enhanced to almost equal extents by 3% and by 15% D_2O, especially at the lower temperature; in the presence of 0.6 M NaCl human serum albumin becomes more rigid, and the transition at 30-35°C becomes degenerate. Polyethylene glycol enhanced the flexibility of human serum throughout the temperature range studied.

These phenomena were interpreted as a change in the rate of relative translational-rotational diffusion of the three domains constituting human serum albumin as a result of the formation and fusion of the short-lived inter-domainal aqueous clusters. The changes in the state of the two cavities between the three domains of the albumin may be responsible for the variation of τ_M around 20 and 35°C. The substitution of DOH and DOD for HOH in the clusters as a result of the addition of D_2O to the protein solution enhances the cooperative nature and stability of these clusters. The clusters are destabilized by 0.6 M NaCl, shifting the A ⇌ B equilibrium to the left, while polyethylene glycol seems to have the opposite effect. The less compact and more hydrated B-conformers of human serum albumin are presumably more flexible than the A-conformers.

---

* This study was carried out by the author in collaboration with S.P. Rozhkov

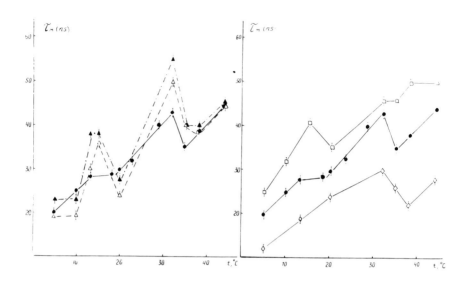

Figure 35. Temperature dependence of correlation time t_M calculated for spin-labeled human serum albumin. (•) in phosphate buffer, pH 7.3; 0.01 M + 0.15 M NaCl; (△) in the presence of 3% D_2O; (▲) in the presence of 15% D_2O; (◇) in the presence of 2% polyethylene glycol, MM = 40,000; (□) in the presence of 0.6 M NaCl. Concentration of human serum albumin: 15 mg/ml. After Käiväräinen and Rozhkov (in press).

Variation in the Degree of Hydration of Human Serum Albumin during Thermally Induced Structural Transitions

The method described in para. 4.1, which involves a combination of proton magnetic resonance with the spin-label methods, revealed that there is a correlation between the changes in the structural flexibility of human serum albumin and the number r of the water molecules which are rigidly bound to the protein. A lower flexibility of the albumin is accompanied by a decrease in the effectiveness of the hydration, while the more flexible molecules are more strongly hydrated (Figs. 36a and 36b). These results confirm the interpretation of the variations in τ_M (Fig. 35) based on the dynamic model: a decrease in the flexibility results from the shift of the A ⇌ B equilibrium of the albumin to the left, and *vice versa*. It is believed that n represents the amount of the water in the cavities whose translational mobility is limited. At high saccharose concentrations (35%) the saccharose seems to act as a perturbing agent, and to alter the properties of the hydrate shell of the protein. The result is a decrease in the hydration of human serum albumin and a change in the overall nature of the relationship between n and t (Fig. 36b).

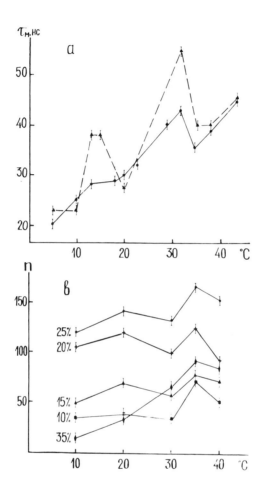

Figure 36. a) (•) correlation time t_M of spin-labeled human serum albumin as a function of the temperature, which reflects the changes in the flexibility of the protein. (▲) correlation time of human serum albumin as a function of the temperature in solution containing 3% D_2O. b) effective hydration $\underline{n}$ of human serum albumin as a function of the temperature, calculated by equation (4.8) at various (10, 15, 20, 25 and 35%) concentrations of saccharose. Concentration of human serum albumin in solution 25 mg/ml. 0.01 M phosphate buffer (pH 7.3) + 0.15 M NaCl. After Käiväräinen *et al.* (in press).

As the saccharose concentration is raised from 10 to 25%, the hydration of human serum albumin increases. It was shown that this effect is connected with the decreased frequency of collisions between the macromolecules as a result of the increased viscosity of the solutions. It may also be the result of the diminished frequency of the A ⇌ B fluctuations in human serum albumin. There are grounds for believing (see below) that the $A \rightleftharpoons B$ type fluctuations of protein cavities affect the structure of the free solvent by increasing T_2^{S+M}. It follows from equation (4.8) that such an effect may alter the effective degree of hydration of the protein.

The average times-between-collisions of human serum albumin under the experimental conditions employed vary between $0.3 \cdot 10^{-6}$ and $2 \cdot 10^{-6}$ sec. This value is much larger than the correlation time τ_M of the mobility of human serum albumin proper, which is between $2 \cdot 10^{-8}$ and $5 \cdot 10^{-8}$ sec as determined by the spin-label method. Thus, collisions between human serum albumin molecules cannot significantly affect the values of τ_M.

The occurrence of protein dehydration during collisions is confirmed by our own PMR experiments performed on frozen solutions of human serum albumin. When the albumin concentration was increased four times — from 25 to 100 mg/ml — the hydration decreased to slightly more than one-third, from 1.7 to 0.6 gm H_2O/gm protein (0.01 M phosphate buffer at pH 6.3 + 0.15 M NaCl). It would appear that as a result of associate formation a part of the water of hydration was displaced from the protein into the free solvent.

For a given frequency of collisions, the effective protein hydration increases as the temperature is raised from 10 to 40°C. This may be explained by an increase in the number of water molecules which become oriented under the effect of r protein groups which vigorously react with water — as the hydrogen bonds in the free solvent become weakened and/or as the destabilizing effect of fluctuations of protein cavities on the free solvent becomes attenuated.

Effect of the Fluctuations of Human Serum Albumin on the Properties of the Free Solvent

The PMR method revealed two extremes of T_2 between 20 and 40°C in human serum albumin solutions (Fig. 37), which correspond to a decrease in the correlation time of water molecules ($T_2 \approx 1/\tau_{H_2O}$). They correlate with changes in the flexibility of the albumin, which is a confirmation of the interdependence of these phenomena.

In the presence of 10% and 20% D_2O these effects become intensified, and their peaks shift towards higher temperatures. These relationships are reversible, and are not affected by the presence of EDTA. This means that paramagnetic impurities (ions of bivalent metals) do not play any part in these effects in protein solutions. An increase in the viscosity of the solution, produced by adding 20% saccharose, resulted in their degeneration. This is explained by a decrease in the frequency of the relative fluctuations of the domains of human serum albumin.

If the variation of T_2 with the temperature in protein solutions is different from that of the control function (for buffer solutions), this may indicate a) changes in the properties of the hydrated shell of the protein alone; b) changes in the properties of the free solvent; c) combinations of effects a) and b).

If, in the presence of 5% - 20% D_2O, which displaces ordinary water from the hydrate shell, the observed deviations of T_2 remain unchanged or increase, this is an indication that the structure of the free solvent is perturbed by the protein (an increase in $T_2 \sim 1/\tau_{H_2O}$ corresponds to an increased mobility of water molecules).

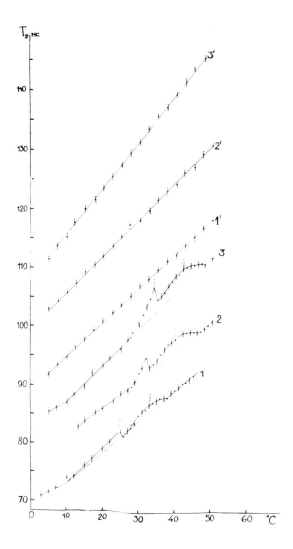

Figure 37. Dependence of T_2 of water protons on the temperature in control solutions (1' − 3') and in 10% solutions of human serum albumin (1 − 3).

1' − 0.01 M phosphate buffer (pH 7.3) + 0.15 M NaCl; 2' − buffer + 10% D_2O; 3' − buffer + 20% D_2O.

1, 2 and 3 correspond to 1', 2' and 3' respectively, and contain human serum albumin in 10% concentration. Peaks I and II represent thermally induced transitions in human serum albumin. Each point represents the average value of a quintuplicate determination of T_2. After Käiväräinen *et al.* (in press).

These results indicate that a rapid exchange between the free solvent and the water contained in the cavities between the domains of human serum albumin during their A ⇌ B fluctuations stimulates fluctuations in the solvent, which destabilize its structure. The effect of these fluctuations is enhanced if HOH is replaced by DOH and DOD.

Compensation Effects in Solutions of Human Serum Albumin*

The application of scanning differential calorimetry to human serum albumin and met-hemoglobin confirmed one of the consequences of the dynamic model — viz., the existence of water-protein compensation effects when the thermodynamic parameters of the macromolecules and of the solvent vary in counter-phase. The resulting change in the C_p of the protein solution as a whole remains constant in the temperature ranges in which the flexibility of the proteins and the mobility of free solvent molecules undergoes significant changes (Fig. 38). It was noted for the first time that low concentrations (5% or less) of D_2O enhance the overall heat capacity of the solution and reduce the thermal stability of proteins (human serum albumin, hemoglobin, papain). Our explanation for it is the destabilization of the tertiary structure of domains and subunits when ordinary water is replaced by heavy water in the vicinity of nonpolar surface segments, because the interaction of DOD and DOH with such segments is less favored thermodynamically than with HOH. Hydrophobic interactions in D_2O are stronger than in H_2O (Lobyshev and Kalinichenko, 1978).

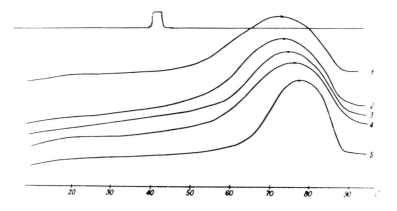

Figure 38. Heat capacity C_p of human serum albumin solutions in the absence and in the presence of D_2O as a function of the temperature: 1 — 3% D_2O; 2 — 5% D_2O; 3 — no D_2O; 4 — 10% D_2O; 5 — 15% D_2O. Concentration of human serum albumin: 10 mg/ml in all cases. 0.01 M phosphate buffer (pH 7.3) + 0.15 M NaCl. After Käivaräinen and Golikova (in press).

* This study was carried out by the author in collaboration with L.I. Golikova.

Higher D_2O concentrations (10-15%) reduce the overall heat capacity of protein solutions as compared to that of D_2O-free solutions, and increase the temperature of denaturation. This last-named fact was known before, and is probably due to the formation of a network of bridging bonds between the polar centers of water sorption on the surface of the protein. The result seems to be a decrease in the number of degrees of freedom of the macromolecule and decreased heat capacity of the solution.

5.3 Ultrasonic Action

Another perturbing factor in the dynamics of biopolymer behavior is the effect of ultrasound. It is reasonable to expect that its effectiveness will be greatest in the zone of frequencies corresponding to the proper fluctuation frequencies (eigen-frequencies) of the proteins themselves, i.e., under specific resonance conditions (El'piner *et al.*, 1970).

The sound wave spreading through the solution disturbs the molecular equilibrium in various ways, and the energy of the sound wave is absorbed. According to Fursov (1971), changes in the hydration of protein molecules are one of several mechanisms by which ultrasound is absorbed. This worker studied the acoustic properties of a number of globular proteins: serum albumin, immunoglobulin G, hemoglobin, casein and RNA-ase. He found that the quenching of sound in solutions of all these proteins decreases with increasing temperature, and increases linearly with increasing concentration of the solution. However, different proteins absorb to different extents under identical experimental conditions. Thus, for instance, serum albumin and hemoglobin, whose molecular masses are approximately equal, show major differences in the magnitude of absorption per molecule: $17.3 \cdot 10^{-34}$ and $23 \cdot 10^{-34}$ $cm^{-1} \cdot sec^2$.

Small detergent concentrations (sodium dodecylsulfate and deoxycholate) enhance the absorption of ultrasound by albumin solutions. Complexing albumin with the acidic mucopolysaccharide heparin in equimolar ratios gives a similar effect; the value of the absorption coefficients of solutions of complex compounds is larger than the sum of the separate absorption coefficients of heparin and albumin solutions separately. The magnitude of this effect decreases to less than one-tenth of its original value when the frequency of the ultrasound is raised from $1.2 \cdot 10^7$ to $6 \cdot 10^7$ cm^{-1}. Unfortunately, no data are available on absorption changes at frequencies below $1.2 \cdot 10^7$ sec^{-1}, and it is thus difficult to say how close it is to the resonance frequency. It may be assumed, on the base of the dynamic model, that the energy of ultrasonic vibrations may be spent, at least in part, on the equalization of the concentrations of A- and B-states of cavities in proteins, just as the resonance absorption of electromagnetic waves results in an equalization of the populations of the different sub-levels. In terms of the theory of resonance absorption, the time T_1 of spin-lattice relaxation will determine the restoration rate of the equilibrium constant $K_{A \rightleftharpoons B} = \exp\left[-(G_A - G_B)/RT\right]$ according to the Boltzmann distribution, by heat exchange between subsystems protein - environmental water.

If it is assumed that the probability of ultrasound-stimulated transitions $P_{A \rightarrow B} \cong P_{B \rightarrow A} \equiv P$, the expression for the absorption of ultrasound will resemble that describing the absorption in an RF field (Carrington and McLaughlin, 1970):

$$dE/dt = \eta_0 \Delta EP/(1 + 2PT_1),\tag{5.24}$$

where $\Delta E = G_A - G_B$; $n_0 = N_A - N_B$ is the difference between the cleft concentrations in the A- and B-states at thermal equilibrium, and P is the probability of induced transitions, which is proportional to the intensity (power) of the sound waves. If the absorption of ultrasound by biopolymers is in fact a resonance process, the saturation effect would be noticeable if the power were sufficiently high. However, this may be difficult to prove experimentally owing to cavitation (generation of voids) and to the destruction of the macromolecule.

Equation (5.24) is a simple interpretation of the increased absorption of ultrasonic energy following the complex formation between the albumin and a detergent or heparin. The formation of specific complexes may produce either a leftward or a rightward shift in the A $\rightleftharpoons$ B equilibrium (Chapter 6); this may be accompanied by an increase in ΔE and n_0 and a corresponding increase in the absorption. It should be borne in mind, however, that as a result of complex formation the frequency of A $\rightleftharpoons$ B transitions may become closer to or move away from the frequency at which the absorption is determined, i.e., the resonance conditions may improve or deteriorate, and thus affect the magnitude of the absorption. In such studies absorption as a function of the frequency must be determined in a wide interval in order to find the ground resonance frequency. This is not a simple task, since a large number of processes, which may be responsible in some measure for the absorption of ultrasound, may take place in biopolymer solutions.

It should be noted in this context that variations of the absorption with frequency which differ from the classical form $af^2 = $ const occur only in solutions of native biopolymers. Fursov (1971) showed that acidic and enzymatic degradations of albumin and hemoglobin considerably reduce the absorption coefficient and restore the quadratic function mentioned above. This means that the relaxation mechanisms, which were operative in the starting solution, now become practically inoperative.

It is interesting to note that the effect of ultrasound on the activity of the enzyme may depend not so much on the carrier frequency as on its modulation frequency (Sarvazyan, 1977). It will be shown (Chapter 7) that the mechanism of the enzymatic action may involve not only the fluctuations of their cavities at frequencies of $\nu_{A \rightleftharpoons B} \leqslant 10^7$ sec^{-1}, but also the relaxational fluctuations of the equilibrium constant $K_{A \rightleftharpoons B}$. If the modulation frequency of ultrasonic waves coincides with the fluctuation frequency of $K_{A \rightleftharpoons B}$, it is easier for the macromolecule to overcome the relaxation-connected activation barriers and thus to accelerate the reaction.

Chapter 6

Dynamic Model of Association and Dissociation of Specific Complexes

In our discussion of the applications of the dynamic model as a description of the properties of biopolymers and their hydrate shells, we have not so far dealt with their functional significance. We shall now attempt to apply the fundamental assumptions which underlie the dynamic model of behavior of proteins in water in order to arrive at the general principles governing the association mechanisms of specific complexes of the type antibody-antigen, enzyme-inhibitor, carrier-ligand, and to discuss the role played by the globule in these processes.

Before a ligand can enter the active site, it must displace water from it, and this requires a certain energy of activation. We shall assume that the sequence of events in the active site which ends in complex formation is as follows (Käiväräinen, 1975, 1978a, 1979a): the ligand collides with the active site in the 'open' (b) state (this event depends on the probability of the collisions and on the lifetime of the active site in the 'open' state); the ligand disturbs the ordered structure of the water in the active site and expels it from the site, partly or completely, the site passing from the b-state to the b*-state; the ligand then induces a transition of the active site from the b*-state to the 'closed' a*-state, when the geometry of the site becomes altered and the specific complex is formed.

In the general case, a transition of an active site to the 'closed' state under the effect of the ligand means that the site becomes less accessible to water and its cavity volume will change to some extent, depending on the size of the ligand. It is quite possible for the volume of the cavity in the a*-state to be almost as large or even larger than its volume in the b-state.

In accordance with the ideas presented in chapter 2, the change in the equilibrium constant $K_{a \rightleftharpoons b} \rightarrow K_{a* \rightleftharpoons b*}$ brought about by complexing with the ligand induces a relaxational process in the protein resulting from the 'mechanical' interactions between cavities; this presupposes a certain structural interrelation between the geometry of the active site and that of the remaining, auxiliary cavities in the protein. The structural relationship between the cavities is determined by the (special) outlined degrees of freedom. It is responsible for the relaxation process, which results in the equalization of the equilibrium constants $K_{A \rightleftharpoons B} \rightarrow K_{a* \rightleftharpoons b*}$. The equalization is mainly due to a change in the value of $K_{A \rightleftharpoons B}$, where A and B are, respectively, the 'open' and the 'closed' state of one of the auxiliary cavities, since the equilibrium a* ⇌ b* becomes stabilized by the ligand (Fig. 39). The relaxation process is accompanied by a dynamic adaptation between the configuration of the ligand and the cavity of the active site; the transition a* → a** takes place and the stability of the complexes increases.

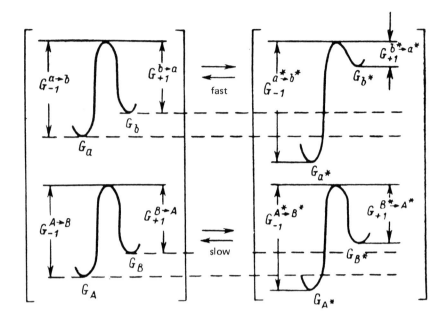

Figure 39. Free energy profiles of the active site and of the auxiliary protein cavity before and after reaction with a ligand. After Käiväräinen (1979a). The diagram shows the free energy levels of the cavities and the free energies of activation of the transitions between the levels.

The ligand does not merely alter the mechanical degrees of freedom in the active site, but also perturbs its electronic structure; the latter effect is particularly prominent in heme-containing proteins, and may constitute an additional contribution to the resulting disequilibrium of the complex. The corresponding perturbations may spread from around the active site throughout the globule, affect the state of the auxiliary cavities and favor the transition of the protein to a new state of equilibrium. The characteristic time of this relaxational process is probably determined by the overall activation energy of elementary rearrangements, which must take place in order to transmit the disturbance from the active site to the auxiliary cavities. This type of interaction we determined as perturbational (Chapter 2).

Auxiliary cavities may interact not only with the active site, on the principle of direct and reverse relationship, but may also react with each other if they are two or more in number. The distinction between mechanical and perturbational interactions between protein cavities is largely arbitrary, since electronic and vibrational motions are interrelated. According to Volkenstein (1975).

"... the distinction between these two types of motion, which may be made, to a sufficient approximation, in the case of simple molecules, is almost impossible in the case of complex molecules. The reason for it is the large amount of stored vibrational energy, which is almost as large as that needed for the excitation of an electron shell."

In constructing our dynamic model of association and dissociation of specific complexes we shall not distinguish between individual types of interactions and their respective relaxational processes, since to do so would complicate the model considerably. However, it will be seen that such distinctions may prove useful in interpreting certain experimental results.

We shall assume, accordingly, that association corresponds to the transition in the state of the active site $b \to b^* \to a^*$, while dissociation corresponds to the opposite process $a^* \to b^* \to b$; the active site interacts with the auxiliary cavity, so that if the latter passes into the 'open' (B) state, the stability of the a^*-state decreases, and the free energy of activation $G^{a^* \to b^*}$ decreases correspondingly. On the other hand, if the auxiliary cavity is in the B-state, the $b^* \to a^*$ transition of the active site reduces the free energy $G^{B \to A}$ of the transition $A \to B$.

In such a case the rate of the transitions $b \to b^* \to a^*$, which result in association, is proportional to the probability of collision between the active site in the b-state and the ligand P_b^{coll}, to the rate of $b \to b^*$ transitions $k_{b \to b^*}$, to the rate of $b^* \to a^*$ transitions $k_{b^* \to a^*}$, and to the lifetime of the auxiliary cavity in the A-state ($t_A \simeq 1/k_{A \to B}$). According to the Eyring-Polanyi equation, the rate constant of the $1 \to 2$ transition is:

$$k^{1 \to 2} = (kT/h) \exp\left(-G^{1 \to 2}/RT\right), \qquad (6.1)$$

where $G^{1 \to 2}$ is the free activation energy of the transition and the remaining symbols have their usual meanings.

The considerations just presented lead to the following expressions for the rate constants of association k_1, dissociation k_{-1}, and for the equilibrium constants of association K_{ass} and dissociation K_{diss} of the ligand with a protein with only one auxiliary cavity (Käiväräinen, 1978a, 1979a):

$$k_1 \sim P_b^{coll} \frac{k_{b \to b^*} k_{b^* \to a^*}}{k_{A \to B}^x} = P_b^{coll} \frac{kT}{h} \exp\left[-\frac{(G^{b \to b^*} + G^{b^* \to a^*}) - xG^{A \to B}}{RT}\right], \qquad (6.2)$$

$$k_{-1} \sim \frac{k_{a^* \to b^*} k_{b^* \to a^*}}{k_{B \to A}^x} = \frac{kT}{h} \exp\left[-\frac{(G^{a^* \to b^*} + G^{b^* \to b}) - xG^{B \to A}}{RT}\right], \qquad (6.3)$$

$$K_{diss} = k_{-1}/k_{+1} \sim (1/P_b^{coll}) K_{b \rightleftharpoons b^*} K_{b^* \rightleftharpoons a^*} K_{B \rightleftharpoons A}^x = (1/P_b^{coll}) K_{b \rightleftharpoons a^*} K_{B \rightleftharpoons A}^x, \qquad (6.4)$$

$$K_{ass} = k_{+1}/k_{-1} = P_b^{coll} K_{b^* \rightleftharpoons b} K_{a^* \rightleftharpoons b^*} K_{A \rightleftharpoons B}^x = P_b^{coll} K_{a^* \rightleftharpoons b} K_{A \rightleftharpoons B}^x, \qquad (6.5)$$

where k are the rate constants of the indexed transitions, G are the free energies of activation of the respective transitions, K are the equilibrium constants between the respective states, and x is the interaction coefficient between the opposite states of the auxiliary cavity and the cavity of the active site. For the sake of simplicity we shall assume that

$$x_{A (b)} = x_{b (A)} = x_{B (a)} = x_{a (B)} \equiv x \; (0 < |x| < 1), \qquad (6.6)$$

even though this condition may not be satisfied in the general case.

These expressions leave out of account the effect of relaxational processes. Since the interaction between the active site and the auxiliary cavity leads to the two opposite states mutually destabilizing each other, the decrease in the free energy of the a-state of the active site ΔG_a during the bonding of the ligand and the increase in G_b by ΔG_b bring about, as a result of relaxation, the following changes $G^{A \to B}$ and $G^{B \to A}$:

$$\Delta G^{A \to B}_{max} = \sigma_1 \Delta G_b \quad (\Delta G_b > 0), \tag{6.7}$$

$$\Delta G^{B \to A}_{max} = \sigma_2 \Delta G_a \quad (\Delta G_a < 0), \tag{6.8}$$

where σ_1 and σ_2 are the flexibility coefficients of the auxiliary cavity of the macromolecule.

If it is assumed that $\sigma_1 \simeq \sigma_2 \equiv \sigma$, it can be readily shown that the values of the equilibrium constants of the auxiliary cavity of the active site before and after relaxation are interconnected as follows:

$$\ln \left(K_{A* \rightleftharpoons B*} / K_{A \rightleftharpoons B} \right) = \sigma \ln \left(K_{a* \rightleftharpoons b*} / K_{a \rightleftharpoons b} \right). \tag{6.9}$$

where

$$\frac{K_{A* \rightleftharpoons B*}}{K_{A \rightleftharpoons B}} = \frac{1 - f_{B*}}{f_{B*}} \bigg| \frac{1 - f_B}{f_B} = \frac{f_{A*}}{f_{B*}} \bigg| \frac{f_A}{f_B} , \tag{6.10}$$

where f_B and f_{B*} are the fractions of time during which the auxiliary cavity is in the 'open' state before and after the relaxational processes, the transition time between each state being much shorter than the lifetime of each state itself.

Since the molecular volumes V_A and V_B correspond, respectively, to the 'closed' and 'open' states of the auxiliary cavity, the time-averaged protein volumes before and after relaxation may be described as follows:

$$\bar{V}_{A \rightleftharpoons B} = f_B V_B + (1 - f_B) V_A, \tag{6.11}$$

$$\bar{V}_{A* \rightleftharpoons B*} = f_{B*} V_B + (1 - f_{B*}) V_A. \tag{6.12}$$

These equations may be used to rewrite (6.10) as follows:

$$K_{A* \rightleftharpoons B*} / K_{A \rightleftharpoons B} = [f_B (\bar{V}_{A* \rightleftharpoons B*} - f_{B*} V_B)] / [f_{B*} (\bar{V}_{A \rightleftharpoons B} - f_B V_B)]. \tag{6.13}$$

If we assume that, under near-physiological conditions $G_a \simeq G_b$, i.e., $K_{a=b} \simeq 1$, equation (6.9) may be written as

$$\ln \left(K_{A* \rightleftharpoons B*} / K_{A \rightleftharpoons B} \right) = -\frac{\sigma}{RT} (G_{a*} - G_{b*}). \tag{6.14}$$

In equation (6.5), K_{ass} refers only to the events involving the active site and the auxiliary protein cavity during the interaction with the ligand. The experimental value of K_{ass} is given by the decrease in the free energy of the entire system protein-ligand-water as a result of complex formation. If the relatively insignificant decrease in the free energy of the system due to dynamic adaptation and to the relaxational changes in the properties of the auxiliary cavity is neglected, it can be shown that the true affinity of the ligand is close to $G_{a*} - G_{b*}$. In fact, the dehydration of the active center and its transition to the 'closed' state do not in themselves alter the free energy of the system if $G_a \simeq G_b$, since the corresponding change in the state of the water is a phase transition of the first order ($\Delta G^{H_2O} = 0$). By definition, G_{a*} corresponds to the free energy of the final state of the active site complex with the ligand. The result is

$$G^{cl} \equiv G_{b*} - G_b \approx G_{b*} - G_a \equiv G^r, \qquad (6.15)$$

where G^{cl} and G^h are the energies of clusterophilic and hydrophobic reactions of the active center, respectively.

The energy of hydrophobic reactions of the ligand with water should be nearly equal to the energy of hydrophobic interactions between the active site in the b*-state and the water (G^h), provided the ligand is complementary to the site. Thus, the overall decrease in the free energy of the system G_{ass} is constituted by the decrease in the free energy of the active site $\Delta G_{ass} = G_a - G_{a*}$, and the decrease in the hydrophobic interactions of the ligand which accompany its transfer from the water to the nonpolar cavity, and which may be expressed as $\Delta G_l = G_{b*} - G_a$. As a result

$$\Delta G_{ass} = \Delta G_{AS} + \Delta G_L = G_{b*} - G_{a*}. \qquad (6.16)$$

Therefore, in equation (6.9), $K_{a*=b*} \simeq K_{ass}$.

We have thus shown that at least a qualitative connection may exist between the K_{ass} of the ligand and the magnitude of the conformational changes in the protein caused by the ligand.

In the general case of a protein containing n auxiliary spaces and $K_{a \rightleftharpoons b} \simeq 1$, the more accurate equation (6.9) will assume the form

$$\ln \left(\frac{K^{(1)}_{A* \rightleftharpoons B*}}{K^{(1)}_{A \rightleftharpoons B}} \frac{K^{(2)}_{A* \rightleftharpoons B*}}{K^{(2)}_{A \rightleftharpoons B}} \cdots \frac{K^{(n)}_{A* \rightleftharpoons B*}}{K^{(n)}_{A \rightleftharpoons B}} \right) = (\sigma_1 + \sigma_2 + \ldots + \sigma_n) \ln K_{ass}. \qquad (6.17)$$

The increase in the equilibrium constants of the auxiliary cavities indicates, in accordance with the dynamic model, that the protein molecule becomes more compact and that its time-averaged linear dimensions decrease.

If the ligand is hydrophobic, it may be expected, in view of the nonpolar nature of the active site, that the free energy of association will to a large extent be determined by the free energy of ligand transfer from water to the nonpolar medium. Since the association reaction is highly specific, the bonding constant may depend on the following factors: the mutually complementary nature of the ligand and the active site, since this is what determines the extent to which water will be displaced out of the active site, and the number of contact points between the ligand and the center; the degree of nonpolarity of the active site; and the size of the ligand and of its nonpolar zones.

The probability of a collision between the ligand and the active site is

$$P_b^{coll} \sim [A] [L] \frac{H^b}{H^L \tau_{L+A}} K_{b \rightleftharpoons a},$$ (6.18)

where $0 < P_b^{coll} < 1$, and $[A]$ and $[L]$ are the respective concentrations of acceptor and ligand; H^b/H^L is the ratio between the cross-sectional areas of the active site in the b-state and of the ligand; $1/\tau_{L+A} = 1/\tau_L + 1/\tau_A$ is the resulting correlation time of the relative Brownian movement between the ligand and the receptor A. According to the Stokes-Einstein Law, $\tau = 4/3 \, \pi a^3 \eta/T$, where a is the effective Stokes radius of the molecule, and η and T are, respectively, the viscosity and the absolute temperature. If the determinant of the ligand which directly participates in the complex formation is small as compared with the size of the ligand as a whole, the dependence of P_b^{coll} on τ_L may strongly affect the value of the association constant.

Hill (1975) discussed the effect of ligand rotation on the diffusion-dependent rate of association between the ligand and the protein from the point of view of Eyring's rate theory. He considered the special cases of spherical and ellipsoidal ligands, and assumed that when the ligand is approaching the protein surface, an effective repulsion potential becomes established, and exerts a restricting effect on the rotation of the ligand. Eyring's gas frequency factor kT/h is replaced by a new factor $D/R\Lambda$, where D is the diffusion coefficient of the ligand, Λ is the so-called de Broglie thermal wavelength, equal to $h/(2\pi mKT)^{1/2}$ (m is the mass of the ligand) and R is the radius of capture around the active site of the protein. Our own expressions for the equilibrium constants will obviously not be affected by this substitution, while in the expressions for the association rate constants the relative thermal motion of the protein and the ligand is, in our view, adequately allowed for by the introduction of P_b^{coll} (e.g. 6.18).

In the case of a protein with three auxiliary cavities (if the relaxation processes are disregarded) k_1, k_{-1} and K_{diss} are given by the following expressions:

$$k_1 = P_b^{coll} \frac{k_{b \to b^*} k_{b^* \to a^*}}{k_{(A \to B)_1}^{x_1} k_{(A \to B)_2}^{x_2} k_{(A \to B)_3}^{x_3}} =$$

$$= P_b^{coll} \frac{kT}{h} \exp\left[-\frac{(G^{b \to b^*} + G^{b^* \to a^*}) - (x_1 G^{(A \to B)_1} + x_2 G^{(A \to B)_2} + x_3 G^{(A \to B)_3})}{RT}\right],$$ (6.19)

$$k_{-1} = \frac{k_{a^* \to b^*} k_{b^* \to b}}{k_{(B \to A)_1}^{x_1} k_{(B \to A)_2}^{x_2} k_{(B \to A)_3}^{x_3}} =$$

$$= \frac{kT}{h} \exp\left[-\frac{(G^{a^* \to b^*} + G^{b^* \to b}) - (x_1 G^{(B \to A)_1} + x_2 G^{(B \to A)_2} + x_3 G^{(B \to A)_3})}{RT}\right],$$ (6.20)

$$K_{diss} = k_{-1}/k_{+1} = (1/P_b^{coll}) K_{b \rightleftharpoons b^*} K_{b^* \rightleftharpoons a^*} K_{(B \rightleftharpoons A)_1}^{x_1} K_{(B \rightleftharpoons A)_2}^{x_2} K_{(B \rightleftharpoons A)_3}^{x_3},$$ (6.21)

where x_1, x_2 and x_3 are the coefficients of interaction of the active site with the first, second and third auxiliary cavity respectively. The relaxation of the protein to a new state of equilibrium is mainly related to the changed equilibrium constants of auxiliary cavities but, since this is also accompanied by a dynamic adaptation of the active site and the ligand, the value of G_{a*} may also change by the small amount ΔG_{a*}.

If the free energies of the states of protein cavities during relaxation vary exponentially during their characteristic times τ, the corresponding changes in $k_1(t)$, $k_{-1}(t)$ and $K_{diss}(t)$ may be represented as follows:

$$k_1(t) = P_b^{coll} \frac{kT}{h} \exp \times$$

$$\times \left\{ -\frac{\begin{array}{l}(G^{b \to b*} + G^{b* \to a*}) - x_1[G_0^{(A \to B)_1} + \sigma_1 \Delta G_b(1 - e^{-t/\tau_1})] + \\ + x_2[G_0^{(A \to B)_2} + \sigma_2 \Delta G_b(1 - e^{-t/\tau_2})] + x_3[G_0^{(A \to B)_3} + \sigma_3 \Delta G_b(1 - e^{-t/\tau_3})]\end{array}}{RT} \right\},$$

$$\hspace{10cm} (6.22)$$

$$k_{-1}(t) = \frac{kT}{h} \exp \times$$

$$\times \left\{ -\frac{\begin{array}{l}[G^{a* \to b*} + \Delta G_{a*}(1 - e^{-t/\tau_0}) + G^{b* \to b}] - x_1[G_0^{(B \to A)_1} + \\ + \sigma_1 \Delta G_{a*}(1 - e^{-t/\tau_1})] + x_2[G_0^{(B \to A)_2} + \sigma_2 \Delta G_a(1 - e^{-t/\tau_2})] + \\ + x_3[G_0^{(B \to A)_3} + \sigma_3 \Delta G_{a*}(1 - e^{-t/\tau_3})]\end{array}}{RT} \right\}, \quad (6.23)$$

where σ_1, σ_2 and σ_3 — in analogy with equations (6.7) and (6.8) — are the flexibility coefficients of the auxiliary cavities. Also

$$K_{diss}(t) = \frac{1}{P_b^{coll}} K_{b \rightleftharpoons b*} K_{b* \rightleftharpoons a*} \exp\left[-\frac{\Delta G_{a*}(1 - e^{-t/\tau_0})}{RT}\right] \times \qquad (6.24)$$

$$\times K_{(B \rightleftharpoons A)_1}^{x_1} \left(\frac{K_{b* \rightleftharpoons a**}}{K_{b \rightleftharpoons a}}\right)^{x_1 \sigma_1(1-e^{-t/\tau_1})} K_{(B \rightleftharpoons A)_2}^{x_2} \left(\frac{K_{b* \rightleftharpoons a*}}{K_{b \rightleftharpoons a}}\right)^{x_2 \sigma_2(1-e^{-t/\tau_2})} \times$$

$$\times K_{(B \rightleftharpoons A)_3}^{x_3} \left(\frac{K_{b* \rightleftharpoons a**}}{K_{b \rightleftharpoons a}}\right)^{x_3 \sigma_3(1-e^{-t/\tau_3})} = \frac{1}{P_b^{coll}} K_{b \rightleftharpoons b*} K_{b* \rightleftharpoons a*} \times$$

$$\times \exp\left[-\frac{\Delta G_{a*}(1 - e^{-t/\tau_0})}{RT}\right] K_{(B \rightleftharpoons A)_1}^{x_1} K_{(B \rightleftharpoons A)_2}^{x_2} K_{(B \rightleftharpoons A)_3}^{x_3} \times$$

$$\times \left(\frac{K_{b* \rightleftharpoons a**}}{K_{b \rightleftharpoons a}}\right)^{x_1 \sigma_1(1-e^{-t/\tau_1}) + x_2 \sigma_2(1-e^{-t/\tau_2}) + x_3 \sigma_3(1-e^{-t/\tau_3})}. \qquad ($$

During protein relaxation, K_{diss} decreases as k_1 increases and k_{-1} decreases, since $\Delta G_b > 0$, while $\Delta G_a < 0$.

The values of τ_1, τ_2 and τ_3, which characterize the rates of relational processes, are determined by the overall free energy of activation of the elementary rearrangements which must take place in protein domains for the disturbance to migrate from the active site to the parts constituting the auxiliary cavity.

If both x_2 and $x_3 \ll x_1$, we obtain the following expression at the end of the relaxation period ($t \to \infty$):

$$K_{diss} = (1/P_b^{coll})\, K_{b \rightleftharpoons b*} K_{b* \rightleftharpoons a*} \exp\left(-\Delta G_{a*}/RT\right) K_{(B \rightleftharpoons A)_1}^{x_1} \left(K_{b* \rightleftharpoons a**}/K_{b \rightleftharpoons a}\right)^{x_1 \tau_1}. \tag{6.25}$$

and, correspondingly

$$K_{ass} = P_b^{coll} K_{b* \rightleftharpoons b} K_{a* \rightleftharpoons b*} \exp\left(\Delta G_{a*}/RT\right) K_{(A \rightleftharpoons B)_1}^{x_1} \left(K_{a** \rightleftharpoons b*}/K_{a \rightleftharpoons b}\right)^{x_1 \tau_1}. \tag{6.26}$$

The relaxation will proceed to completion if the times needed for the establishment of the a*-state of the active site are shorter than those required for the protein to relax back into its initial, ligand-less state. Accordingly, the smaller the value of P_b^{coll}, which depends on the ligand concentration, the smaller will be k_1 and the smaller the probability that the relaxation, including a corresponding increase in affinity, will proceed to completion. The condition for the relaxation to proceed to completion may be written as

$$k_1 \gg 1/\tau, \tag{6.27}$$

where τ is the characteristic time of reverse relaxation of the protein.

When the concentration of the ligand becomes so low that $k_1 \leqslant 1/\tau_1$, a smooth $(A* \rightleftharpoons B*) \to (A \rightleftharpoons B)$ transition should be observed. If k_1 is known, τ may be experimentally determined in this way. It may also be expected that the dissociation rate constant K_{-1} and the equilibrium dissociation constant K_{diss} will increase with decreasing ligand concentration as the effect of relaxational processes becomes weaker.

It follows from equation (6.18) that $P_b^{coll} \to 1$ when the concentrations of the ligand and of the receptor increase and/or the resulting correlation time τ_{L+A} decreases — for example, as a result of a decrease in the Stokes radius of the ligand.

The magnitude which is experimentally determined by such methods — equilibrium dialysis and ultracentrifugation — is K_{dss}, which is diffusion-dependent. In order to compare the trans-globular effects induced in a given protein by different ligands, the working conditions must be such that P_b^{coll} is the same in all cases.

6.1 Effect of Complex Formation on the Stability of the Protein

It was shown in Chapter 5 that the overall, time-averaged free energy of a protein containing two nonpolar fluctuating cavities may be represented by the equation:

$$\bar{G} = G_0 + f_a G_a + (1 - f_a) G_b + f_A G_A + (1 - f_A) G_B, \tag{6.28}$$

where f_a, f_A and $(1 - f_a)$, $(1 - f_A)$ are the fractions of time during which the active center and the auxiliary cavity are in the states a, A, b and B respectively; G_a, G_b, G_A and G_B are the free energies of the states indicated by the indexes, and G_0 is the free energy of the rigid molecular matrix, which is independent of the equilibrium constant of the fluctuating protein cavities.

As before (eqs. (5.14) and (5.17)), if the relaxational process and dynamic adaptation are taken into account, we obtain

$$\overline{\Delta G\ (t)} = \Delta G_0 + [RT \Delta \ln f_a - (G_b - G_a)\Delta f_a] + \Delta G_a^0 + \Delta G_{a*}\left(1 - e^{-t/\tau_0}\right) +$$

$$+ [RT \Delta \ln f_A\ (t) - (G_B - G_A)\Delta f_A\ (t)] + \Delta G_A\left(1 - e^{-t/\tau_1}\right). \quad (6.29)$$

Since it is assumed that the change of state of the cavities by the effect of the ligand does not affect $\overline{G_0}$, we have $\overline{\Delta G_0} = 0$. It follows from the model that both ΔG_a and $\Delta G_A < 0$, so that $\overline{\Delta G_0}(t) < 0$ if

$$G_b - G_a \text{ и } G_B - G_A \geqslant RT. \quad (6.30)$$

Thus, it is predicted by the model that when a ligand is bound by a protein, the free energy of the protein rapidly decreases due to a change in the state of the active site. If relaxational changes of a similar type are thereby induced in the behavior of the auxiliary cavities, the overall free energy of the protein will decrease at a corresponding rate. It should be borne in mind that $\Delta G_{a*} \ll \pm\Delta G_a^0$. In the general case, the overall change in the free energy of a protein as a result of binding a ligand will depend on its affinity, on the number of auxiliary cavities and on their flexibility coefficients.

It is seen that in accordance with the dynamic model of association and dissociation of specific complexes the role played by the globule is determined by the interaction between the active site and the auxiliary cavities which, as a result of relaxational processes, enhance the ligand-binding constant and improve the stability of the protein molecule.

6.2 Effect of Complex Formation on Dynamic Properties of Proteins

The fluctuation frequency of any protein cavity may be represented as

$$^v A \rightleftharpoons B = \frac{1}{t_A + t_B} = \frac{1}{(k_{A \to B})^{-1} + (k_{B \to A})^{-1}} = \quad (6.31)$$

$$= \frac{h/kT}{\exp\left(G^{A \to B}/RT\right) + \exp\left(G^{B \to A}/RT\right)}.$$

If, on forming a complex with the protein, the ligand produced an increase of ΔG_b in the free energy of its active site in the 'open' state, and a decrease of ΔG_a in the free energy of its 'closed' state, the fluctuation frequency of the active site is

$$\nu_{a* \rightleftharpoons b*} = \frac{1}{t_{a*} + t_{b*}} = \frac{h/kT}{\exp\left[(G^{a \rightarrow b} + \Delta G_a)/RT\right] + \exp\left[(G^{b \rightarrow a} - \Delta G_b)/RT\right]}$$

$$(6.32)$$

Following the completion of relaxational processes, the fluctuation frequency of any one of the auxiliary cavities of the protein may, in view of equations (6.7) and (6.8), be represented as follows:

$$\nu_{A \rightleftharpoons B} = \frac{h/kT}{\exp\left[(G^{A \rightarrow B} + \sigma_A \Delta G_b)/RT\right] + \exp\left[(G^{B \rightarrow A} - \sigma_B \Delta G_a)/RT\right]} \quad (6.33)$$

where σ_A and σ_B are the 'flexibility' coefficients of the auxiliary cavity in states A and B respectively.

If $|\Delta G_b| > |\Delta G_a|$ or $\sigma_A > \sigma_A$, the increase in the lifetime of the A-state of the auxiliary cavity will be larger than the decrease in the lifetime of the B-state and $\nu_{A=B}$ will decrease, in accordance with equation (6.33). However, since clusterophilic interactions are more sensitive to cavity changes than hydrophobic interactions, the case $\sigma_B > \sigma_A$, with an increase in $\nu_{A=B}$, is the more probable.

A shorter lifetime of an auxiliary cavity in the 'open' state means that the clusterophilic interactions have weakened, and that the structure of the water molecules in the 'open' state is less orderly. Consequently, the entropy change ΔS which accompanies the migration of these molecules into the environmental medium as a result of B → A transitions will also be smaller. The perturbing effect of protein fluctuations on the environmental water will weaken (Chapter 5).

These processes may be reflected as changes in the IR and NMR spectra of the water in protein solutions. We shall now consider a number of examples, in order to demonstrate that the available experimental data correspond to the proposed mechanism of formation of specific complexes and its implications.

6.3 Immunoglobulins

As a result of the work carried out in numerous laboratories, in particular by the research teams of Porter (1973) and Edelman (1973), the fundamental structure of antibody (AT) molecules could be established. The molecules of all five main classes of immunoglobulins – the monomeric IgG, IgD and IgE, and the polymeric IgM and IgA – contain heavy chains, which are specific to each class, and light chains of κ- or λ-types, which are common to all classes (Putman, 1969; Nezlin, 1972). The structure of the active site of the molecule, i.e., the antigen-bonding sites, includes the amino acid residues of the so-called variable zones of both chains – about 110 residues counting from the amino end of the chain; the remaining biological functions of antibody molecules – complement bonding, bonding to cell membranes etc. – are performed by the remaining, larger zones in the chains, which are constant by primary structure and are specific to the given class of immunoglobulins (Gally, 1973; Nisonoff *et al.,* 1975). The structure and the functions of immunoglobulins have been reviewed in several monographs (Käivaräinen and Nezlin, 1978; Cathou, 1978; Metzger, 1978).

Immunoglobulins G are T-shaped or Y-shaped molecules, with a molecular mass of 150,000, consisting of identical Fab-subunits (molecular mass 50,000), each one containing one antigen-binding site; they are connected through flexible polypeptide chain segments with an Fc-subunit (molecular mass 50,000). Each subunit comprises four domains, which are stabilized by internal disulfide bonds (Fig. 40). X-ray diffraction studies indicate that the variable and the constant domains of Fab- and Fc-subunits are situated at some distance (1.0 - 1.5 nm) from one another. The domain pairs of immunoglobulins are interconnected by thin polypeptide chain segments which are flexible, so that the interdomain cavities may vary in size.

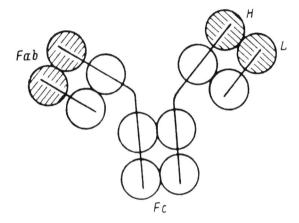

Figure 40. Multiglobular structure of immunoglobulin G. After Metzger (1978).

It is not very likely that the signal transmission from the active site of the antibody, formed by the variable domains of Fab-subunits, to the complement-bonding segments, located on the remote Fc-subunit, takes place through these flexible segments. We believe that it is the water in the protein cavities which takes part in this process.

The behavior dynamics of the macromolecules is determined by the structure and the size of their elements (domains), which are relatively independent and relatively rigid. We shall accordingly consider the structure and the interactions of these domains, as indicated by X-ray diffraction data obtained on immunoglobulins.

Domain Structure

All the domains studied had a similar three-dimensional structure. As could be expected from the data on their primary structure, homologous variable and homologous constant domains resemble each other more closely than do variable and constant domains. The principal structural feature of a domain is the antiparallel β-folded layer, which consists of stretched segments of a polypeptide chain, interconnected by curved 'joints'. Two such layers lie next to each other, much like a sandwich, antiparallel to the chain segments. The contiguous segments are interconnected by hydrogen bonds. Such a two-layered structure is especially evident in constant domains, in which it is more regular. Variable domains contain a supplementary loop, which contains a second, hypervariable region.

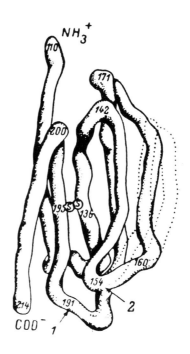

Figure 41. Spatial model of a constant domain of λ-type polypeptide chain of human immunoglobulin. After Poljak (1975). Dotted line shows location of the supplementary loop in the variable domain. Arrows show the locations of antigenic zones of Or (1) and Kern (2).

It is known from studies of the primary structure (Wu and Kabat, 1970) that residues in the variable region may be subdivided into hyper-variable residues, which are responsible for the qualitative differences between the chains, and the non-hypervariable or carcass residues. The hypervariable residues are accessible to solvents, and may be coiled in different manners in different variable domains. Thus, for instance, the first hypervariable light-chain zone of REI protein is a stretched chain, while in NEW and Mcg proteins it is coiled (spiral-shaped). In McPC603 protein this site contains an insert of six amino acid residues, which is stretched into a loop. On the other hand, the layouts of the carcass residues are very much alike in all domains, except for small variations. This similarity in the tertiary structures of the variable domains indicates that nonvariable residues constitute a rigid carcass (matrix), with the hypervariable loops built, as it were, into it.

One Fab-fragment is constituted by four domains — two variable and two constant ones, subject to the following rules.

1. The two constant domains interact more closely and, as a result, the constant region as a whole is more compact than the variable region. The reason for it is that constant domains interact by way of four-segment layers, forming a large contact surface, while the variable domains interact by three-segment layers.

2. The principal interactions between domains are lateral, i.e., the constant domains interact with each other, and the variable domains interact with each other, while interactions between variable and constant domains are less evident.

3. Both heavy-chain domains – V_H and C_H – are in closer contact than the two light-chain domains and cause the Fab-fragment as a whole to be somewhat bent. A similar bend is also present in the light-chain diner (Fig. 42), which consists of two light-chain monomers with an identical structure. Nevertheless, one of the monomers forms a 70° angle between the domains, while the other forms a 110° angle, which determines the curvature of the molecules as a whole.

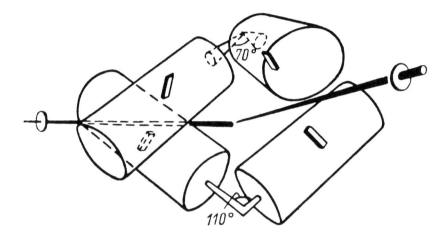

Figure 42. A schematic representation of Bence-Jones protein, modelling the Fab-subunit, based on X-ray diffraction data. After Schiffer *et al.* (1973). *To the right* – variable domains; *to the left* – constant domains.

The structure of the Fc-fragment has no such curvature. It is seen (Fig. 43) that both the C_H3-domains are in close contact with each other, much like the C_L- and C_H-domains in the Fab-fragment, while the C_H-domains are out of contact with each other (except for the covalent bond in the 'joint' region) and form a large cavity.

Structure and Mobility of the Active Site of Antibodies

The Fab-fragments of two myeloma proteins – human IgG and mouse IgA – were studied by high-resolution X-ray technique. It was found that the former protein effectively reacts with a number of compounds, in particular with the hydroxylated Vitamin K_1

Figure 43. A schematic representation of the position of α-carbon atoms of the polypeptide chains of the Fc-fragment of myeloma IgG protein. After Huber (1976).

(K$_1$-OH). The latter — McPC603 protein synthesized by a myelomatous mouse — forms precipitates with several natural antigens, including pneumococcal polysaccharides, and reacts with posphotylcholine, which is the active part of this antigen. Both these proteins are very suitable models for studies of antigen-antibody complexes, since it is possible to prepare crystals of their Fab-fragments with their respective haptens.

Atomic models of Fab-fragments, constructed in conformity with X-ray diffraction data, show that variable domains of light and heavy chains are interconnected in a pattern where hypervariable segments form a single surface in contact with the antigen. The shapes and sizes of the active site cavities were different in the two proteins. The active site of McPC603 protein formed a fairly deep pit, about 1.5 nm wide and 1.2 nm deep, while that of the NEW protein formed a shallow groove, about 1.5 nm long, 0.6 nm wide and 0.6 nm deep. The reason for these differences are the inserts in the three hypervariable segments of the former protein. The active site zone of the NEW protein, which contained the K$_1$-OH hapten, was formed by 23 residues of hypervariable zones of light and heavy chains. The naphthoquinone ring of the hapten was in close contact with the phenolic Tyr90 ring of the light chain, with another two residues of the same chain, and with one residue of the heavy chain. The phytyl part of the hapten interacts with four heavy chain residues and with the side groups of three other residues of the hypervariable zone. Accordingly, this hapten reacts with 10 residues in all, i.e., only with some of the active site residues.

The active site cavity of the McPC603 mouse protein is larger and consists of 35 residues of hypervariable zones of both light and heavy chains. It is seen in Fig. 44 that phosphorylcholine is located inside a deep pit. The choline part reacts with the residues of both chains, while the phosphate reacts with the residues of the heavy chain alone. As before, hapten fills only a part of the active site cavity; the remaining part is probably capable of reacting with the hapten carrier and helps in stabilizing the mobility of the site.

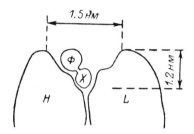

Figure 44. Location of phosphorylcholine hapten in the active site of myelomatous mouse protein McPC603. After Padlan *et al.* (1974).

Φ — phosphate; X — choline; H and L — heavy and light polypeptide chains.

Studies of dimers of the variable domains of Bence-Jones protein, whose structure resembles that of the Fab-subunit of IgG, revealed that the inter-domain contact zone contains not only the quasi-active site cavity, but also a small nonpolar cavity, containing three water molecules (Epp *et al.*, 1975). One H_2O molecule only, which was fairly firmly bonded, could be identified on the very bottom of the active site cavity. It would appear that the mobility of the other water molecules located in this cavity is fairly high, so that no sharp X-ray pattern can be obtained. This study revealed yet another interesting phenomenon — to wit, that the intra-domain S-S bonds are capable of existing as two stereo-isomers in roughly equal concentrations. Differential Fourier analysis also revealed two possible locations of Glu100, thus indicating that variable domains display a dynamic behavior in solution.

The V_L-dimers of the myelomatous MOPS315 protein can also exist in two pH-dependent conformations. In the more 'open' conformation at pH 8 the V_L-dimers interconnect two haptens of dinitrophenyllysine, while at pH 6 only one hapten is bound, but the affinity is 3-4 times higher (Givol, 1978).

High-resolution NMR (270 MHz) was applied to the study of the active site of the variable domain (Fv-fragment) of dinitrophenyl-binding myeloma immunoglobulin G (MOPS315) (Dower *et al.*, 1977). The differential spectra obtained after the addition of various dinitrophenyl (DNP) modifications to Fv, showed that the active site is conical-shaped, and that its dimensions are 0.8x1.1 nm at the top and not less than 0.03x0.08 nm at the bottom, and that it is strongly saturated with aromatic amino acids (Sutton *et al.,* 1977).

The absorption bands in the NMR spectra, which correspond to certain amino acids, become broader in the presence of hapten. This effect is most probably due to a change in the mobility of phenylalanine or tyrosine residues which, as was shown by Campbell *et al.* (1975), fluctuate around their symmetry axes at 10^3 - 10^4 sec^{-1}. At such rates the resonance of two *ortho*- and two *meta*-protons becomes identical. If the motion of the amino acids slows down, or if the difference between the chemical shifts of the *ortho*- and *meta*-states increases, the absorption line may broaden considerably. Since the Fv-fragment is distinguished by small chemical shifts, the broadening is most probably due to the stabilization of tyrosine or phenylalanine residues of the active site as a result of hapten binding.

In the course of our studies (Käiväräinen *et al.,* 1980a) of affinity by heterogeneous pig anti-DNP antibodies and their subunits, using spin-labeled DNP-SL haptens, it was noted that the EPR spectrum of the AT + DNP-SL complex conserved the form which indicated that DNP-SL can exist in two different environments, which have, respectively, a larger and a smaller micro-viscosity, up to a molar ratio [hapten] : [active site] = 0.35. These two states of spin-labeled hapten were interpreted as resulting from the fluctuations of the active site between 'open' and 'closed' states, at a frequency of $\leqslant 10^7$ sec^{-1}. The results thus obtained indicated that the stronger the affinity of the active site to the spin-labeled hapten, i.e., the lower the value of the [hapten] : [active site] ratio, the larger the equilibrium shift towards the 'closed' state, and the more rigid becomes the complex. This is manifested by the redistribution between A- and B-components of the EPR spectrum and by the increase in the resulting correlation time τ_{R+M} of the AT+DNP-SL complex (Fig. 45).

Hsia and Piette (1969) studied the cross-reactivity and structural heterogeneity of rabbit anti-DNP with the aid of spin-labeled haptens and found that the EPR spectra of spin-labeled cross-haptens differ from the homologous spectra by the presence of components indicating a more rapid rotation. This may also result from the equilibrium between two orientations of spin-labeled cross-haptens in the active site – possibly as a result of cavity fluctuations. Homologous spin-labeled haptens produce a large shift in the equilibrium between the two states of the active site towards the more compact state, and for this reason the narrow components disappear almost entirely.

Zav'yalov *et al.* (1975) noted a reversible conformational transition in the zone of antigen-binding site of Fab-fragments and in the variable parts of Bence-Jones protein between 25 and 35°C at physiological pH-values. This transition may be recorded by dispersion of optical rotation, differential perturbation spectrophotometry and electrochemical iodination. The temperature dependence of the optical rotation is S-shaped, which is typical of conformational transitions between two states. A comparison between differential temperature-perturbation spectra of proteins at low and high temperatures reveals a diminished accessibility of tyrosines in the nonpolar cleft of the active site to water, ethylene glycol and iodine. Since the β-structure, which forms the rigid matrix of the V-domains, remains unchanged, this transition may be caused by a changed conform-

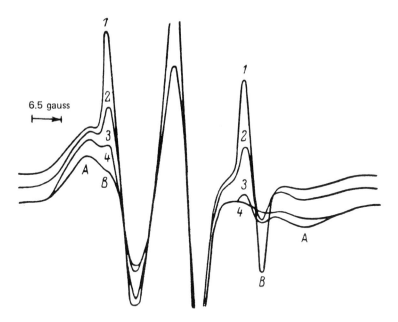

Figure 45. EPR spectra of complexes with spin-labeled haptens (Ab + DNP-SL) at various [hapten] : [active site] antibody ratios. After Käiväräinen (1980).
1 − [1.2] : [1]; *2* − [0.9] : [1]; *3* − [0.7] : [1]; *4* − [0.35] : [1]. 0.01 M phosphate buffer.
(pH 7.3) + 0.15 M NaCl. 22°C.

ation of hypervariable segments and/or by a change in the relative locations of variable domains. In our opinion, the changes which were noted could result from a shift in equilibrium between two short-lived states of the fluctuating active-site cavity towards the 'closed' state. Here, the temperature acts as an agent which perturbs the structure of water cluster and enhances hydrophobic interactions in accordance with the mechanism described in Chapter 5. Agents such as glycerol or salts affect the amplitude of the conformation transition, which may be interpreted as the result of their additional perturbing effect on the stability of the low-temperature 'open' state. A similar explanation may be offered for the effect of a sufficiently high proton concentration at pH ≤ 6.5.

Relative Mobility of the Domains Located in the Subunits

The above description of the properties of active sites in antibodies included a discussion of the experimental data indicating that the variable domains constituting the active site cavity are relatively mobile. However, it would appear that pairs of variable and constant domains of Fab-subunits are also capable of executing relative fluctuations.

Käiväräinen *et al.* (1973b) noted a surprising resemblance between the EPR spectra of various spin-labeled IgG, their Fab-fragments and light-chain dimers, whose structure is analogous to that of Fab-subunits (Fig. 46). The spectrum actually observed is the result of superposition of two spectra, corresponding to the slower (A-component) and the faster (B-component) rotation of the 2,2,6,6-tetramethylpiperidine-4-amino-N-dichloro-triazine label.

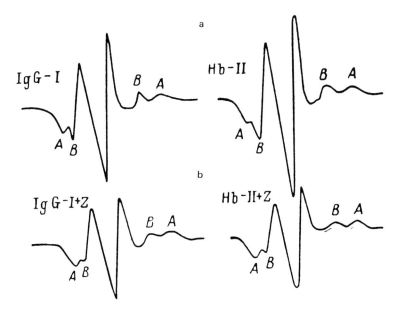

Figure 46. EPR spectra of spin-labeled rabbit IgG (IgG-I) and human hemoglobin (Hb-II) (a), (b) in the presence of $6.4 \cdot 10^{-2}$ M potassium ferricyanide. After Käiväräinen *et al.* (1973b). 0.1 M Medinal buffer; pH 7.4; 20°C.

Since under the experimental conditions employed one label was bound to each Fab-fragment and each light-chain dimer, it was natural to assume that the EPR spectrum of spin-labeled proteins is produced owing to the ability of the label to exist in two states relative to the protein surface. There was no link between the label and the Fc-subunit. The similarity between the EPR spectra of various spin-labeled immunoglobulins and their subunits seems to be due to the uniform structure of elements such as domains and their mutual interactions.

Since the primary structure of the variable part of the various spin-labeled proteins varied strongly, whereas the constant, C-domain-forming part remained essentially unchanged, the unchanged form of EPR spectra indicates that the label is bound to the (probably histidine) residue of the constant domain.

The qualitative aspect of the spectra thus obtained resembled the spectrum of hemoglobin, spin-labeled at the β-93 SH group. It had been conclusively proved that the EPR spectrum of this protein was in fact generated by the fluctuation of the label between two states, and that the equilibrium constant between A- and B-states of the label reflects the equilibrium between two conformations of the β-chains (McConnel *et al.*,

1968; Moffat, 1971). In the A-state the label replaces TyrHC2(145)β and becomes immobilized in a small-sized location in the protein — the tyrosine 'pocket' — whereas in the B-state it is displaced from the 'pocket' by the tyrosine, becomes exposed to the solvent and rotates freely.

In order to confirm the similarity between the EPR spectra of the spin-labeled Fab-fragments of IgG and Hb, the reactions taking place between the N-O groups of the labels of these proteins and the broadening ions were compared. Likhtenshtein's (1974) method for localizing the N-O group of the label with respect to the protein surface is based on the fact that the broadening of the line (ΔH) in the EPR spectrum of the label interacting with the paramagnetic complex depends on the rate constant of exchange relaxation:

$$\Delta H = KC/1.52 \cdot 10^7 \text{ gauss} \tag{6.34}$$

where C is the molar concentration of the paramagnetic complex and K varies with the accessibility of the N-O groups of label for the complex.

It is seen from Fig. 46 that a given concentration of potassium ferricyanide (broadener) results in practically the same changes in both spectra. The broadening and the decrease in the intensity of the B-component indicates that the N-O groups of the labels in the B-state are located on the surface of the protein and are sufficiently accessible to the action of broadener ions. The N-O group in the A-state seems to be located in the pits of IgG and of hemoglobin molecules, and is not readily accessible to broadeners, since the presence of a broadener in solution of a spin-labeled IgG does not significantly alter the A-components of EPR spectra. It would seem that the nature of the spectra of spin-labeled oxyhemoglobins and immunoglobulin subunits is quite similar but, judging by the results of X-ray diffraction, the Fab-domains do not contain a zone analogous to the tyrosine 'pocket', and have no free C-end capable of displacing the label from such a cavity.

It would seem, accordingly, that a change in the micro-environment of the label's N-O group results from changes in the compactness of Fab and light-chain dimers, caused by thermal fluctuations. According to the hypothesis of Käiväräinen et al. (1972a, 1973), which was also subsequently arrived at by Huber (1976) on the strength of X-ray diffraction data, C- and V-domain pairs may approach or move away from each other owing to the flexibility of the polypeptide chain segments by which they are interlinked. Jump-like transitions of the cavities formed by the domains between the 'open' and the 'closed' states result.

If this hypothesis is accepted, it may be expected that a proteolytic digestion of the fragments interlinking V- and C-domain pairs will result in the disappearance of the broad A-components of the EPR spectrum, corresponding to the 'closed' state of the cavity. We could in fact observe this effect in the case of spin-labeled light-chain dimers (Fig. 47). Physicochemical (Björk et al., 1971), immunochemical (Vengerova et al., 1972) and X-ray diffraction (Epp et al., 1975; Poljak, 1975) studies showed that the spatial structure of the domains remains unchanged after they had been separated by proteolysis.

This hypothesis has also been confirmed by other workers. It was noted in Edmunson's laboratory that Bence-Jones proteins are capable of crystallizing in two forms — trigonal and needle-shaped, corresponding to the different conformers of the protein. The trigonal conformer passes into the other one when acted upon by ligands. The reverse

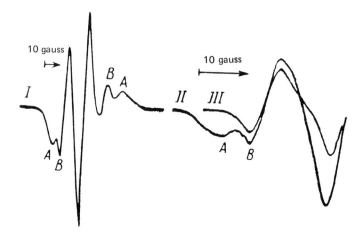

Figure 47. EPR spectra of spin-labeled (at histidine residues) light-chain dimers. After Käiväräinen and Nezlin (1976a).

I — complete EPR spectrum of intact light-chain dimers; II — the left-hand part of the spectrum drawn to a different scale; III — the same part of the spectrum after tryptic digestion of spin-labeled light chains into domains.

transition may be effected by reducing the S-S bond in the C-end region of the chain (Ely, 1978).

Simple hydrodynamic assumptions were used to calculate the probable correlation times of the respective fluctuations of variable and constant Fab domains, and of thermal rotation of Fc relative to $F(ab')_2$; the calculated values were $6 \cdot 10^{-9}$ and $3 \cdot 10^{-8}$ sec (McCammon and Karplus, 1977). The value of $3 \cdot 10^{-8}$ is in agreement with experimental studies of flexibility of immunoglobulins, but $6 \cdot 10^{-8}$ is close only to the correlation time of isolated Fv-fragments representing V_H-V_L domain pairs of IgA (Dwek *et al.,* 1975a, b).

Pilz *et al.* (1973, 1974) showed, using small-angle X-ray scattering and X-ray diffraction data, that the structural mobility of immunoglobulin subunits is indicated, *inter alia,* by the fact that their sizes in the crystal and in solution are different. The subunits assume a less compact form in solution. The respective volumes of Fab- and Fc-subunits in solution are 75 and 91 nm^3. Since the two subunits have practically the same molecular mass (50,000), the much larger volume of Fc indicates that this subunit is more mobile. This conclusion is confirmed by a study of the accessibility of tyrosines and tryptophans to the solvent (Isenman *et al.,* 1975), which revealed that most of the exposed Fc chromophores are localized in C_H2 domains, and that the number of the exposed tyrosines in the isolated C_H2- and C_H3-domains is only 5% larger than in the intact Fc subunits. When the domains are isolated, the number of exposed tryptophans also increases to a small extent: from 1.0 in the intact Fc to 1.4 in $pF'c + C_H2$.

The method of fluorescence polarization revealed that the intramolecular mobility of Fc is larger than that of Fab (Dudich *et al.,* 1978). Thus, for instance, in the case of dansylated Fab from human IgG the relaxation time was 65 nsec, while being 49 nsec for Fc. The authors noted an interesting effect — viz., a fast decrease in the correlation time of the dansylated Fc at concentrations between 2.5 and 2 μm. No such effect was noted for Fab. The effect noted for Fc was attributed to its partial dissociation. Similar effects are sometimes encountered in very dilute solutions of oligomeric proteins.

Willan *et al.* (1977) and Nezlin *et al.* (1978) made promising attempts to selectively modify carbohydrates by Fc spin label following their oxidation by periodate. It was confirmed by the spin-label method that the internal mobility of Fc is larger than that of Fab (Timofeev *et al.*, 1978). The correlation time obtained for Fc was only one-half as large as the value calculated on the assumption that this IgG subunit is a solid particle.

We studied the temperature dependence of conformational mobility of intact IgE molecules, labeled at the carbohydrate components with 4-amino-2,2,6,6-tetramethyl-piperidine-1-oxyl spin label after previous periodate oxidation. Analysis of experimental correlation times led to the conclusion that both in IgE and IgG molecules the intramolecular mobility of the Fc-fragment is higher than that of Fab under normal conditions. The mobility of spin-labeled carbohydrates with respect to the protein surface increases with the temperature.

In another study (Käiväräinen *et al.*, in press), involving the use of spin-labeled DNP-SL, intramolecular mobility of anti-DNP Fab subunits was noted. When the temperature decreased from 20 to 5°C, the calculated correlation time of Fab decreased from 20 to 12 nsec. The labilization of the relative mobility of Fab domains is probably caused by the weakening of hydrophobic interactions, which stabilize the more compact A-conformer of Fab. Very probably, rotary motions through a certain angle around the long axis of Fab take place — in addition to the relative translatory-rotary motion of 2V- and 2C-domains of Fab, which is responsible for the creation of A- and B-conformers, and whose frequency is relatively low (about 10^7 per second). If in the less compact B-conformer such movements have significantly shorter correlation times than in the A-conformer (for example, 7 and 25 nsec respectively), then the increased flexibility of Fab at lower temperatures may result not merely in the destabilization of the A-conformer, but also in a shift of the A $\rightleftharpoons$ B equilibrium to the right (see eq. 6.60). It is interesting to note that in the presence of 5% D_2O in the Fab + DNP-SL solution, intramolecular flexibility of this protein and of Ab + DNP-SL (Fig. 50) cannot be noted even at 5°C.

The properties of the Fc-fragment are of special interest, because its C_H2-domains determine the complement-binding functions. In addition, the fragment may form complexes with special receptors of immuno-competent cells. The fixation and activation mechanism of the complement system is still unclear, but it is certain that in the intact molecule there is a connection between the conformational state of its $F(ab')_2$ part and the complement-fixing activity of Fc. Thus, for instance, the native IgGI is active, while the reduced and alkylated IgGI is not, but an isolated Fc-fragment of the latter molecule is again active. The digestion of Fc off the inactive IgG also converts it to its active form (Isenman *et al.*, 1975). There are two possible explanations for the activation effect: (i) exposure of complement-fixing Fc sites as a result of conformational changes in $F(ab')$ or complete removal of this part of the molecule; (ii) generation of fixation sites during the structural changes in Fc, induced by changes in $F(ab')_2$ or by the cleavage of $F(ab')_2$.

That the conformation of Fc is in fact changed during the digestion of F(ab′)₂ is indicated by the fact that it becomes capable of forming complexes with lymphocyte membranes or of causing their proliferation *in vitro*. A similar effect can be produced in antibodies in the IgG class by thermal aggregation and formation of sufficiently large-sized
complexes with the antigen.

Small thermal IgG aggregates, consisting of 2 or 3 molecules, are active towards the
complement as well. As the degree of aggregation increases under the effect of heat, so
does the activity. However, alkaline treatment also yielded highly active units which were
no larger than dimers (Füst *et al.*, 1978).

The Shape of Antibody Molecules and the Relative Mobility of their Subunits

In the view of Huber (1976), who studied the X-ray diffraction patterns of the integral
molecule of myeloma immunoglobulin G, that molecule possesses three types of flexibility. Firstly, the Fab subunits may rotate with respect to one another, depending on the
heavy chain fragments between the subunits and on the disulfide bond between the heavy
chains. Secondly, the flexibility of the chain fragments between the variable and the constant domains of Fab-fragments allows the domains to move (Fig. 48). Thirdly, the type of
flexibility which is responsible for the mobility of the Fc-fragment with respect to both
Fab-fragments, and for the low electron density in the 'hinge' zone of the myeloma IgG.

According to our own data (Käiväräinen *et al.*, 1973b; Käiväräinen and Nezlin, 1976a,
b), it is the second flexibility type which is probably responsible for the existence of A-
and B-conformers of Fab. Fab may yet display a fourth type of mobility, viz., reorientation of V- and C-domain pairs around their interconnecting longitudinal axis; this accounts for the flexibility of Fab at low temperatures (about 5°C) (Käiväräinen 1980).

The classical study of Valentine and Green (1967) deals with soluble complexes of
antibodies with dinitrophenyl group and bivalent hapten. The microphotographs obtained
in this study showed ring structures of different shapes consisting of di-, tri-, tetra- and
pentamers of antibody molecules interconnected by hapten. It is noteworthy that the
angle between the Fab-fragments was different in aggregates of different types, thus indicating that it might be affected by the flexibility of the polypeptide chain segments interconnecting the Fab-fragments. In another series of experiments (Feinstein and Munn,
1969), the flexibility of IgM was confirmed by electron microscopy.

The method of fluorescence depolarization is an important tool in the study of the independent rotation of immunoglobulin subunits. Studies of immunoglobulins included
comparisons between experimental correlation times of immunoglobulins and the correlation times calculated on the assumption that immunoglobulin is a rigid molecule. Thus,
for instance, if it is assumed that IgG molecule is a rigid ellipsoid of rotation, with axis
lengths in the ratio of 1 to 2, and a molecular mass of 150,000, its correlation time
should be 73 nsec, whereas the experimental values are usually very much lower − 20 - 40
nsec (Zagyansky *et al.*, 1969; Nezlin *et al.*, 1973; Cathou, 1978). This indicates that IgG
fragments are capable of independent relative displacements, i.e., that their interconnecting polypeptide segments are flexible. Data obtained with the aid of spin labels were in

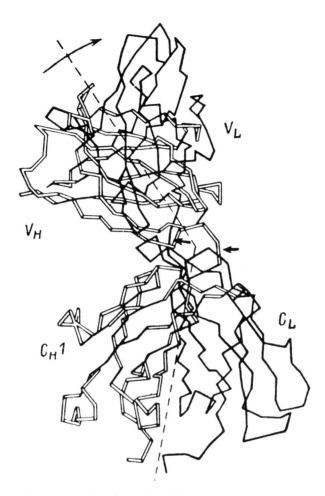

Figure 48. Diagramatic representation of α-carbon skeleton of the Fab fragment of NEW protein. After Poljak (1975).

V_H and C_HI are the heavy chain domains, V_L and C_L arc the light chain domains. The two short arrows indicate chain segments between the domains; the long arrow shows the possible direction of rotation of variable with respect to constant domains.

agreement with these conclusions. Thus, Dwek et al. (1975b) found 44.4, 23 and 6.5 nsec as the respective correlation times of spin-labeled dinitrophenyl, localized in the active site of IgA (MOPS315), and its Fab- and Fv-fragments.

Käiväräinen *et al.* (1973a, 1974) were the first to prepare rabbit antibodies directly against the 2,2,6,6-tetramethylpiperidine-4-amino-(N-dichlorotriazine) spin label involved in the immunization by hapten. As a result of formation of specific AT complexes and their pepsin and papain fragments with a spin label, the latter became firmly immobilized (Fig. 49). Mixing appropriate amounts of solutions of nonspecific rabbit IgG and label under similar conditions merely caused the EPR spectrum of the free label to appear. The correlation times were calculated by the method proposed by Shimshick and McConnel (1972). Around the isoelectric point (pH 6.3) they were found to be 32, 30 and 18 nsec for complexes of spin label with antibody, $F(ab')_2$ and Fab, respectively.

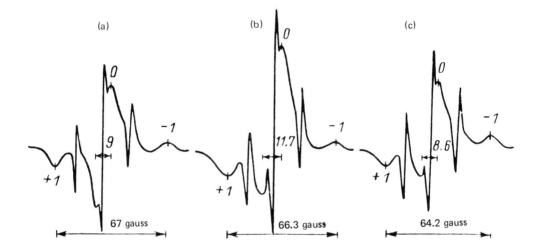

Figure 49. EPR spectra of solutions of complexes of spin label hapten with rabbit antibodies (a), their $F(ab')_2$ subunits (b) and monovalent Fab subunits (c). After Käiväräinen *et al.* (1973a). 0.1 M phosphate buffer (pH 6.3); 25°C.

The differences between the correlation times of the isolated Fab-fragment and as a constituent part of $F(ab')_2$ and the intact antibody indicate a limitation on the freedom of rotation of Fab in composition of $F(ab')_2$ and IgG. It was noted that at pH 6.3 (around the isoelectric point of Fab-fragments) the center line of the EPR spectrum of the complex between the label and the bivalent pepsin hydrolyzate is much wider than the center lines of the spectra of two other complexes (Fig. 49). This may be caused by dipole-dipole interaction between the spins of unpaired label electrons, located in the active sites of neighboring Fab-fragments.

When the pH was changed from 4.5 to 5.5 and from 7.5 to 8.0, the dipole-dipole broadening of the spectral lines of this complex was not noted. In the case of intact antibody molecules with a spin label, the dipole broadening of EPR spectral lines failed to be observed throughout the 4.5 to 8.0 pH range. Thus, cleavage of the Fc-fragment reduces the distance between the Fab-fragments around the isoelectric point only, indicating that the Fc-fragment and the electrostatic repulsion both stabilize the native, Y-shaped structure of IgG. A very important consequence of these results as concerns the understanding of the complement-binding effect of the antibody, is that a reverse relationship — viz., between the angle formed by Fab-subunits and the conformation of the Fc-subunit — is also possible.

Humphries and McConnel (1976) confirmed that it was possible to obtain an antibody against the spin label used for hapten immunization; this technique is now being used with great success in studies of various immunological reactions.

The approach described in section 3.6 for determination of τ_R and τ_M was adopted by Käiväräinen (1980) in their study of temperature dependence of the flexibility of pig's anti-DNP antibodies with the aid of spin-labeled DNP-SL; this approach was necessary, since in this case the N-O group of the label is itself free to rotate about the hapten which is fixed inside the active site.

It was noted that standardized values of τ_M increase with the temperature; this reflects the decreased flexibility of the molecule. In the presence of 3% D_2O in solution the antibody molecule was stabilized at 5°C, while in the presence of 10% D_2O the protein could be stabilized even at 10°C. At 20 and 30°C the D_2O effect was absent even at D_2O concentrations of 15%. At 5°C the effect rapidly attained saturation (Fig. 50).

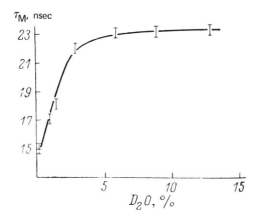

Figure 50. τ_M of antibody + DNP-SL complex as a function of D_2O concentration at 5°C. After Käiväräinen (1980). 0.1 phosphate buffer (pH 7.35) + 0.15 M NaCl. τ_M — calculated for standard conditions (water, 25°C; $\eta/T = 3 \cdot 10^{-5}$ P/K).

The quenching of the D_2O effect with increasing temperature is probably due to the intensification of the exchange processes taking place between the D_2O molecules in the hydrate shell of the protein and the H_2O of the solution.

In another series of experiments the flexibilities of immunoglobulins G and E were compared, both by the method described in section 3.6 and by the method of fluorescence depolarization (Nezlin *et al.*, 1973). The correlation times determined by both methods were found to be similar — 35 and 22 nsec for IgG, 60 and 55 nsec for IgE. This result indicates that IgE molecules are less flexible than IgG molecules. It would appear that the relative shift of IgE fragments is smaller than that of IgG fragments, signifying that there are additional interactions between IgE subunits, but not between IgG subunits.

Sykulev *et al.* (1980a, b) attempted to estimate the contribution of disulfide bonds to the stabilization of the intramolecular mobility of IgE. At 21°C no differences were observed between the flexibilities of the intact and the reduced molecules; this indicates the important part played by noncovalent interactions between immunoglobulin chains and in the stabilization of their native structure. However, at 35°C τ_M of intact IgE molecules remains unchanged, while increasing almost 1.5 times in reduced molecules. This may be interpreted either as the intensification of hydrophobic interactions between the subunits of the reduced IgE at higher temperatures or as an increase in the effective Stokes radius of the reduced molecule as a result of cleavage of the non-covalent bonds between the subunits and the C_H2-domains of the Fc-fragment.

The large τ_M-value of reduced IgE (as compared with the native protein) at 35°C is in agreement with the data reported by Cathou (1978) for anti-DNP of rabbit antibodies by the method of nanosecond fluorescence depolarization.

Thus, there is convincing evidence for the existence of relative thermal displacements of individual subunits of immunoglobulins, and for intramolecular mobility of the subunits themselves. It may be concluded that this type of molecular flexibility has the important function of ensuring the optimum course of the reaction between antibodies and antigens, the latter having differing spatial structures. Flexibility is very important, in particular, in the reaction between both active sites of the antibody and one antigen molecule.

Conformational Changes of Antibodies During Formation of Specific Complexes With Antigens

The consequences of the dynamic model theory can best be compared with experimental results by kinetic studies of complex formation. A study (Levison *et al.,* 1975) of the variation rate of the intensity and polarization of fluorescence during the reaction between fluorescein hapten (H) and a homologous antibody Ab under static conditions, and by the method of stop-flow revealed that the reaction proceeds in two stages:

$$\text{Ab} + \text{H} \underset{k_{-1}}{\rightleftharpoons} \text{Ab} \ldots \text{H} \overset{k_2}{\rightarrow} \text{Ab} - \text{H}. \qquad (6.34)$$

The formation of the intermediate complex Ab...H can be described by a second-order rate constant, and is completed within 0.3 sec after the reagents have been mixed. The reaction Ab...H $\overset{k_2}{\rightarrow}$ Ab-H can be described by a first-order rate constant, and is fully completed only after 60 minutes. The authors interpret the second stage of the reaction as a slow conformational change in the antibody, which may be accompanied by desorption of water. This reaction also proceeds in two stages with isolated Fab-subunits. Neutral salts enhance the association rate constant k_1 in accordance with Hofmeister series. This effect follows from equation (6.19), since the presence of ions reduces the value of $G^{b \to b*}$ by perturbing the ordered water structure in the active site in the b-state.

Thus, salts produce an equilibrium shift between the 'open' and 'closed' states of protein cavities towards the 'closed' state. Accordingly, less water is displaced out of the active site and out of the auxiliary cavities during the binding of the ligand. It may be expected that the positive enthalpy and entropy values, which characterize the association reaction, would decrease correspondingly. This decrease in fact occurs, and proceeds in accordance with the Hofmeister series.

Lancet and Pecht (1976), who studied the kinetics of the reaction between the myeloma monomer of IgA (MOPS 460) and DNP-lysine by the temperature jump method, came to the conclusion that the protein exists as two conformers, both in the free state and when bound by a ligand. Binding the hapten merely alters the equilibrium constant between the conformers. The isomerization process was detected by following the variation in the fluorescence of the protein. The calculated difference between the free energies of these conformers was about 3 KJ/mole.

It was found that the association rate constant increases linearly with increasing hapten concentration, whereas the dissociation rate constant decreases until it has reached a certain value. The reason for it may be the increased probability of completion of relaxational processes in the protein with increasing hapten concentration, in accordance with equation (6.27).

A similar study was performed by Vuk-Pavlovic *et al.* (1978), who studied the reactions of two other myeloma immunoglobulins — XRPC-24 and J-539 — with oligogalactan haptens. The relaxation spectrum also revealed the existence of a fast process, related to the association of the hapten, and of a slow process, which is the isomerization of the proteins. The kinetic behavior of intact proteins and their Fab-subunits was similar, indicating that the effect of the interaction between Fab-subunits and Fc is weak. An analysis of temperature relationships made it possible to calculate the activation processes in all reaction stages, the reaction being satisfactorily described by the following scheme:

$$H + T_0 \underset{k_{-T}}{\overset{k_T}{\rightleftarrows}} T_1$$

$$k_0 \downarrow\uparrow k_{-0} \qquad k_1 \downarrow\uparrow k_{-1}$$

$$H + R_0 \underset{k_{-R}}{\overset{k_R}{\rightleftarrows}} R_1,$$

where $K_0 = k_0/k_{-0}$, $K_1 = k_1/k_{-1}$, T and R are two conformational states of the protein with different affinities to the ligand ($k_R > k_T$).

The difference between the free energies of bonding $\Delta(\Delta G) = \Delta G_R - \Delta G_T$ acts as the driving force for the shift of the conformational equilibrium $T \rightleftharpoons R$ to the right, when $k_0 \rightarrow k_1$. In terms of the dynamic model this process corresponds to a change in the equilibrium constant of the auxiliary cavity $(A \rightleftharpoons B) \rightarrow (A^* \rightleftharpoons B^*)$ due to the hapten-induced change of state of the active site ($b \rightarrow b^* \rightarrow a^*$). The rate constant of this process is between 1 and 10 sec^{-1} at 25°C, which corresponds to the free energy of activation G++ = 67.71 KJ/mole. The isomerization is accompanied by evolution of heat ($\Delta H = 75.4 - 92.2$ KJ/mole) and by an increase in entropy, and becomes 3-4 times faster at 35°C. This is an exact thermodynamic description of the accumulated distortions in the geometry of the cavities between variable and constant domains, destabilization of the ordered water structure in its B-state, and of the consequent shift in the equilibrium constant $K_{A \rightleftharpoons B} \rightarrow K_{A^* \rightleftharpoons B^*}$, accompanied by displacement of water, K_{ass} and the stability of the protein may also be expected in increase at the same time.

Table 8 presents a synopsis of the kinetic parameter ratios.

TABLE 8. Kinetic parameters of the reaction between intact XRPC-24 and hapten at 25°C

i	K_i	k_i, c^{-1}	k_{-i}, c^{-1}	ΔG_i, KJ/mole
T	$4.5 \cdot 10^4$	$4.5 \cdot 10^5$	10.0	–26.5
R	$2.7 \cdot 10^5$	–	–	–30.9
0	0.5	1.05	2.09	1.7
1	3.0	0.3	0.1	– 2.7
Total	$1.2 \cdot 10^5$	–	–	–28.9

If, continuing our analogy with our model, we assume that $K_0 \equiv K_{R \rightleftharpoons T} \sim K_{A \rightleftharpoons B}$ and that $K_1 = K_{R_1 \rightleftharpoons T_1} \sim K_{A* \rightleftharpoons B*}$, then, by equation (6.17) above

$$\ln(K_{A* \rightleftharpoons B*}/K_{A \rightleftharpoons B}) = \sigma \ln K_{ass}$$

it is possible to calculate the 'flexibility' coefficient of the auxiliary cavity of the Fab-subunit.

If the numerical values for XRPC-24 — $\ln(3.0/0.5) = \sigma \ln 1.2 \cdot 10^5$ — are substituted in the equation, we obtain $\sigma = 0.15$; in the case of J-539, $\ln(3.67/0.93) = \sigma \ln 1.2 \cdot 10^5$, and $\sigma = 0.12$.

These values of flexibility coefficients are reasonable. It follows from equations (6.7) and (6.8) that the ultimate changes in the free energies of 'open' and 'closed' states of the auxiliary cavity of Fab subunits constitute 12-15% of the change in the free energies of the states of the active site, induced by the reaction with hapten.

Zav'yalov et al. (1977b), who studied the reaction between anti-dansyl rabbit antibodies and hapten by the method of differential perturbational spectroscopy, also came to the conclusion that dansyl-lysine stabilizes one of the conformers of the antibody, with similar free energies.

These results are in agreement with our own data (Käiväräinen et al., 1973b; Käiväräinen and Nezlin, 1975, 1976a,b), obtained with the aid of a spin label localized on Fab-subunits in a manner which ensures that the EPR spectrum reflects the A ⇌ B equilibrium between more and less compact conformers of the Fab-subunit. It was noted that the bonding of spin-labeled antibodies with antigens invariably resulted in a shift of the A ⇌ B equilibrium to the left (Fig. 51). According to the model, such a shift may result in the stabilization of the protein. This is confirmed by experimental results indicating an enhanced stability of antibodies and their Fab-fragments to proteolysis (Grossberg et al., 1965), to the action of denaturing agents (Cathou and Werner, 1970), and a slower rate of deuterium exchange (Ashman et al., 1970). Calculations based on kinetic curves of H-D exchange of rabbit antibodies against pneumococcal polysaccharides show that the free energy of the protein decreases by 8.2 - 12.5 KJ/mole. This finding is in agreement with its enhanced thermostability (Zavodzky et al., 1978).

Very interesting data were obtained by the dilatometric method during a study of the reaction of rabbit anti-dansyl antibodies with DNP-lys and the macromolecular multi-

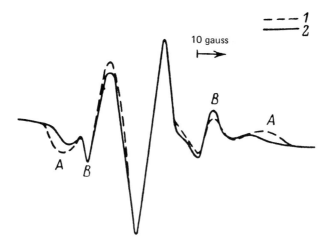

Figure 51. EPR spectra of spin-labeled rabbit antibody after formation of specific complexes with hemoglobin (1), and a control spectrum of mixed solutions of spin-labeled rabbit IgG and hemoglobin (2). After Käiväräinen *et al.* (1973b).

valent antigen (DNP-IgG). The system monovalent Fab + DNP-lys was also studied (Ohta *et al.*, 1970). Antibodies of differing affinities were prepared from three groups of rabbits. In all cases the volume of the solution increased as a result of formation of immune complexes. On the basis of one mole, the volume increase accompanying the formation of DNP-lys complex with the Fab-subunit was about half as large as during complex formation with the intact antibody. This means that any changes in the interaction between antibody subunits and conformational rearrangements in the Fc-subunits make only a small contribution to the increase in the volume of the solution.

The correlation noted between the volume change of the solution as a result of the reaction and the logarithm of K_{ass} is of interest. If the volume increase ΔV is a manifestation of transglobular conformational transitions in the antibody molecule, similarly to $K_{A^* \rightleftharpoons B^*}/K_{A \rightleftharpoons B}$ in equation (6.17), $\ln \Delta V$ should vary linearly with $\ln K_{ass}$, as is in fact the case (Fig. 52). For the sake of comparison, studies were carried out on reactions between lysozyme, RNA-ase and their low-molecular competing inhibitors – N-acetyl-d-glucosamine and cytidine-2'(3')-monophosphate respectively. In such cases complex formation brings about an increase in the volume of the solution (Table 9), which proves that ligand-induced conformational transitions in proteins are governed by common relationships.

The increase in the volume of the solution may be caused by several different factors, including conformational changes occurring in proteins, neutralization of charges or intensified intramolecular hydrophobic interactions. Ohta *et al.* (1970) reviewed all these factors in the context of the system studied, and concluded that the principal contribution to the volume increase is made by hydrophobic interactions. It should be remembered in this connexion that these interactions result from the penetration of the non-

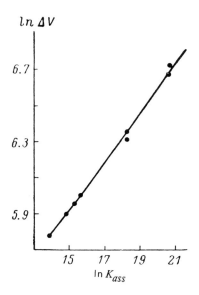

Figure 52. Logarithm of the volume change of the solution accompanying the reaction between antibody and hapten as a function of K_{ass}. After Ohta *et al.* (1970).

TABLE 9. Volume changes during reaction between antibodies and antigens (after Ohta *et al.*, 1970) and between enzymes and inhibitors

Protein	Ligand	K_{ass}	$\ln K_{ass}$ $(-\Delta G_{ass}/RT)$	ΔV, ml/mole	$\ln \Delta V$
Antibody	DNP-lys	$3.1 \cdot 10^6$	14.9	365	5.90
		$6.6 \cdot 10^6$	15.7	410	6.01
		$9.0 \cdot 10^7$	18.3	580	6.36
		$1 \cdot 10^9$	20.7	800	6.68
		$6.6 \cdot 10^6$	15.7	400	5.99
		$9.0 \cdot 10^7$	18.3	550	6.31
2Fab	,,	$\sim 1 \cdot 10^9$	20.7	850	6.74
		$1 \cdot 10^6$	13.8	324	5.78
		$4 \cdot 10^6$	15.2	392	5.97
Lysozyme	N-acetyl-d-glucosamine	20–50	3.0–3.9	48	3.87
RNA-ase	Cytidine-2'-(3')-monophosphate	$2 \cdot 10^3$	7.6	23	3.13

Note: $\ln K_{ass}$ and $\ln \Delta V$ were calculated by the author.

polar hapten into the active site, and that their effect becomes intensified by the equilibrium shift between the 'closed' and 'open' states of auxiliary cavities towards the 'closed' state, which becomes stabilized by hydrophobic interactions. In our view, it is chiefly this latter process which should correlate with the K_{ass}-value of the protein with the ligand.

One consequence of transglobular conformational changes in antibodies induced by the reaction with the antigenic determinant is the appearance of their effector functions, i.e., the ability to fix and to activate a complement. This presupposes a change in the protein conformation in the zone of the Fc-subunit in which the bonding segment of the first component of the complement is localized, and/or an interaction between the IgG subunits. Fc is found at a distance of 8-10 nm from the active site. Studies involving the use of haptens and monovalent antigens, which do not precipitate on reacting with antibodies, deserve special mention. Such a study was carried out on the system IgM — rabbit antibody + phenyl-β-lactosidase hapten and monovalent macro-antigen, representing a complex of hapten with RNA-ase (Brown and Koshland, 1975). It was found that a hapten with $K_{ass} = 1.6 \cdot 10^5$ is incapable of inducing conformational changes in IgM, such as are required to bind the complement, whereas the monovalent antigen RNA-ase with $K_{ass} = 4.7 \cdot 10^4$ does have this capacity.

It thus follows that a ligand with a lower affinity produces more extensive conformational rearrangements in the protein. The authors interpret this effect by postulating that some of the interaction energy is spent on effecting structural changes in Fc, which are responsible for the fixation of the complement. In our view, the contradictions involved in the interpretation of the highly interesting results are due to the imperfect determination of the K_{ass} of the monovalent macro-antigen. Our previous results indicate in fact that any comparison of the transglobular effects exerted on a given protein by ligands with differing affinities must be carried out under conditions in which the values of P_b^{coll} of these ligands with the active site are the same. According to equation (6.18), P_b^{coll} is inversely proportional to the correlation time of the ligand: $\tau_L = 4/3\pi a_L \eta/T$, where a_L is the effective Stokes radius of the ligand.

Determinations of K_{ass} of free hapten (molecular mass 500) and of hapten conjugated with RNA-ase were conducted at similar $[A] [L] \cdot H^b/H_L \cdot K_{b=a}$ and $1/\tau_A$ values for both types of ligands. The concentration of the active sites of the IgM was $\cong 0.5 \cdot 10^{-1}$ M, while that of the ligand was $0.5 \cdot 10^{-5} - 5 \cdot 10^{-5}$ M.

Nevertheless, the values of $1/\tau_L + 1/\tau_A$ varied with the molecular mass and with the effective Stokes radius of the ligand. Since the molecular mass of IgM (900,000) is much larger than that of RNA-ase and, *a fortiori*, much larger than that of hapten ($1/\tau_A \ll 1/\tau_L$), we may write, in view of (6.18):

$$\frac{P_b^{coll} \text{ (hapten)}}{P_b^{coll} \text{ (antigen)}} \approx \frac{\tau_L \text{ (antigen)}}{\tau_L \text{ (hapten)}} \approx 30. \qquad (6.40)$$

It then follows from equation (6.21) that it would be possible to compare K_{diss} and K_{ass} for two types of ligands if the concentration of monovalent antigen were about 30 times higher than that of the hapten. In actual fact they were practically equal, which means that the true value of K_{ass} for a monovalent antigen is larger than the experimental value, and is very probably larger than the K_{ass} of the free hapten with the active site of the antibody. This is confirmed by the fact that a 2500-molar excess of the low-molecular ligand is required to displace the monovalent antigen from the active site by the hapten.

There are thus reasons to conclude that, both in these experiments and in those described above, the high-affinity antigenic determinant induces larger conformational changes in the protein, in conformity with equation (6.17). In the case of a large monovalent antigen these changes suffice for IgM to be able to fix the complement.

Koshland obtained interesting results on the monovalent antigen of papain, modified at its only sulhydryl group by the ester M: azophenyllactoside-N-chloracetyltyrosine (LacTyr). The papain carrier was hydrolyzed by trypsin into fragments of various sizes, which contained the hapten LacTyr, after which the capacities of the fragments to induce complement fixation in hapten-homologous antibodies of IgM class were compared. It was shown that this capacity is determined by the size of the hapten carrier. Five or six amino acid residues are needed to ensure minimum fixation, while about 120 are required for maximum fixation (Koshland, 1978). Unfortunately, data on the K_{ass}-values of these monovalent antigens are lacking. According to Koshland, his results indicate the existence of two types of signals, which are needed for the manifestation of the effector functions of IgM — one induced by the specific reaction of the active site with antigenic determinants, while the other originates from the non-specific effect of the carrier.

According to Koshland, the non-specific effect of the bulky carrier may be to restrict the freedom of rotation of Fab-units or to enhance the effect of the hapten by completely filling the active site. In terms of the dynamic model, an increase in carrier size is tantamount to an increase in the effective Stokes radius of the variable domains, the corresponding decrease in their diffusion rate relative to the constant domains and frequency of the A $\rightleftharpoons$ B fluctuations. This may result in a redistribution of the interactions between the domain pairs $(V_L H_H - C_L C_H 1)$, and also between the subunits of the antibody, and favor the exposure of the complement-binding segments on Fc.

In recent years several studies of conformational changes of antibodies were performed by the relatively new method of circular fluorescence polarization (CFP), which is the emission analog of the circular dichroism method. Large antibodies not containing tryptophan were studied to follow the changes in the CFP spectra: RNA-ase, polyanyl-polylysine polypeptides and the 'loop--like' lysozyme polypeptide (Schlessinger *et al.*, 1975 a, b). In all the antibodies studied, the CFP spectra underwent major changes, which were all similar, after having reacted with the corresponding antigen. Changes originating from the reactions between the antigen and isolated Fab-fragments affected other parts of the spectra.

It is conceivable, in principle, that the differences in the CFP spectra are due to some impairment of the Fab structure during their proteolytic digestion. However, the results of kinetic, dilatometric and other studies quoted above, which indicate that the properties of Fab are almost unaffected by isolation, make such an interpretation unlikely. We are thus justified in concluding that it is precisely the rearrangements of the Fc-subunits which are responsible for the changes in the CFP spectra which are observed for integral antibody molecules, but are absent in the case of isolated Fab-subunits.

A study (Jaton *et al.*, 1975) of anti-polysaccharide antibodies yielded similar results. The antigens employed were oligosaccharides of increasing length, including tetra-, hexa- and octasaccharides, and a 16-membered oligomer. As in the experiments described above, the reaction with the antigen resulted in changes in CFP spectra, which were less manifest in the case of isolated Fab- or $F(ab')_2$-fragments. These studies also showed that the Fab- and Fc-parts of the IgG molecule affect one another, and that changes occur in

Fc after it has bound a sufficiently large-sized antigen. The small-sized tetrasaccharide hapten failed to induce any conformational changes in the antibody, thus confirming the importance of the size of the antigenic determinant. This may also be connected with the decrease in the affinity of the anti-polysaccharide antibodies with decreasing oligosaccharide length.

It was observed during a study of the complement-activating capacity of antibodies in the IgC class that, unlike the case of IgM, it cannot be induced by the conformational changes of Fab- and Fc-subunits which, according to CFP data, accompany the binding of monovalent antigen. However, the reaction between the IgG antibody and double-loop lysozyme peptide, which is a bivalent antigen, imparted to the antibody the complement-activating capacity. Special experiments conducted in an analytical centrifuge revealed that such immunity reaction, just as in the experiments with monovalent antigens, is not accompanied by the formation of large aggregates. It is believed that bivalent antigen is needed to produce the required angle ratio between Fab and Fc, which would result in the exposure of complement-binding segments or in the formation of new ones. This conclusion was confirmed by Jaton *et al.* (1976) by another series of experiments. These workers studied the complement-binding capacity of rabbit antibodies of IgG class after reaction with antigens, as a function of the size of the latter. The antibodies were prepared by immunization of the rabbits by pneumococcal polysaccharides. The antigens represented a set of different oligosaccharides isolated from the initial polysaccharide.

The complement-binding activity of immune complexes was manifested only if the sugar oligomer contained 21 or more sugar residues. It was shown by sedimentation analysis that the aggregates formed by antigen-antibody complexes then contain four antibody molecules or more. The angles between the Fab-subunits of IgG in such clusters varied between 90 and 180°, while in complement-inactive complexes the angle between the Fab-subunits is about 60°. Parallel experimental work carried out by the CFP method revealed a correlation between the size of the oligosaccharide hapten and the magnitude of conformational changes in the Fc.

Thus, the activation of complement-binding parts of IgG is due to the combined effect of antigen-induced conformational changes in the Fc-subunits of IgG and to the increase of the angle between the Fab-fragments to more than 90°. In the case of antibodies of IgM class the latter condition may have been met prior to combining with the antigen, so that aggregation of these molecules need not necessarily take place. In terms of the dynamic model, the angle between Fab-subunits must be regarded as a factor which influences the state and the mobility of the cavity between $F(ab')_2$- and Fc-subunits, and also the distribution of the respective states of Fc domains, which determines the exposure of the complement-binding segment.

Experimental results obtained by the circular dichroism (CD) method revealed differences in CD spectra of three homologous anti-pneumococcal antibody polysaccharides before and after their interaction with specific hexasaccharide ligands (Jaton *et al.*, 1975). CD spectra were also employed to study homologous and hybrid recombination products of H- and L-chains under the effect of hapten. The CD-spectrum of homologous antibody recombination products was identical with that of the native antibodies, despite the absence of the interchain S-S bonds. Changes in CD spectra occurred as a result of hapten binding in all three native antibodies, in their Fab fragments and in molecules resulting from homologous recombination. However, no evidence of conformational changes could be detected as a result of the interaction between hapten and heterologous antibody recombination products. These results are confirmed by the change in the quantum yield of antibody fluorescence, indicating that inter-chain interactions in immunoglobulins are highly specific.

A similar conclusion was arrived at on the strength of the results obtained by Preval and Fougereau (1976), who studied the reassociation of the H- and L-chains of immunoglobulins by the method of competing hybridization. These workers studied the interaction between two monoclonal L-chains — the autologous L_A and the heterologous L_H chains — and the monoclonal heavy chain H_A. Twelve human proteins were used for this purpose, which made it possible to study the behavior of 44 different chain combinations by the method of competing hybridization. It was found that in 80% of the cases it is the autologous L- and H-chains which reassociate. The preferential reassociation seems to be the result of antigen-dependent selection processes, which are observed during the differentiation of antibody-forming cells.

Since the geometry of the nonpolar cavities of immunoglobulins is ultimately formed following the recombination of L- and H-chains, it may be assumed that the association of the chains with cavities which are best adapted to the quasi-crystalline structure of water in the 'open' state is thermodynamically favored. This may be a hitherto neglected factor which determines the course of self-organization of proteins.

Hapten-induced Changes in the Flexibility of Antibodies and in the Intramolecular Mobility of their Fc-subunits

The problem of signal transmission from the active site of the antibody to the remote parts of the molecule has been studied for a long time, in the context of the initiation of its effector functions. Numerous data on conformational changes of antibodies during their binding with antigenic determinants were quoted above. It was found, by methods of small-angle electron scattering (Pilz *et al.*, 1973, 1974) and neutron scattering (Cser *et al.*, 1977) that the effect of the hapten is to considerably reduce (by up to 10%) the radius of gyration of the antibody, owing to the contraction in the effective volume of the macromolecule and/or its increased flexibility. NMR data indicate a changed interaction between the subunits of the antibody (Vuc-Pavlovic *et al.*, 1979).

Tumerman *et al.* (1972) used the method of fluorescence depolarization of dansyl-lysine, both on a specimen bound in the active site by the anti-dansyl antibody and on one covalently conjugated with the antibody outside the active site, to calculate the correlation times of this protein; his results were 37 and 30 nsec respectively. This was interpreted as a decrease in the flexibility of the antibody caused by the binding of hapten. However, in this case the difference between the correlation times may have been caused by the different locations of the fluorescent labels with respect to the antibody molecule, and by the different environments and lifetimes in the excited state. Conclusions about conformational changes in the Fc-subunits of the antibody caused by reaction with hapten, based on CFP data (Jaton *et al.*, 1975) are also unclear (Metzger, 1978) and shed no light on their nature.

Accordingly, there is as yet no conclusive answer to the problem of the changes in the conformation and dynamic properties of antibodies as a result of formation of a specific complex with hapten.

We studied this problem by the method of separate determination of τ_M and τ_R described above (Section 3.6) (Käiväräinen *et al.*, 1981). We used pig antibodies against 1-dimethylaminonaphthalene-5-sulfonyllysine (DNP-lys). The antibody was labeled with spin-label I - 2,2,6,6-tetramethyl-N-1-oxylpiperidine-4-amino-(N-dichlorotriazine) - in 0.05 M phosphate buffer (pH 7.3) during 3 days at 4°C (Ab–I) and with spin-label II - 4-amino-2,2,6,6-tetramethylpiperidine-1-oxyl - as described above (Ab–II) (Nezlin *et al.*, 1978). I preferentially reacts with histidine residues, while II can only bind carbohydrates under these conditions. The excess label was removed by exhaustive dialysis and gel-filtration on Sephadex G-25. The ϵ-dansyl-lysine hapten was added to the antibody solution in fourfold molar excess. In a number of experiments, $1/\tau_{R+M}$ as a function of T/η was plotted in the presence of 3% D_2O.

Under the experimental conditions employed, two Ab–I labels were bound to each molecule of the antibody. In view of the symmetry of the IgG molecule — two identical Fab-subunits and absence of Ab–I on Fc of rabbit antibodies (Käiväräinen *et al.*, 1973b) — it may be concluded that AT-I here becomes localized on Fab-subunits.

Changes in the EPR spectra of Ab–I after hapten binding (Fig. 53) are in agreement with those previously noted by Timofeev *et al.* (1979). It is significant that the A- and B-components of the EPR spectrum, which have been attributed by ourselves to the more compact A-conformer and the less compact B-conformer of the Fab-subunits, become narrower and their resolution improves.

The τ_{R+M} determined from the A-components of the EPR spectrum is the result of superposition of two types of motion: label rotation relative to the protein (τ_R) and rotation of the protein as a whole (τ_M). In order to acquire a better understanding of the physical meaning of changes in these magnitudes, the latter were determined at several temperatures (Table 10). The effect of 3% D_2O on the correlation time was studied at 5°C. All τ_M-values were reduced to standard conditions (water, 25°C). The error was $\Delta\tau_M = \pm 2$ nsec; $\Delta a_M^{eff} = \pm 0.1$ nm; $\Delta\tau_R = \pm 0.2$ nsec.

TABLE 10. Values of the parameters τ_M, a_M^{eff} and τ_R before and after Ab–I binding by hapten at various temperatures

SLI protein	5°			24°			30°		
	τ_M nsec	a_M^{eff} nm	τ_R nsec	τ_M nsec	a_M^{eff} nm	τ_R nsec	τ_M nsec	a_M^{eff} nm	τ_R nsec
			in the absence of D_2O						
Ab–I	20	2.7	10.2	43	3.5	8.3	50	3.7	7.9
Ab–I + H	30	3.1	9.1	33	3.2	8.5	27	3.0	8.2
			in the presence of 3% D_2O						
Ab–I	22	2.8	9.8	43	3.5	8.3			
Ab–I + H	30	3.1	9.1	33	3.2	8.5			

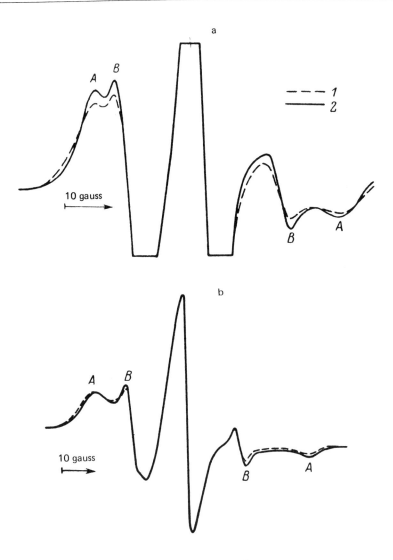

Figure 53. EPR spectra of pig antidansyl antibodies at 30°C, I-labeled (a) and antibody at 5°C, II-labeled (b) at the carbohydrates, before (1) and after (2) reaction with hapten. After Käiväräinen *et al.* (1981). 0.01 M phosphate buffer (pH 7.3) + 0.15 M NaCl.

It is seen that, in the case of Ab-I + H, τ_m and τ_R are practically independent of the temperature; in the case of the intact Ab-I antibody, τ_M increases considerably (from 19 to 50 nsec), while τ_R decreases from 10.2 to 7.9 sec.

The presence of 3% D_2O in solution produced a slight increase in the τ_M of the intact antibody, but did not affect the τ_M of the complex. The increase in the resolution between the A- and B-components of the EPR spectrum of AB-I after hapten binding (Fig. 53) means — as was proved by comparison with theoretically calculated spectra — a decrease in the frequency $\nu_{A \rightleftharpoons B}$ of fluctuations between the A- and the B-states of the label, i.e., between two Fab conformers. Since the τ_R values determined by ourselves for the A-state of the label are of the order of 10^{-8} sec (Table 10), it follows from the theory of the spin-label method (Antsiferova *et al.*, 1977) that $\nu_{A \rightleftharpoons B} = 1/(t_A + t_B) \leqslant 10^7$ sec^{-1}. $\nu_{A \rightleftharpoons B}$ decreases with increasing lifetime t_A of the A-conformer and/or that of the B-conformer (t_B) under the effect of hapten. The activation barrier separating the A- from the B-state may be enhanced not only by stabilization of variable domains, but also by an increase in their effective Stokes radius after reacting with a large antigen.

A comparison of the ratio between the A- and B-components of the EPR-spectra of Ab-I with the calculated values (Antsiferova *et al.*, 1977) makes it possible to effect an approximate estimate of the equilibrium constant between A- and B-states of the label: $k_{A=B} \simeq 4$. According to the Boltzmann distribution, this means that the free energy of the A-conformer is lower by 1.7 KJ than that of the B-conformer.

At 24 and 30°C the τ_M of Ab-I decreases after hapten binding from 43 to 33 nsec and from 50 to 27 nsec respectively. At the same time τ_R increases only from 8.3 to 8.5 nsec and from 7.9 to 8.2 nsec (Table 10). These results indicate that a slow (during 4-5 minutes) increase in the amplitude of A- and B-components of EPR-spectra of Ab-I resulting from their contraction, which was previously observed by Timofeev *et al.* (1979), is an indication of the increased flexibility of the antibody during complex formation with hapten. According to the dynamic model of formation of specific complexes (Käiväräinen, 1979a), the reduced interaction between Fab- and Fc-subunits of the antibody may be caused by a relaxation process which stabilized the tertiary structure and the relative fluctuations of Fab-subunit domains. The rate of the relaxation process, which is induced by ligand binding, is determined by the overall free energy of activation of elementary rearrangements in the protein domains of Fab, which must take place if the nature of their fluctuations is to be changed.

It was demonstrated in the case of chymotrypsinogen (Brandts *et al.*, 1967; quoted according to Aleksandrov, 1975) that the protein stability, which is determined by hydrogen bonds, remains unchanged between 0 and 70°C, while the stabilizing effect of hydrophobic interactions is intensified by almost 30% when the temperature is raised from 0 to 30°C. In view of the above, the differences in the variation of τ_M of Ab-I and Ab-I + H with the temperature (Table 10) may be explained by postulating that the relaxational changes in the properties of Fab reduce the contribution of hydrophobic interactions between the F(ab')$_2$ and the Fc-subunits.

New results were obtained with carbohydrate-II-labeled antibodies. Under the experimental conditions employed one II-label was bound to each antibody molecule. Since carbohydrates are localized mainly on the Fc-fragment of IgG (Nezlin, 1972), there is reason to believe that label II is mostly bound to Fc.

The EPR spectra of Ab-II and Ab-II + H were practically identical (Fig. 53). At room temperature, in the absence of saccharose, the EPR spectra consisted of three narrow lines, corresponding to a rapid rotation of the N-O group of the label, and there are no wide A-components, but the latter can be made to appear by decreasing the temperature or increasing the viscosity of the solution. The variations of $1/\tau_{R+M}$ with T/η for Ab-II and Ab-II + H (Fig. 54) reflect the changes brought about by hapten and by D_2O (sec. 3.1). If these data are evaluated (Table 11), it is seen that at both temperatures (1°C and 5°C) the hapten increases the mobility of Fc-domains, but the mobility of the labeled carbohydrate with respect to the protein, as indicated by τ_R, decreases. The presence of 3% D_2O results in a decrease of τ_M and a_M^{eff} and in a marked increase in τ_R for both free Ab-II and the complex Ab-II + H. However, conformational changes of Fc, induced by the hapten in the presence of D_2), are no longer noted. The changes of τ_M and τ_R of Ab-I and Ab-II at 5°C (Tables 10 and 11) differ in sign and magnitude, which confirms that labels I and II are localized on different subunits.

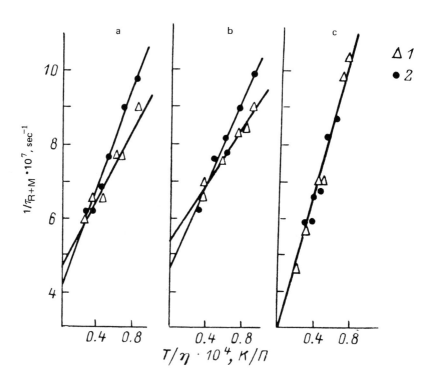

Figure 54. $1/\tau_{R+M}$ as a function of T/η for carbohydrate-labeled Ab-II (1) and Ab-II + H (2) at 0°C (a), at 5°C (b) and in the presence of 3% D_2O (c). After Käiväräinen *et al.* (1981). 0.01 M phosphate buffer (pH 7.3) + 0.15 M NaCl.

The EPR spectra of carbohydrate-labeled Ab (Ab-II) remained practically unchanged following the binding of hapten. Similar observations made at an earlier date (Wilan *et al.*, 1977; Timofeev *et al.*, 1978) led to the conclusion that conformational rearrangements in Fc are absent. However, it became clear, as a result of separate determinations of τ_M and

τ_R (Table 11), that the reason for the constancy of the EPR spectrum of Ab-II is that the decrease in τ_M is compensated by the increase in τ_R during complex formation.

TABLE 11. Values of the parameters τ_M, a_M^{eff} and τ_R before and after AT-II - hapten binding at different temperatures

Protein SL II	1°			5°		
	τ_M nsec	a_M^{eff}, nm	τ_R, nsec	τ_M nsec	a_M^{eff}, nm	τ_R, nsec
	in the absence of D_2O					
Ab–II	6	1.8	22	7.5	1.9	18
Ab–II + H	4	1.6	28	5.0	1.7	22
	in the presence of 3% D_2O					
Ab–II	–	–	–	3	1.5	33
Ab–II + H	–	–	–	3	1.5	33

The presence of A- and B-components in the EPR spectrum of Ab-II may be due to two Fc conformers, as was the case for Fab, or to the ability of the spin-labeled carbohydrate to fluctuate between two different orientations relative to the protein, as in the case of oxyhemoglobin, spin-labeled at the SH-groups of the β-chains (Moffat, 1971). The X-ray model of Fc suggests that the carbohydrates are mainly localized of C_H2 - C_H2 domains of Fc, separated from one another by a large space, which may be filled with water (Deisenhofer *et al.,* 1976).

The weak interaction between these domains is probably responsible for the fact that Fc is more flexible than Fab, as manifested by the low values of τ_M and a_M^{eff}. The very low values of $\tau_M \simeq 3$ nsec, and $a_M^{eff} \simeq 1.5$ nm, noted in the presence of 3% D_2O, are close to those which may be expected for isolated C_H2-domains ($a_M^{eff} \simeq 1.1$ - 1.2 nm).

According to the dynamic model, the interaction between water and the cavities in the antibody molecule affects the relative mobilities of the domains and of whole subunits. D_2O forms stronger hydrogen bonds and has a much larger constant of association with proteins K_{ass} than H_2O (Lobyshev and Kalinichenko, 1978). It is not surprising, therefore, that even small D_2O concentrations can produce major effects (Table 11). Our own results (Tables 10 and 11) indicate that the presence of 3% D_2O affects the flexibility of Fc more strongly than the interaction between the subunits. As was stated in the case of pig anti-DNP antibodies, the effects produced by 3% D_2O are not due to hydrogen exchange, since they rapidly disappear when the temperature is increased.

We are accordingly led to the following conclusions.

1. The results obtained with the aid of spin label I indicate that hapten induces a relaxational process, resulting in the stabilization of the tertiary structure and a diminished frequency of the relative fluctuation of Fab domains.

2. As a result of this process, changes take place in the interactions between the subunits of the antibody. At near-physiological temperatures the freedom of rotation of Fab-subunits relative to Fc is greatly enhanced by the hapten effect.

3. Data on Fc-induced conformational changes have been obtained for the first time using spin-label carbohydrates. At 1°C and 5°C these changes become manifested as a limited increase in domain mobility and intensification of the interaction between labeled carbohydrates and domain surfaces.

4. The presence of 3% D_2O in Ab-II solution produces a major effect, which is in qualitative agreement with the effect of the hapten, but the signal transmission from active sites to Fc is blocked.

Relationship between the Association Constant of the Antibody – Hapten Complex and the Changes in the Dynamic Properties of the Antibody.

The results of dilatometric studies carried out on antibodies and their Fab-subunits (Ohta *et al.*, 1970) and of CFP studies (Jaton *et al.*, 1975; Schlessinger *et al.*, 1975 a,b) indicate the existence of a direct relationship between the magnitude of conformational rearrangements in the antibody caused by the binding of an antigen determinant or a hapten, and the association constant of the respective complex. Dilatometric studies indicate that such rearrangements are mostly localized in Fab-subunits and are connected with the intensification of intramolecular hydrophobic interactions in Fab.

In order to obtain more detailed information about the nature of the hapten-induced structural changes in pig anti-DNP antibodies and their Fab-subunits, we used a spin-labeled DNP-SL hapten (Käiväräinen *et al.*, 1980c). The DNP-SL was added to the antibody solution and its Fab-fragments, which had differing affinities to the hapten, and it was assumed that if the ratio [H] : [active site] is small, the antibodies involved in the reaction will be those with the large binding constant (competition by antibody and the reversibility of the reaction Ab + H); as this ratio approaches unity, the reacting antibodies were expected to have a smaller average binding constant. It follows that the value of K_{ass}, averaged for each ratio, decreases as [H] : [active site] → 1. Magnitude-averaged conformational changes in proteins may be expected to correspond to the averaged K_{ass}-values.

Table 12 shows the values of τ_M and τ_R, calculated in this way. τ_R reflects the mobility of the N-O group of the DNP-SL, which is known to be higher than that of the DNP-hapten itself when the active site is in the 'closed' a-state, since the spin label is free to rotate around the rigidly fixed DNP.

TABLE 12. Values of τ_M and τ_R, nsec, of antibody complexes and their Fab-subunits with spin-labeled hapten at two [H] : [active site] ratios and at different temperatures

t, °C	Ab + DNP-SL				Fab + DNP-SL			
	[H]:[active site] = 1		[H]:[active site] = 0.35		[H]:[active site] = 1		[H]:[active site]=0.35	
	τ_M	τ_R	τ_M	τ_R	τ_M	τ_R	τ_M	τ_R
5	15	10.2	33	10.2	12	10.2	20	10.2
15	30	8.3	33	9.1	–	–	–	–
20	50	7.8	43	8.3	20	8.7	20	9.1
30	60	7.1	43	8.0	20	8.0	–	–

Note: τ_M-values are accurate to within ±2 nsec; τ_R-values are accurate to within ±0.2 nsec.

It is seen from Table 12 that the values of τ_M and τ_R for Ab + DNP-SL vary more strongly with the temperature when the [H] : [active site] ratio is unity than when it is 0.35. This is probably due to the higher thermostability of the stronger immune complexes. A comparison of τ_M-values for Ab + DNP-SL with Fab + DNP-SL shows that there is a correlation between the stabilization of the complex antibody - spin-labeled hapten and a decrease in the intramolecular mobility of the Fab complex, which is most conspicuous at 5°C. As the temperature increases and hydrophobic interactions in proteins become intensified, τ_M of both Ab + DNP-SL and Fab + DNP-SL increases. The τ_M-value of the Fab + DNP-SL complex increases only if [H] : [active site] = 1, since if this ratio is 0.35, the complex behaves as a rigid particle even at 5°C. This follows from the Stokes-Einstein Law, according to which the time period of 20 nsec is close to the correlation time to be expected for a rigid particle of the size of Fab.

The increase in the thermostability of the relative mobility of antibody subunits during the formation of stable complexes with spin-labeled hapten is in agreement with the results obtained with the aid of a spin label linked by covalent bond to the Fab-subunits of antidansyl antibodies (see above). It would appear that at physiological temperatures the Fab-subunits which constitute antibody complexes with a large K_{ass} rotate more freely around Fc than when forming part of Ab + DNP-SL with a smaller K_{ass}.

The flexibility of Ab + DNP-SL and Fab + DNP-SL complexes at 5°C and at [H] : [active site] = 1 and at 5°C is strongly diminished in the presence of 3% D_2O. This is manifested as an increase in the τ_M-value from 15 to 22 nsec and from 12 to 19 nsec respectively. It is important to note that there is no longer any D_2O-effect at [H] : [active site] = 0.35; this effect is also absent in both types of complexes at 15°C and above.

It is not very likely that the changes in τ_M are caused by changes in the fluctuation frequency between A- and B-conformers of Fab (the 'closed' and 'open' states of the cavity formed by 2V- and 2C-domains), since EPR spectra of Ab labeled at their Fab-subunits show that the transition frequency $\nu_{A \rightleftharpoons B}$ does not exceed 10^7 sec^{-1} even at 5°C, when τ_M of Fab + DNP-SL decreases to 12 nsec ($1.2 \cdot 10^{-8}$ sec).

The stabilizing effect of hapten on the flexibility of Fab may be explained as follows: We shall assume that the relative reorientation of V- and C-domain pairs constituting the B-conformers is faster than that of the pairs constituting the A-conformer. Thus, say, the correlation time of the B-conformer may be 6-7 nsec, while that of the A conformer is of the order of 22-24 nsec. If the hapten shifts the A $\rightleftharpoons$ B equilibrium to the left, so that $K_{A=B}$ increases from 0.7 to 0.9, the resulting correlation time τ_M will change within the same limits as those within which it changes in the Fab+DNP-SL complex when [H] : [active site] ratio varies between unity and 0.35 at 5°C (see eq. 6.60).

The stabilization of the fluctuation domains of Fab-subunits by hapten weakens the interaction between these subunits and the Fc-subunit; this is manifested by lower τ_M-values at temperature close to the physiological (Table 12).

The increase in the relative mobility of antibody subunits under the effect of hapten may be explained by shift in the equilibrium between the two states of the central cavity towards that corresponding to a larger number of degrees of freedom and a faster rate of rotation, or by diminishing of the hydrophobic interaction between subunits.

It is not very likely that an antibody with a higher affinity differs from an antibody

with a lower affinity of the same population in its general structural and dynamic properties. The difference in K_{ass} is most probably due to small differences in the hypervariable
zone, which render the 'closed' active site more complementary to the hapten. Judging by
the EPR spectra of DNP-SL localized in the active center (Fig. 45), the equilibrium b-
tween the more compact state *a* and the less compact state *b* at [H] : [active site] = 0.35
is shifted towards the former state more strongly than if that ratio is unity. The variation
of τ_R with the temperature (Table 12) indicates that the environment of DNP-SL in the
a-state is more rigid when this ratio is 0.35 than when it is unity.

These results confirm one of the postulates of the dynamic model of complex formation, viz., that an interaction, varying with the association constant between the active
site and the ligand, is operative between the states of the active site and auxiliary cavities
of the protein.

Heterogeneity of Ab Hydrate Shell

Our studies of antibody + DNP-SL complexes at [H] : [active site] ratio of unity at subzero temperatures yielded results which are important to the understanding of the mechanism of protein hydration. Changes in the active site were followed by following the
shift of the low-field (+1) A-component of EPR spectra. The shift of the A-component,
which reflected the immobilization of the label, increased rapidly between –8 and –13°C.
This is the first stage in the changes taking place in the active site. The second stage was
noted between –20 and –40°C; thereafter, the shift showed a continuous increase up to
its maximum value at –80°C. In our view, the three-stage nature of this variation is due to
the successive freezing of different types of water, viz., the water fractions near the nonpolar group and, at lower temperatures, of the water near the polar and the charged zones
and in cavities. When the initial solution contained 3% D_2O, the freezing curve underwent
an upward shift of 3-5°C; moreover, between –10 and –40°C the shape of the curve was
considerably changed and the transitions became sharper and more cooperative.

The presence of a perturbant (0.5 M NaCl and 1 M urea) resulted in the disappearance
of the sharpness of the stages, eliminated the effect of D_2O and produced a small shift
towards slower motions. These results indicate that perturbants are capable of disturbing
the structure of the water shell of the proteins.

In our experiments, conducted on oxyhemoglobin and IgG solutions in the presence of
5% D_2O, the proportion of the water exchanged against D_2O and HOD at –10°C was 10
and 15% respectively.

We may now conclude that the differences between the flexibility of antibodies,
produced by added D_2O, are mainly due to the interaction of D_2O and HOD with the
nonpolar cavities of these proteins (see Section 4.1). The hydration of the surface zones
could affect the stability of the tertiary domain structure, but not the relative fluctuations of the domains.

Dehydration of Antibodies during their Reaction with Hapten

It follows from the dynamic model that the perturbation of the properties of subunits
under the effect of hapten or a large antigen determinant may alter the geometry and the

stability of the central cavity between F_{ab} and F_c by antibody subunits. This results in the destabilization of the water cluster in this cavity, and shortens its lifetime in the 'open' B-state. As a result, the A $\rightleftharpoons$ B equilibrium of the central cavity becomes shifted towards the left. This shift may be expected not merely to decrease the effective volume of the antibody molecule, but also to reduce its degree of hydration by displacing water out of the cavity.

Up till now no data on changes in the degree of hydration of Ab were available.

We attempted to answer this question by the technique of frozen solution NMR (para. 4.1) on the system: pig anti-DNP-antibodies + H.

The NMR lines of the water of hydration of intact A antibody and Ab + H differed both in integral intensity, which is proportional to the number of mobile water molecules in the vicinity of the protein, and in their half-width ($\Delta \nu$ Hz). The quantitative determination of the hydration (in gm H_2O per gm of protein) was carried out by comparing the surface areas of the signals given by frozen protein solutions at different temperatures and of the signal given by the frozen control salt solution (0.01 M Mn^{++} + 4.5 M LiCl), in which the concentration of the water observed at $-35°C$ is 50 ± 1 M (Kuntz *et al.,* 1969).

The results of the processing of PMR spectra are shown in Figs. 55a and 55b. It is obvious that conformational changes brought about in the antibody by its reaction with DNP-hapten in fact bring about its dehydration. The amount of the water bound by the antibody at $-18°C$ (0.35 gm H_2O/gm protein or $2.9 \cdot 10^3$ moles H_2O per mole of the antibody) is more than 35% larger than that observed for the Ab + H complex (0.20 gm H_2O/gm protein or $1.7 \cdot 10^3$ molecules of H_2O per molecule of the antibody). However, at higher temperatures ($6-10°C$) this dehydration is practically imperceptible against the background of the large amount of the non-frozen water in the fractions of the hydrate shell which have remained unaffected by the reaction. As these fractions gradually freeze, this difference becomes more marked (Fig. 55a).

The effect of displacement of water from the active center by hapten with a molecular mass of about 300 is much weaker (not more than 50 mol. H_2O during the B → A transition of the active center. The evolution of the half-width of the line of water in the hydrate shell $\Delta \nu$ with the temperature (Fig. 55b) indicates that the change is not only in the amount of the bound water, but also in its properties. The broadening of the PMR line with decreasing temperature indicates a decrease in the time of spin-spin proton relaxation (since $\Delta \nu = 1/T_2 \pi$) and an increased correlation time of the water molecules.

The decrease in $\Delta \nu$ between $-10°C$ and $-13°C$ may be caused by the freezing out of the water fractions in the vicinity of the non-polar segments of the protein (Kurzaev *et al.,* 1975) which are least mobile, i.e., which give the broadest lines. The water in the cavities and the water next to the polar and to the charged groups freezes at lower temperatures and is more mobile than the water of solvation of nonpolar surface groups at the same temperature. The increase in $\Delta \nu$ of the intact antibody when $t°$ is reduced from $-13°C$ to $-18°C$ may reflect the immobilization of the water in the large central cavity between the antibody subunits. In the case of Ab + H this cavity contains a smaller amount of water, whose structure is more resistant to freezing. For this reason the increase of $\Delta \nu$ between $-13°C$ and $-18°C$ is insignificant (Fig. 55b).

It is thus seen that the experimental results are compatible with the assumption that

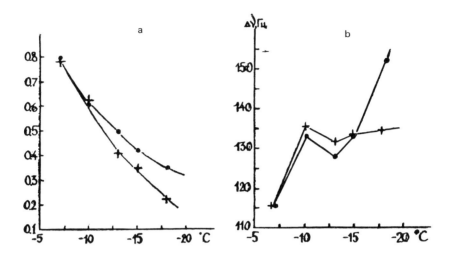

Figure 55. Hydration (a) and half-width ($\triangle \nu$ Hz) of PMR lines (b) of frozen solutions of antibodies (•) and their hapten complexes (+). 0.01 M phosphate buffer (pH 7.3). Antibody concentration: 20 mg/ml. After Käiväräinen *et al.* (in press).

the dehydration of the antibody as a result of the reaction with hapten is principally due to the leftward shift of the A $\rightleftharpoons$ B equilibrium of the central cavity of the antibody, accompanied by the displacement of the corresponding number of H_2O molecules into the solvent.

Effect of Antibodies on the Water Environment

One non-trivial consequence of the dynamic model is the capacity of the fluctuating protein to perturb the surrounding medium by exchange between the ordered water in the cavities and the free water of the solution. This perturbance must be a function of the changes in the water-protein interactions in the cavities brought about by specific and nonspecific agents. Such effects can in fact occur in albumin solutions under the effect of heat (para. 5.2).

Taking the reaction Ab + H as an example, we shall see that similar phenomena also take place in protein solutions during specific conformational changes. Pig antibodies with hapten 8-DNP-5, 8-azo-4-octanoic acid, were used, as well as their monovalent Fab- and Fc-subunits (Käiväräinen *et al.*, 1981).

We studied the temperature dependence of the shift of the peak of the deformational-librational (associative) absorption band of water in antibody solutions, and in their Fab- and Fc-subunits, before and after reaction with hapten between 2100 and 2140 cm^{-1} — an interval in which the protein itself does not absorb (Fig. 56a, b). The shift of this peak towards lower frequencies indicates a weakening of the hydrogen bonds in the water or a decrease in their number. Studies of the temperature dependence of the IR spectra of

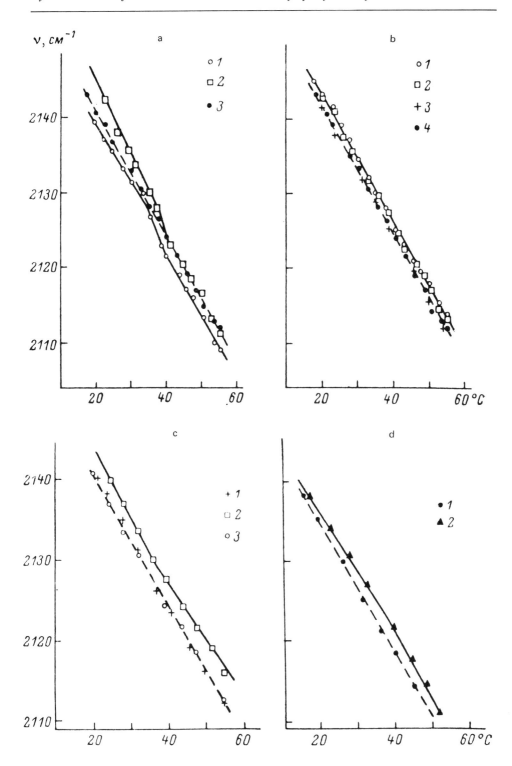

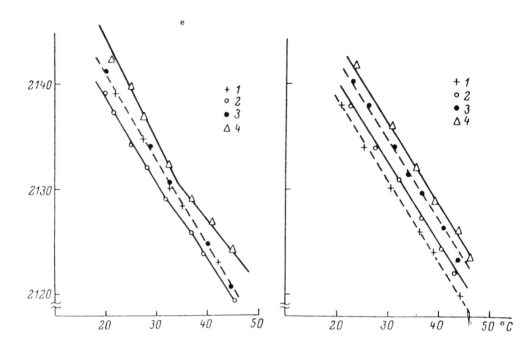

Figure 56. Temperature dependence of the shift in the peak of the associative absorption band of water in various solutions. After Käiväräinen *et al.* (1981a).

a) 1 — Ab; 2 — Ab + H; 3 — control: 0.01 M phosphate buffer (pH 7.30) + 0.15 M NaCl. b) 1 — Fab-subunits of the Ab; 2 — Fab-subunits + H; Fc-subunits; 4 — 0.01 M phosphate buffer (pH 7.30) + 0.15 M NaCl. Concentration of Ab, M: [Ab] = $1 \cdot 10^{-4}$; [Fab] = $3 \cdot 10^{-4}$; [Fc] = $3 \cdot 10^{-4}$. Molar ratio [H] : [active site] = 1.25:1. c) 1 — 0.01 M phosphate buffer (pH 7.30) + 0.15 M NaCl; 2 — Ab in this buffer in the presence of 20% saccharose; 3 — 0.01 M phosphate buffer (pH 7.30) + 0.15 M NaCl in the presence of 20% saccharose. d) 1 — 0.01 M phosphate buffer (pH 7.30) + 0.5 M NaCl; 2 — Ab in this buffer. Ab concentration: 15 mg/ml. e) 1 — 0.01 M phosphate buffer (pH 7.30) + 0.15 M NaCl; 2 — oxidized cytochrome C (20 mg/ml) in this buffer; 3 — this buffer + 20% saccharose; 4 — oxidized cyto-chrome C (20 mg/ml) in buffer with 20% saccharose; f) as e), for cytochrome $\bar{C}$ reduced by Fe++ (EDTA) complex; this complex, at the concentration of $1.7 \cdot 10^{-2}$ M, was present in all buffer and pro-tein solutions in Fig. e).

water show that, as the water is heated from 6 to 72°C, the fraction of the molecules not containing hydrogen bonds decreases from 31 to 18%, while the fraction of molecules with a single hydrogen bond remains approximately constant at 42% (Buijus and Choping, 1963; quoted from Eizenberg and Kautsman, 1975).

The effect of protein on the water environment may be of two kinds. Firstly, protein fluctuations and exchange between the bulk water in solution and the hydrate shell perturbs the environment and reduces the number and the strength of hydrogen bonds; secondly, the interaction between water molecules and the numerous polar and nonpolar groups on the protein surface stabilizes their thermal motion. Thus, the observed shift of the absorption peak of water is the result of superposition of opposite effects.

In the case of isolated Fab-fragments, the stabilizing effect of the polar groups is stronger than the destabilizing effect, and the temperature variation of the peak absorption of water in Fc-solution is the same as that of the control buffer (Fig. 56), within the limits of the experimental error. Thus, the stabilizing effect of Fc on water is fully compensated for by the destabilizing effect of intramolecular mobility of Fc. However, in solutions of the intact antibody there is a shift of the absorption peak towards lower frequencies than those of the control. A juxtaposition of these results leads to the conclusion that the largest contribution to the perturbance of the hydrogen bonds of the surrounding solvent is made by cavity fluctuations between the antibody subunits and by the exchange between the cavity water and the external medium. These experimental results are compatible with a possible effect of the changes in the dynamic properties of Fc, which is a constituent part of the antibody, on the strength of the hydrogen bonds in water under the effect of hapten.

Special control experiments conducted on low-molecular (molecular mass 13,300) proteins – cytochrome C, hemoglobin and albumin – confirm the dual nature of the interaction between protein and water.

Ferrocytochrome C, whose heme cavity is in the 'closed' state, has a stabilising effect on water at concentrations of 20 mg/ml. This is manifested as a high-frequency shift of the absorption maximum by 1.5 ± 0.2 cm^{-1} at 30°C. An increase of its concentration from 20 to 150 mg/ml results in an increase of the shift to 8 cm^{-1}.

During the oxidation of ferrocytochrome C its effect on water became reversed, and in the presence of ferricytochrome C at a concentration of 20 mg/ml, the water showed incipient destabilization. This result is in agreement with our own theories, according to which the heme cavity of the oxidized ferrocytochrome C may fluctuate between 'open' and 'closed' states, unlike its reduced form.

Ferrihemoglobin (MtHb) behaves in a similar manner. The presence of human and bovine MtHb in solution at a concentration of 20 mg/ml at 30°C causes the absorption band of water to shift by 2 cm^{-1} towards lower frequencies relative to the control (0.01 M phosphate buffer (pH 7.3) + 0.15 M NaCl). The band shift curve of hemoglobin solution in the oxidized form coincides with the control curve.

In order to demonstrate the stabilizing effect of the charge and of the polar protein groups on the water environment, we studied the pH-dependence of the stabilizing effect of human serum albumin, present in a concentration of 200 mg/ml, at 30°C. The positive shifts of the absorption peak of water in the presence of human serum albumin at pH 5.8, 7.4 and 7.8 were 2 ± 0.2, 3.5 ± 0.2 and 6 ± 0.2 cm^{-1} respectively. All determinations were conducted in 0.01 M phosphate buffer + 0.15 M NaCl. It is seen that as the pH values move further away from the isoelectric point of human serum albumin (pH 4.9), i.e., as the total charge on the protein increases, its stabilizing effect on the water is enhanced.

On the other hand, human IgG in the same buffer, at pH 7.3, had a destabilizing effect on water even when present in a concentration of 150 mg/ml.

The magnitude of the water-destabilizing effect exerted by intact IgG and by met-hemoglobin varies slowly and nonlinearly with the protein concentration. However, the stabilizing effect increases almost proportionally to the concentration of ferrocytochrome C and albumin in solution; this indicates that the two effects are different in nature.

In order to find out if the hapten-induced changes in the solution of the antibody are or are not caused by the increased charge on the protein, we carried out comparative disk electrophoresis on the antibodies and their hapten complexes. No differences in the electrophoretic mobilities of the proteins could be observed.

We shall now discuss the possible reasons for the observed destabilization of water by proteins. We may say at once that the Brownian motion of the antibody molecule as a whole does not affect the average mobility of the hydrogen bonds, or the vibration spectrum of the water environment, since the presence of Fab-units in water, which have a higher mobility than the intact antibody, fails to produce a low-frequency shift of the band with respect to the buffer control.

The lifetime of hydrogen bonds in water (10^{-11} - 10^{-12} sec), which is determined by the frequency of jump-like reorientations of water molecules, can only be affected by processes taking place at a comparable rate. In protein solutions such processes may comprise local fluctuations in the state of the solvent (density, concentration, pressure), which accompany the 'displacement' or the 'sorption' of the water molecules during $A \rightleftharpoons B$ transitions of protein cavities. These involve sudden changes in enthalpy and entropy, when

$$\Delta G_{A \rightleftharpoons B} = n\Delta H^{H_2O} - Tn\Delta S^{H_2O} \approx 0, \tag{6.41}$$

where n is the number of water molecules whose state undergoes a major change during $A \rightleftharpoons B$ transitions. The water molecules are ejected, as it were, out of the cavities into the external environment, altering the properties of the latter. The sudden changes in the orientations of water molecules and their translational motions during their expulsion from the cavity must be accompanied by a reorientation of the molecules of the surrounding solvent, and thus involve overcoming a certain potential barrier.

Obviously, the larger the values of $n\Delta H^{H_2O}$ and $n\Delta S^{H_2O}$, the stronger the disturbance which is introduced into the aqueous medium surrounding the protein. The magnitude of the effect must also depend on the frequency of the $A \rightleftharpoons B$ transitions of the cavities. Judging by the results described above, the isolated Fab behaves above 20°C as a rigid particle ($\nu_{A \rightleftharpoons B}^{Fab} \leqslant 10^7$ sec^{-1}). The central cavity of IgG is probably larger and may act as a more effective generator of anisotropic fluctuations.

A change in the orientation of only a small number of water molecules is sufficient for the anisotropic fluctuations to significantly affect the properties of the liquid. Thus, for instance, if only one out of 10,000 polar molecules of the liquid is oriented in a definite manner, while the distribution of the remaining molecules is isotropic, an electric moment corresponding to a field intensity of about 100 V/cm is generated (Shakhparonov, 1976).

The fluctuations in the water surrounding the protein may take place not only during the transitions B → A or A → B, but also during the lifetime of the cavities in the 'open' B-state. Here they may take place owing to the fast exchange between the ordered water in the cavity and the free water in the solution. Such a process may in fact take place, because the lifetime of the B-state is several orders longer than that of hydrogen bond in free water.

If these ideas are correct, it may be expected that the effect of the agents which disturb water structures in protein cavities and reduce the value of ΔS^{H_2O} and frequency of A ⇌ B transitions, should result in a strengthening of the hydrogen bonds of the water. Studies of the effect of 20% saccharose and 0.5 M NaCl on the shift of the water absorption peak in Ab solutions show that in both cases there is a high-frequency shift relative to the absorption band of the control solution, similarly to the shift produced by the effect of hapten (Figs. 56b and 56c). The saccharose reduces the value of ΔS^{H_2O}, and enhances the viscosity of the solution; as a consequence, the frequency of the relative fluctuations of Ab subunits – i.e., $\nu_{A=B}$ – decreases.

Methemoglobin and ferricytochrome C present in a concentration of 20 mg/ml in 0.01 M phosphate buffer (pH 7.3) in the presence of 20% saccharose (Fig. 56e) also showed an incipient stabilization effect on water. These results indicate that the interaction between protein cavities and water always proceeds by the same mechanism.

The fact that the saccharose effect disappears after the cytochrome C has been reduced by Fe^{++} (EDTA) complex is convincing evidence in favor of our interpretation of the observed effects. In fact, the positive shift of the absorption peak of water in ferrocytochrome C solutions relative to the buffer control remains the same both in the presence and in the absence of saccharose (Fig. 56f). This is in fact what may be expected, since it is indicated by the data of X-ray diffraction analysis that during the reduction of ferrocytochrome C its heme-containing cavity is screened from the solvent by Phen82, and water is displaced from the cavity.

We used the method of proton magnetic resonance in order to verify that proteins are in fact capable of reducing the average strength of hydrogen bonds. The location of the water absorption band in NMR spectrum is a sensitive parameter which reflects its diffusion-averaged (D) structure. It was found that in 3% oxyhemoglobin solution at 25°C its site coincides with that of the absorption band of buffer control, while in methemoglobin solution it is shifted by 2.3 Hz towards low frequencies, which corresponds to a weakening of the hydrogen bonds of water. This result is in full agreement with IR spectroscopic results. The half-width of the line, which was directly proportional to the correlation time of water molecules, was very small (2.00 Hz) in methemoglobin solution, was larger (2.10 Hz) in the case of the buffer, and largest in the case of oxyhemoglobin. Other data concerning interaction of Hb with water may be seen in Section 6.6.

The perturbances introduced in the solvent by the fluctuating protein impair not only the V-structure of the water, as recorded by the method of IR-spectroscopy, but also its D-structure.

It is of interest to compare the results obtained by IR spectroscopy with the data of refractometric studies performed on a solution of porcine anti-DNP antibody before and after the reaction with hapten (Fig. 57). Despite the fact that IR-spectroscopy yields in-

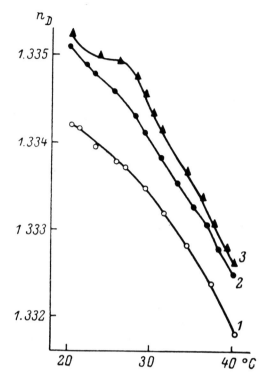

Figure 57. Refractive index of 0.01 M phosphate buffer (pH 7.3) + 0.15 M NaCl (1), of the solution of antibody (10 mg/ml) in this buffer (2) and of solutions of antibody complexes with DNP-hapten (3) as a function of temperature.

formation on the averaged vibrational V-structure of water, while the refractive index yields information on the diffusional-averaged D-structure (Sec. 1.1), it can be seen that the temperature variations of ν (Fig. 56a) and of n_D (Fig. 57) are very similar.

The experimental refractive index of a protein solution is

$$n\,(t) = n_0 + n^P\,(t), \qquad (6.42)$$

where n_0 is the refractive index of the buffer and n^P is the refractive index of the protein. n_0 and n^P are directly related to the molar polarizability a of the respective components of the solution:

$$n_0 - 1 = 4\pi N_0 a_0, \qquad (6.43)$$
$$n^P - 1 = 4\pi N^P a^P, \qquad (6.44)$$

where N is the number of molecules in 1 ml of the solution.

The refractive index is a measure of local polarizability, produced by the deformation of the electron configuration around the nucleus by an electromagnetic field; it contains no direct information about the extended structure of the macromolecules, but is a description of the averaged parameters of their individual chemical bonds. The variation of n_0 with increasing temperature is caused by the perturbed D-structure of water due to the decrease in the average number of undisrupted hydrogen bonds. The redistribution of electron density in H_2O molecule during the rupture of its hydrogen bonds affects its polarizability a_0. This effect may be either enhanced or weakened by the presence of the protein. Were the contribution of the protein to the relationship $n(t)$ constant, this function would be parallel to $n_0(t)$. However, this is not the case (Fig. 57). It may be said that hapten enhances the stabilizing effect of the antibody on water, since in its presence the n-value is higher than in the presence of the same concentration of intact protein.

These results may be interpreted in the same manner as those obtained by IR spectroscopy. However, in our case the conformational rearrangements of the protein itself, which are accompanied by a disturbance of its chemical bonds, may also contribute to the change in the refractive index of the solution. We also noted changes in the refractive index of a solution of human serum albumin, which correlate with the observations made by the method of IR spectroscopy, after the molecule has bound three molecules of methyl orange.

It should be borne in mind that a change in n, induced by the binding of ligands in protein solutions, means that the dielectric constant ϵ of the medium changes as well, since $n^2 = \epsilon$. This factor may significantly affect reactions taking place inside membranes and the accompanying structural rearrangements of the latter, and in intra-cellular processes involving electrostatic interactions.

On the strength of our experimental results obtained by the spin label method, IR spectroscopy and refractometry, we may conclude that the reaction Ab + H proceeds in the following sequence of events: the hapten, which has penetrated into the active site cavity, produces a shift in the equilibrium between the 'open' and 'closed' states of the cavity in favor of the 'closed' state, which is thereby stabilized; the relative rotation and stabiliziation of variable domains reduce the mobility of the $V_L V_H$ and $C_L C_H 1$ domains which constitute the Fab-subunits; the intensified interaction between the $V_L V_H$ and $C_L C_H 1$ domain pairs alters the interaction between the domains forming the large central cavity between IgG subunits; as a result, the intramolecular mobility of the domains which constitute the Fc-subunit changes as well.

The fact that the tendency to generate anisotropic fluctuations of anti-DNP-Ab decreases after complex formation with hapten may be due to the perturbance of the cooperative system of the water molecules in the central cavity of the antibody. The fusion of the water cluster is probably the result of a deformation in the geometry of the cavity and a change in its dynamic properties, caused by the events occurring in the active site. That water plays an important part in hapten-induced conformational changes of the antibody is confirmed by the fact that these changes are strongly affected by small additions of D_2O.

The accessibility of the 'hinge' segment between $F(ab')_2$ and Fc to the attack by proteases changes only insignificantly following the reaction between Ab and hapten (Jaton *et al.*, 1978). This is not in contradiction with our interpretation of its conformational changes, which are related to changes in the dyanmic properties of central cavity. Nevertheless, Jaton's results constitute a weighty argument against the model of a simple transition from a less compact to a more compact structure of the antibody under the effect of ligands.

According to our model, the inflexions on the curves in Figs. 56a and 56b around the physiological temperatures indicate that temperature-induced conformational transitions taking place in the antibody cause changes of ΔS^{H_2O} in the cavities between the constituent subunits of the molecule. It is very probable that the changes in the geometry of the segments which interlink the subunits and subunit domains of immunoglobulins are not arbitrary, but are determined by individual degrees of freedom, the latter being in turn determined by specific stereochemical features of these segments. Their conformational changes may also contribute to the dynamics of the protein.

The sum total of the data available on the conformational properties of the antibody, their isolated subunits and on the specific features of the interaction between these proteins and water may be regarded as a confirmation of the fundamental assumptions underlying the dynamic model of behavior of proteins in water. We shall now illustrate the dynamic model by taking a small protein with one subunit — cytochrome C — as an example.

Effect of Perturbing Agents on the Hydrate Shell and on the Flexibility of Immuno-globulins

The effect of stabilization of the flexibility of proteins by low D_2O concentrations (Fig. 50) was employed in the study of the action of perturbing agents on the hydrate shell.

The presence of 3% D_2O in solutions of F_{ab} + DNP-SL and AT + DNP-SL resulted in an increase of τ_M from 12 to 15 nsec and from 15 to 21 nsec respectively. The addition of perturbing agents — 0.5 M NaCl, 1 M urea and 10% butanol — fully eliminated the isotopic effect. The presence of 0.5 M NaCl not only eliminated the difference between the protein stability in the presence and in the absence of D_2O, but also raised the τ_M of F_{ab} and Ab at 5°C to 17 and 27 nsec respectively.

At 20°C no effect of 3% D_2O or of the perturbing agents on the τ_M of these proteins was noted, which indicated that the effects of the temperature and of the perturbing agents are additive.

The effect of perturbing agents on the flexibility of proteins was also studied on subfractions of spin-labeled IgG of mink. This effect was most conspicuous as the changes of flexibility of the first (electrophoretically slow) subfraction of mink IgG (Table 13). This table also shows the freezing temperatures of buffer and of perturbing agents solutions.

The fact that there is an obvious correlation between τ_M and t_f^0 shows that, while protein may affect the properties of the solvent, the reverse effect is also possible. The t_f^0-values indicate that the lower the activity of the water, the less flexible is the protein. This confirms one of the most important consequences of the dynamic model of behavior of proteins in aqueous media. An increased activity of water may be expected to shift the equilibrium between the 'closed' state A and the 'open' state B towards the more highly hydrated state B. Since the corresponding protein conformers are more flexible, this is experimentally manifested as a shorter correlation time. A decrease in the activity a of water will result in the opposite effect. This may be demonstrated as follows. The activity of water is directly connected with its chemical potential:

$$\mu = RT \ln a + \mu_0$$

TABLE 13. The values of τ_M for IgG-SL subfractions of mink on the absence and presence of perturbing agents at $5°C$, and freezing points $t_f^°$ of the buffer and of the perturbing agents solutions in the absence of proteins

	pure buffer*	1M urea in buffer	0.5M NaCl in buffer	10% butanol in buffer
τ_M (nsec)	33 ± 2	35 ± 2	40 ± 2	50 ± 2
$t_f^°$	-0.35 ± 0.05	-1.00 ± 0.05	-1.50 ± 0.05	-2.70 ± 0.05

* 0.01M phosphate buffer, pH 7.3 + 0.15M NaCl

The similarity between the state transitions of the water in cavities as a result of the fluctuation process $A \rightleftharpoons B$ and phase transitions of the first kind is caused by the difference between the chemical potentials of water in the open cavities μ^B and of the free water in solution μ^F. One of the reasons for the change in the conformation and in protein dynamics as a result of their association with ligands is the shift in the $A \rightleftharpoons B$ equilibrium, indicated by the change in μ^B. However, both similar and opposite changes in protein properties may be produced by altering the activity of the free water, i.e., by altering μ^F.

According to the Boltzmann distribution, the equilibrium constant $K_{B \rightleftharpoons F}$ describing the exchange between the free fraction F of the water in solution and the water bound to the 'open' protein cavities B is:

$$K_{B \rightleftharpoons F} = \exp\left(-\frac{\eta^B - \eta^F}{RT}\right)$$

It follows from the above equation that a decrease in μ^F should shorten the lifetime of the flexible B-conformers of the protein and shift the $A \rightleftharpoons B$ equilibrium towards the left, while an increase in μ^F should have the opposite effect and enhance the resulting flexibility of the protein.

6.4 Cytochrome c

Cytochrome c is a heme-containing protein with a molecular mass of 12,300, whose function is to transfer the electron in the membranes of mitochondria from the cytochrome-reductase to the cytochrome-oxidase enzyme complex. The cleft accommodating the heme is formed by a polypeptide chain of 16 nonpolar residues, of which more than half were identical in all the cytochromes studied. The iron of ferricytochrome and ferro-cytochrome is bound on one side of the heme to the Met80 sulfur atom, and on the other to the imidazole groups of His18. Up to pH 3, the heme iron in both the valent states is predominantly present in the low-spin state. It has been shown by Genkin *et al.* (1977) that between pH 1.5-2.0 and pH 5 the chaotropic ions (NaCl and $NaClO_4$) change the equilibrium between the low-spin and the high-spin states of ferricytochrome C towards the low-spin form of the protein.

Mammalian cytochromes C do not react with O_2 or NO_2 in a neutral medium. Their absorption spectra are typical of ferri- and ferro-derivatives of complexes with their fifth and sixth coordinate sites occupied by strong-field ligands (Harbury and Marks, 1978).

Results of X-ray studies indicate that the prosthetic group of equine cytochrome C is so positioned that only its edge can be in contact with the surrounding medium (Dickerson and Timkovich, 1975). The bottom of the heme cleft comprises nonpolar zones only, which border on the surface with clusters of positively charged lysine residues, in conformity with the distribution of polar and nonpolar cavity-forming residues in general (sec. 4.2).

The globule of ferricytochrome c also contains two channels interconnecting the heme-containing cavity with the outside environment. The left-hand channel, which is not accessible to the solvent, is constituted by the 55-57 chain, which contains the hypothetical 'electron path' to Met80 (which is the sixth ligand of Fe^{+++}), formed by the aromatic rings Tyr74, Tyr59 and Tyr67, disposed in parallel. During the reduction of ferricytochrome C the place of the Tyr74 ring becomes 'inverted', and the 'electron chain' is broken. The right-hand channel, formed by the 19-25 residues, is accessible to the solvent molecules.

Dickerson *et al.* (1972) assume, just as we do, that the active site cavity of ferricytochrome C fluctuates between states which are 'open' or 'closed' to the solvent. The actual state of the molecule depends on the positioning of the phenylalanine residue 82. According to diffraction patterns, the 'open' state, corresponding to ferricytochrome, may be stabilized by the Cl^--anion. This conclusion is probably justified by the fact that in the absence of Cl^--ions the oxidized form of the protein does not crystallize, while the reduced form may crystallize both in the presence and in the absence of Cl^--ions.

Judging by the electron density distribution in the active site zone, the site occupied by Cl^- in ferricytochrome is identical with that occupied by phenylalanine ring in ferrocytochrome. It may be expected, accordingly, that the increase in the concentration of Cl^- in solution will result in a shift of the equilibrium between Fe(III) ('open') and Fe(II) ('closed') states towards the former.

A number of hypotheses have been advanced as to the mechanism of the electron transfer from cytochrome reductase to the Fe(III) of the cytochrome; however, since none is as yet recognized as conclusive, they will not be discussed further. From our point of view, it is the conformational changes which accompany this process or are consequent upon it (Fig. 58) which are of interest. Phen82, which is exposed to the solvent in ferricytochrome, blocks the opening of the heme cavity in the ferro-form and thus converts it into the 'closed' state (Fig. 58, heavy arrows).

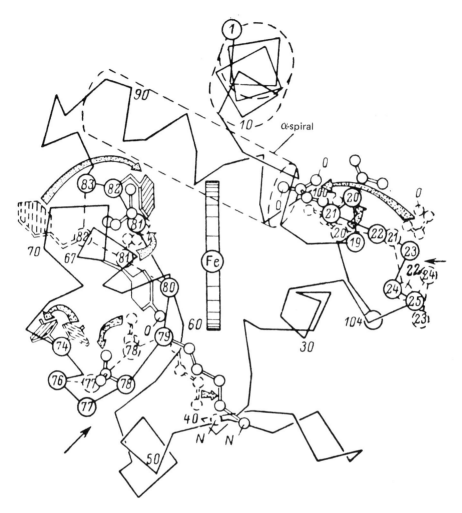

Figure 58. Principal structural changes during the reduction of ferricytochrome **c**. After Takano *et al.* (1971). Thin arrows indicate the left-hand and the right-hand channels.

The existence of the mechanical link between the states of the heme 'pocket' and the right-hand channel of cytochrome **c**, which was directly observed by Takano *et al.* (1973), is especially important in the context of our theories of dynamic interaction between the fluctuating cavities in protein molecules. When the active center passes into the 'closed' state, this link ensures the displacement of the residues, resulting in the blockage of the channel. The molecule as a whole becomes much more compact. These conformational changes were observed on comparing heart ferricytochromes of horses and tuna fish, but X-ray data indicate that there is no difference between the ferri- and ferro-form of cytochrome C in the heart of tuna fish. This may result from the conditions under which the crystallization took place, when the same iso-form of the protein is always stabilized, irrespective of the ligand condition.

Numerous available data indicate that ferrocytochromes c have a more stable and more compact conformation than mammalian ferricytochromes in solution; they display a slower rate of proteolysis (Yamanaka *et al.,* 1959) and of hydrogen exchange (Kägi and Ulmer, 1968), and are more resistant to heat (Butt and Kellin, 1962) and to guanidine and urea (Myer, 1968). This phenomenon is explained not by random conformational changes which are detectable by X-ray methods, but by dynamic changes, i.e., changes in the amplitude and frequency of structural fluctuations of cytochrome c (Salemme, 1977). However, changes in these parameters reflect a rise in the activation barriers between the conformers of the protein, and shift the equilibrium a $\rightleftharpoons$ b to one or another side.

We were able to show (Figs. 56, e, f) that oxidized cytochrome c is capable of inducing anisotropic fluctuations in the surrounding aqueous medium, but reduced cytochrome C is not. This effect is compatible with an interpretation of X-ray diffraction data as a sharp shift in the equilibrium between the 'open' and the 'closed' states of the heme-containing cavity towards the 'closed' state following the reduction of cytochrome c. We shall accordingly assume in what follows that the ferri- and ferro-forms of cytochrome c are conformationally different.

In terms of the dynamic model, the transition process of cytochrome c or another heme-containing protein from the ferri- to the ferro-form may be described as follows: bonding an electron results, in the first stage, in a fast ($t \leqslant 10^{-9}$ sec) change in the oxidation state of the iron [Fe(III) $\overset{k_0}{\rightarrow}$ (Fe(II)], after which the electron state of the heme varies more slowly, at a rate constant of k_1. In the cases of methemoglobin and hemoglobin the iron passes from the low-spin to the high-spin state at the same time. This is followed by an even slower variation in the behavior of the active center site itself, with a shift in the equilibrium between 'open' and 'closed' states towards the latter: (a $\rightleftharpoons$ b) $\overset{k_2}{\rightarrow}$ (a* $\rightleftharpoons$ b*). Since the mechanical link between the active site and the right-hand channel (or, in the general case, with auxiliary cavities) is not perfectly rigid, the corresponding shift in the equilibrium between their states A and B is somewhat delayed. As a result of the corresponding relaxational process (A = B) $\overset{k_3}{\rightarrow}$ (A* $\rightleftharpoons$ B*), the geometry of the cavity of the active site becomes dynamically adapted to the new electronic state of the heme, which may also affect its spectral characteristics. We may add that, in accordance with the dynamic model of association and dissociation of specific complexes, K_{ass} may be expected to increase as a result of this relaxational process.

The experimental work of Blumenfeld *et al.* (1974 a, b, c; 1977 a, b) seems to confirm the sequence of events postulated above. That the electronic states of the heme in cytochrome c, methemoglobin, hemoglobin, and also the electronic state of the active sites in heme-free ferrous proteins, change at a slower rate than their states of oxidation, and that they depend on the conformation of the protein globule, was proved during the reduction of their active sites by electrons thermolyzed by γ-irradiation in frozen water matrices at 77°K. The authors noted that spectroscopic and EPR characteristics of the active sites of all these proteins in conformationally non-equilibrated states may be very different from the corresponding characteristics at equilibrium. Blumenfeld *et al.* (1977a) studied the reduction of heme- and iron-containing proteins in aqueous solutions by the method of pulse radiolysis, in order to verify the existence of similar conformationally non-equilibrated states in solution and to determine their kinetic parameters. The solutions were irradiated with a high-energy electron pulse lasting for 10^{-6} sec. The kinetics of the processes were followed spectrophotometrically by following the change in the optical absorption at 500-600 nm, which is the absorption range of the active site.

The reduction Fe(III) $\xrightarrow{k_0}$ Fe(II) is realized within a time shorter than the duration of the pulse ($\simeq 10^{-6}$ sec); accordingly, the experimental values reflect only those disequilibrium-induced relaxational processes in the active center and in the globule residual after 10^{-6} sec have elapsed after the reduction.

In agreement with the prediction of the dynamic model, three relaxation state constants are observed for cytochrome c, methemoglobin and hemoglobin under physiological conditions, with $k_1 > k_2 \gg k_3$; according to the dynamic model, k_1 is connected with the change in the electronic state of the heme, k_2 with the changed dynamics of behavior of the active site cavity, while k_3 describes the change in the dynamics of auxiliary cavities and dynamic adaptation of the active center.

Yet another consequence of the dynamic model — the stronger affinity of the protein-ligand complex at equilibrium as compared with its initial state — was also confirmed by Blumenfeld *et al.* (1977a). The rate of oxidation of ferrocytochrome by potassium ferricyanide $K_3 Fe(CN)_6$ was chosen as the measure of the affinity protein-ligand for the case of cytochrome c- electron complex. The rate of this reaction was determined immediately after the reduction of the active center by solvated electrons. The oxidation kinetics were recorded spectrophotometrically at 550 nm. Pulse radiolysis was carried out directly in the presence of this oxidant.

In all the cases studied, the oxidation rate constant prior to the termination of conformational relaxation of the protein from the ferri- to the ferro-form was higher than after relaxation process is finished. This means that the equilibrium form of the complex cytochrome C - electron is more resistant to the action of the oxidant than it is just after being formed. Accordingly, relaxational processes induced by the binding of the ligand, in fact result in an increase of K_{ass} (eq. 6.24).

Not only the low-molecular potassium ferrocyanide, but also protein-plastocyanin or ferricytochrome C, may function as the oxidizing agent in these experiments. In such cases the transport mechanism seems to be much more sensitive to the conformation state of the ferro-form, since the period of time during which the rate constants were different from their equilibrium values was much longer than in the case of similar processes taking place in the presence of ferricyanide.

A systematic analysis of the influence of various factors on the capacity of cytochrome C to reduce nicotinamide adenine dinucleotide-H was carried out by Tabagua (1977) by the NMR method. The pH-dependence of the reduction rate of ferricytochrome C by this nucleotide is very interesting. In low-pH regions, in which the heme cleft is exposed to a lesser extent, the reaction is about twice as fast as at alkaline pH values, when it is more exposed. This is connected with the exchange of the sixth ligand of heme iron with Met80 against Lys79 due to the ionization of the Tyr67 hydroxyl group in alkaline pH-range. In our view, these results may also be interpreted as the consequence of pH-dependent variation in the equilibrium constant between two short-lived states of the heme-containing protein cavity with various ligands in position six, which differ in their activation energy of reduction of ferric nicotinamide adenine dinucleotide-H and the affinity to the accepted electron. That the ferro-form of cytochrome C is more stable (i.e., has a higher affinity to the electron) at pH 7 than at alkaline pH values is indicated by the fact that electron exchange between ferro- and ferri-cytochrome C is inhibited when pH decreases to 5.

Important results were obtained when studying the interaction between ferricyto-chrome c and the polyamino acids blocking the charged zones on the protein surface, which may be responsible for the reactions with reductase and oxidase. Tabagua (1977) noted their effect on the state of the heme, which alters the rate of reduction of cyto-chrome C by nicotinamide adenine dinucleotide-H. This is yet another confirmation of the fact that the states of the globule and of the active site are interconnected.

Another consequence of the dynamic model of protein behavior is possible intra-molecular water-protein compensation effects (Chapter 5) which may accompany con-formational changes. This is indicated by the experimental work carried out on cyto-chrome c. Dawson *et al.* (1972) studied the temperature variation of the reduction rate of cytochrome C by $Fe(EDTA)^{2-}$ and by chromium ions by the stop-flow method, and obtained differing results: activation enthalpy ΔH = 19.7 and 73 KJ/mole, activation entropy ΔS = 58 and 75.4 J/mole·°K and free energy of activation ΔG = 46.1 and 50.3 KJ/mole respectively. These data may be interpreted as follows. In the former case the increase in the H and S of the solution caused by displacement of water molecules from the cavities when the protein passes into the ferro-form is compensated by decreased 'looseness' and conformational mobility of the protein as a result of the shift of equilib-rium between the states of its active site and the right-hand channel towards the 'closed' state. In the latter case there is no such compensation, since the chromium ion may still form part of the complex with heme after the reduction of cytochrome c, just like Cl^-, while competing for accommodation in the cavity with Phen82. This prevents the active site from 'collapsing' and the protein molecule as a whole cannot pass into the more com-pact state, with accompanying decreases in the entropy and enthalpy.

Thus, cytochrome c proved to be a very suitable object on which to illustrate all the principal consequences of the dynamic model of protein behavior in water. It will be shown below that the results of studies carried out on myoglobin confirm to a large extent the conclusions arrived at in the case of cytochrome c.

6.5. Myoglobin

Myoglobin — the oxygen carrier in muscles — contains one polypeptide chain with 153 residues and one heme. The nitrogen atom of the imidazole ring of histidine interacts with iron in the fifth coordination site. Oxygen and other ligands may become coordina-ted to iron in the sixth site. Myoglobin (Fig. 59) comprises eight spiral-shaped regions of different lengths, from the N-end to the C-end (*A* to *H*).

Heme is localized in the cleft between the spirals *E* and *F*, the proximal HisF8 residue being located in the *F*-spiral (Kendrew *et al.*, 1960). The border of the active site cavity contains polar propionyl residues, whereas its main part, which accommodates the heme, is distinctly nonpolar. Ligands such as O_2, CO or NO are bound to Fe(II) only, the metal passing from the high-spin to the low-spin state at the same time.

The iron of ferromyoglobin is incapable of coordinating water, but becomes capable of doing so after oxidation to Fe(III), when the myoglobin passes into the met-form. In addition to water, the protein in this form can also bind OH^-, F^-, N_3^-, CN^-, SCN^- and imidazole (Rifkind, 1978). The difference between the low-spin and the high-spin energy states is small for many metmyoglobin complexes, and the activation barrier separating the two states from one another is sufficiently low for the concentrations of the two spin states to be approximately equal at room temperature. The absorption spectra in the visible range are mostly due to $\pi \to \pi^*$ transitions in the porphyrin. They depend both on the spin state and on the ligand state of the iron. The properties of myoglobin and isolated hemoglobin subunits are similar in many respects.

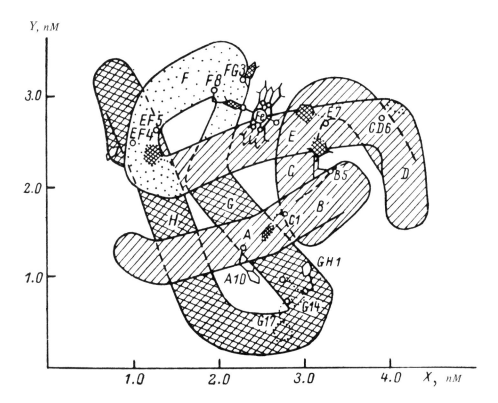

Figure 59. Spatial structure of the sperm whale myoglobin in *XY*-projection. After Artyukh *et al.* (1977). *Thin shading* indicates histidine residues inside the molecule which are not accessible to the solvent. HisA10 and HisGH1 are located in the site of active contact between myoglobin and cytochrome C during the electron transfer process.

X-ray diffraction studies of myoglobin in the presence and in the absence of specific ligands failed to reveal any significant conformational rearrangements in the protein. A possible reason for it is that in both states — ligand and ligand-less — only one of the possible conformers of myoglobin becomes stabilized during crystallization. Nevertheless, numerous data, obtained by various methods, indicate that the function of myoglobin in solution is related to changes in its conformational state. These data include the results of a comparative study of myoglobin and oxymyoglobin by the method of proton NMR, which show that the state of the protons is changed not only in the zone of the active site, but also in segments distant from the heme (Patel *et al.*, 1970; Schulman *et al.*, 1970). The conformational changes are also indicated by the large decrease in entropy which occurs when O_2 and CO are bound by myoglobin (-251.4 and -275.3 J/mole·°K respectively) (Keyes *et al.*, 1971).

The spin-label method proved to be very useful in studies of conformational properties of myoglobin. The analysis of the mobility of spin labels selectively bound to various protein segments, and its changes consequent upon ligand binding and pH variation, are in agreement with the structural model of myoglobin proposed by Atanasov (1970), as presented by Artyukh *et al.* (1975, 1977). According to this model, the myoglobin molecule consists of three independent parts or blocks: *ABCDE, F* and *GH*, which are interconnected by other mobile, nonhelical fragments of the polypeptide chain, and are capable of thermal shifting with respect to each other (Fig. 59). If we assume that the *F*-fragment is immobile, then the two extremal locations of the other fragments correspond to four possible conformers of myoglobin. According to Artyukh *et al.* (1977), they are determined by the interactions at the C- and N-ends of the molecule. A change in the identity of the ligand brings about conformational changes resulting in a rearrangement of the salt cluster at the N-end of the molecule. This is manifested by a change in the pK of group ionization, and indicates that the charge on the N-end of the molecule is affected by the electronic state of the heme.

The labels on HisA10 and HisG19 at the contact sites of the mobile *ABCDE* and *GH* fragments proved to be the most sensitive to conformational changes. The HisE5 label on the *ABCDE* fragment was only slightly affected by conformational changes, which means that it had shifted as one whole. The TyrG4 and HC2 labels also reflected the stability of the *GH* segment as a whole.

It is important to note that not only the forward (heme-globule) but also the reverse (globule-heme) interrelation is present (Artyukh *et al.*, 1977); an example is the dependence of the pK-transition of myoglobin from the met-form (with H_2O as ligand) to the hydroxy-form (OH⁻ as ligand) on the manner in which the globin is modified by the spin label. If the label was bound to HisA10 and to the α-NH_2-group of the valine at the N-end, the pK of transition decreased from 8.87 to 8.83 and 8.5 respectively, which indicates that the equilibrium between the high-spin and the low-spin conformers of met-myoglobin was shifted towards the latter. Absorption spectra, and the fact that during the transition from the H_2O- to the OH⁻-ligand the pK remains unchanged, seem to show that the modification of the tyrosine residues of methemoglobin have no influence on the state of the active site.

The correlation times of spin labels conjugated with various parts of globins were used as indexes of conformational changes induced by changes in pH. The conformational transitions of the protein as a result of the ionization of the α-NH$_2$-group of the N-terminal valine and the ϵ-NH$_2$-group LysH9 were reflected as angles in the curves of correlation times as a function of pH (between 6.0 and 11.6). Both these groups participate in the formation of the N-terminal salt cluster.

Several workers (Atanasov, 1970, 1971; Atanasov *et al.*, 1972; Likhtenshtein, 1974) presented convincing evidence that the binding of a CN$^-$-ligand and an increase in the temperature from 15 to 50°C produce an equilibrium shift, at least between the two conformers of metmyoglobin. The normalized variations of increasing mobility of spin-labels bound to myoglobin and the degree of saturation of metmyoglobin with CN$^-$ as a function of the logarithm of CN$^-$-concentration were found to be symbatic. Moreover, the variations of heat capacity, rate of spin-lattice relaxation ($1/T_1$) and a number of other parameters of metmyoglobin and cyanomyoglobin proved to be symbatic (Fig. 60).

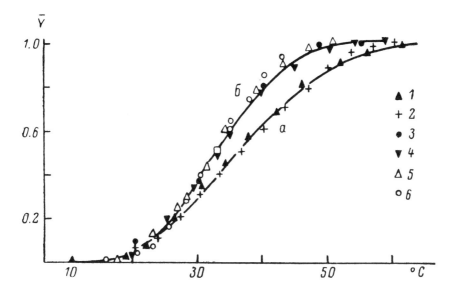

Figure 60. Normalized temperature variations ($\overline{V}$) of heat capacity (1) and optical density at 245 nm (2) of cyanomyoglobin solutions (a), rate of spin-lattice relaxation of water protons (3), heat capacity (4), activation energy of rotation of nitroxyl spin labels (5) and partial semi-saturation pressure of myoglobin by oxygen (6) and of metmyoglobin solutions (b). After Atanasov *et al.* (1972).

Since metmyoglobin is stabilized by CN$^-$, the normalized curves for this substance are shifted towards higher temperatures. The increased rate of proton relaxation (T_1^{-1}), i.e., the decrease in T_1, means, in terms of the theory and experimental results obtained by Blombergen, that the correlation times of water molecules in solution have increased (Fig. 8). This result may be explained either by an increased accessibility of water to the coordination sphere of the heme iron or, as was shown above (para 6.3), by a lower effectiveness of anisotropic fluctuations of the solvent due to the a $\rightleftharpoons$ b transitions of the active site cavity, owing to the destabilization of the water cluster in this cavity and a decrease of $\Delta S^{H_2 O}$ with increasing temperature (equation 6.41)). In the former case the equilibrium between the 'open' (b) and the 'closed' (a) state of the active site (a $\rightleftharpoons$ b) would be shifted to the right, while in the latter case it would be shifted to the left. The second explanation is the more likely, since a leftward shift of the a $\rightleftharpoons$ b equilibrium may be expected to result in an increased heat capacity of the solution due to the displacement of some of the water from the protein cavities into the environmental medium (Fig. 60). The 5-7% change in the extinction of the Soret band as the temperature increased from 20 to 45°C shows that the environment of the heme has in fact changed (Brunori *et al.*, 1966). The van't Hoff enthalpy and entropy of the temperature transition of metmyoglobin are, respectively, 209 KJ/mole and 628.5 J/mole·°K. It follows that the difference in the free energies $\Delta G = \Delta H - T\Delta S$ between the myoglobin conformers is only about 8.4 KJ/mole. However, caution should be exercised in using results obtained with the aid of the van't Hoff equation (Chapter 5).

EPR spectra of a spin-labeled single crystal of myoglobin show that the N-O group of the isocyanate label, localized in the heme-containing cavity, is mobile, and that the correlation time is about $2 \cdot 10^{-8}$ sec. According to the X-ray model, the label will not penetrate into the cavity unless the distance between the cavity walls increases to 0.3 to 0.4 nm. Nevertheless, it seems that the cavity may open up even more in order that a reorientation of the spin label might become possible (Likhtenshtein, 1974).

Magonov *et al.* (1978b) demonstrated the existence of non-equilibrium states of ferrimyoglobin and its complexes with OH^-, F^-, N_3^-, CN^- and imidazole, which are formed immediately following the reduction of iron. These workers studied the absorption spectra and the magnetic circular dichroism of myoglobin and its complexes in the states generated during the reduction of water-glycerol solutions of oxidized forms of protein by thermolyzed electrons at 77°K. Under such conditions iron is already present in its ferrous form, while the protein globule is frozen in the conformation typical of the oxidized form of the protein. In these circumstances the reduction of metmyoglobin and its fluoride complex results in the formation of low-spin and high-spin ferrous states. The reduction of the azide and cyanide complexes is accompanied by the formation of low-spin ferrous forms, whose spectroscopic characteristics reflect the retention of CN^- and N_3^- in the coordination sphere of the heme iron. When the temperature is increased to 178°K, the absorption spectrum of hydroxymyoglobin displays changes indicating an increased content of the high-spin ferrous form. It is interesting to note that in hemoglobin this form does not yet show signs of increasing under similar conditions (Magonov *et al.*, 1978a). The spectroscopic characteristics of the low-temperature and equilibrium-reduced forms of isolated heme and its dimidazole complex are identical. This fact also confirms the ability of the globule to stabilize the non-equilibrium states of heme at low temperatures.

Studies of reduction of ferrimyoglobin by solvated electrons at room temperature (Blumenfeld *et al.*, 1977a) and also of cytochrome **c**, revealed that conformation relaxation proceeds in three, and sometimes even in four stages. The first stage, according to our model, involves a change in the electronic state of the heme after the Fe(III) $\rightarrow$ Fe(II) transition; in the case of myoglobin this is accompanied by a transition of the low-spin to a high-spin Fe(II). In the second stage there is a change in the dynamics of the active site $(a \rightleftharpoons b) \rightarrow (a^* \rightleftharpoons b^*)$; the third stage, which is the slowest stage of all, is the result of the dynamic adaptation of the active site to its new state (heme + ligand) during the relaxation of the protein in the equilibrium conformation. Since, according to our ideas, there are two types of interactions — mechanical and perturbational — between the active site and the auxiliary cavities of the protein, the third and fourth stages may reflect the respective relaxational processes. It should be borne in mind, however, that the generation of the 3rd stage, which is intermediate between the 2nd and the 4th stages, may be merely due to a rather arbitrary subdivision of the kinetic relaxation curve into independent exponential segments. During the relaxation to the state of equilibrium, the ferro form of myoglobin becomes more stable to oxidation by potassium ferricyanide and by protein oxidants.

Noteworthy data on the dynamics of reassociation of myoglobin with CO and O_2 were obtained by Austin *et al.* (1975) by the method of pulse photolysis. These workers studied the kinetics of reassociation following a pulse lasting for $2 \cdot 10^{-6}$ seconds between 40 and $350°K$ in water, water-glycerol solutions and in polyvinylethyl matrix by following the variation of the absorption in the Soret range. In order to interpret the kinetic effects, these authors propose the following mechanism of interaction with ligands. There are four activation barriers which must be overcome prior to the reduction of the complex: the first barrier corresponds to the collision between the ligand and the hydrate hull in the active site region, and depends on the concentration of the ligand; the second barrier is determined by the amino acid residues which interfere with the entry of the ligand into the cavity of the heme; the third barrier is constituted by the steric hindrances to the motion of the ligand in the cavity of the heme; and the fourth barrier is related to the direct approach of the ligand to the iron atom. These stages can be distinguished from each other experimentally only at sufficiently low temperatures. Above $280°K$ they become so fast that they merge with one another, and the reassociation process is described by a single exponent.

The large number of hindrances to reassociation may be due to the fact that, having photo-dissociated under conditions of limited mobility, the myoglobin molecule does not manage to return to its original, ligandless state, so that the active site is present in its relatively inaccessible 'closed' form.

Subsequent studies performed in this laboratory led to the important conclusion that the recombination of myoglobin with the ligand is accompanied by major fluctuations in the protein structure (Beece *et al.*, 1980). It was found that the height of the first three out of the four activation barriers in the approach of CO towards the heme increases with increasing viscosity of the solution. The viscosity was varied by the addition of glycerol, saccharose and ethylene glycol in the $200\text{-}300°K$ temperature range. The results were in conformity with the Kramers equation, according to which the rate constant of overcoming the activation barrier is inversely proportional to the viscosity of the medium. The authors concluded by accepting the dynamic model of myoglobin, according to which the penetration of the ligand into the active center is conditional on the cavity of this center having effected the transition from the 'closed' to the 'open' state.

Data which indicate that myoglobin and hemoglobin are capable of binding small non-polar molecules outside the sixth coordination site of the heme are relevant to the problem of interaction between protein cavities. One of the segments which bind xenon and cyclopropane, recorded by X-ray diffraction technique, is situated in a nonpolar cavity on the proximal side of the heme between HisF8 and the pyrrole ring of the heme (Schoenhorn, 1969, 1972); this worker also detected a second xenon-binding center, at some distance from the heme between the segments *AB* and *GH*. Studies of the binding of xenon, pentane and other alkanes with myoglobin and hemoglobin showed that at least two molecules are cooperatively bound (Wishnia, 1969). If they are absorbed in different cavities, then this constitutes direct evidence of their interaction. This is confirmed by the existence of at least two xenon-binding centers, which may enhance the affinity of myoglobin to CO (Keyes and Lumry, 1968).

Since myoglobin is a satisfactory model of the properties of hemoglobin subunits, we may immediately proceed with the discussion of the special features of hemoglobin, which are a consequence of the quaternary structure.

6.6 Hemoglobin

The principal function of hemoglobin molecules which are inside erothrocytes, is the direct transport of oxygen from the lungs of the animal to its internal organs, and the transport of CO_2 in the opposite direction. Hemoglobin (molecular mass 64,400) consists of two pairs of myoglobin-like subunits ($\alpha_2\beta_2$). The four subunits together form what is practically a regular tetrahedron, 6.0 x 5.5 x 6.9 nm in size. The oxygen molecule reacts with the heme, using the sixth coordinate bond of Fe^{++}, just as in the case of myoglobin. According to the available data, free α-amino groups at the N-ends of hemoglobin participate in the binding of CO_2 (Rossi-Bernardi and Roughton, 1967).

Inside the tetramer there is a cavity running through the entire molecule at a height of 5 nm (Fig. 61). The cavity has the aspect of two boxes, one on top of the other, turned perpendicularly to each other. Each such 'box' is about 2.5 nm tall, 2.0 nm long and 0.8 - 1.0 nm broad. The width of the boxes corresponds to the distance between identical subunits.

High-resolution X-ray diffraction analysis made it possible to identify the location of water molecules which are the most firmly bound to the deoxy- and met-forms of hemoglobin (Fermi, 1975; Ladner *et al.,* 1977). A number of water molecules are localized at the contact sites between the subunits and form bridge bonds, thus additionally stabilizing the tetramer.

At least four molecules are noted to be present in the $\alpha_1 - \beta_2$ contact region. In the deoxy- and in the met-forms of hemoglobin these subunits are shifted slightly with respect to each other. For this reason the systems of bridge bonds in these forms of hemoglobin show differences as well. According to Perutz (1977) water molecules make an important contribution to the interaction between hemoglobin subunits.

Large numbers of water molecules on the surface of the subunits were observed near the polar groups of hemoglobin and methemoglobin subunits (Takano, 1977). Only about 90 water molecules were recorded on electron density charts; this is less than 10% of the hydrate hull of hemoglobin. The remaining part is highly mobile and is not responded to by X-ray diffraction methods.

The process of addition of oxygen to the hemoglobin tetramer is cooperative. Here, the meaning of the word 'cooperative' is that the addition of the first few oxygen molecules facilitates the addition of the remaining ones. Hemoglobin may be considered as a protein which models the allosteric properties of enzymes. Perutz (1970) summed up the results of his X-ray diffraction studies and, as a result, proposed a stereochemical mechanism of heme-heme interaction. During the conversion of hemoglobin from the oxy- to the deoxy-form it is the quaternary protein structure which is changed most of all. These changes are triggered by a 0.075-nm shift of the iron atom with respect to the plane of the porphyrin ring of one of the subunits when an O_2 molecule is bound to the sixth site of Fe(II). At the same time the Fe(II) atom passes from the high-spin to the low-spin state and becomes located in the plane of the heme. The iron atom 'pulls', as it were, the proximal histidine residue towards the heme. This results in a shift of the F spiral towards the center of the molecule, and TyrC2(140) is displaced from the cavity between the F and the H spirals. Subsequent events result in a stepwise rupture of the salt bridges which stabilize the quaternary structure of the deoxy form of hemoglobin, and in the transition of hemoglobin to the oxy-form.

At the same time the subunits are shifted along the $\alpha_1 - \beta_2$ and $\alpha_2 - \beta_1$ contacts by about 0.7 nm; this is accompanied by a small (about 0.1 nm) shift of the subunits along the contacts $\alpha_1 - \beta_1$ and $\alpha_2 - \beta_1$. It would appear that the forces which alter the quaternary structure of hemoglobin affect mainly the $\alpha_1 - \beta_2$ and $\alpha_2 - \beta_1$ contacts.

When hemoglobin passes into the ligand state, the α-chains rotate slightly with respect to each other. The distance between their hemes increases slightly — from 3.49 to 3.60 nm. On the other hand, the distance between the hemes of the β-subunits shows a marked decrease — from 3.9 to 3.3 nm (Perutz and Lehmann, 1968). Since this is accompanied by an 8% decrease in the volume of the tetramer (Muirhead and Perutz, 1963), the central cavity as a whole contracts.

The tertiary structure of hemoglobin, i.e., the conformation of α- and β-subunits, also changes when a ligand is bound. Thus, for instance, the width of the protein fold of the so-called tyrosine pocket in the α-chain decreases by 0.13 nm, while that of the β-chain decreases by 0.20 nm. This is accompanied by a change in the conformational equilibrium of the C-ends of hemoglobin chains. In the oxy- and in the met-forms of hemoglobin the C-end residues of all four chains are partly free and are shifted, in the sense that they remain in the bound state only a part of the time. In the deoxy-form of hemoglobin, on the contrary, each C-terminal residue is fixed by two salt bridges. The diagram (Fig. 62) which represents a possible sequence of the reaction between hemoglobin and oxygen reflects the multi-stage nature of the reaction.

The transition of hemoglobin from the deoxy- to the oxy-form is accompanied by the rupture of six salt bridges, with the liberation of protons (Bohr effect). The process is accompanied by a 25-50 KJ/mole decrease in the free energy of interaction between the subunits. This is roughly equal to the energy of interaction of four O_2 molecules with hemoglobin. Basing himself on his experimental results, Perutz described the allosteric

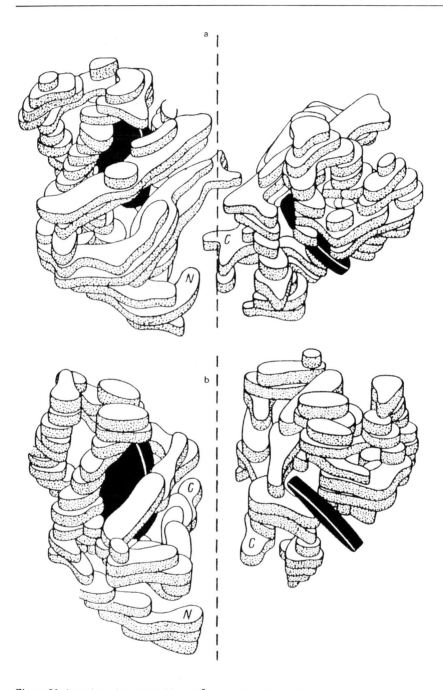

Figure 61. Location of α-chains (a) and β-chains (b) of hemoglobin in the tetramer. After Perutz *et al.* (1960). The pairs lie perpendicularly to each other. To obtain the complete model, two α-chains should be turned upside down and placed above the two β-chains.

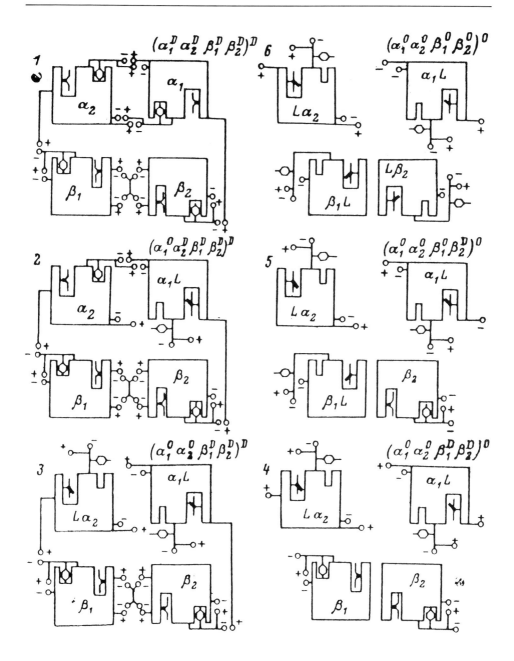

Figure 62. A possible sequence of the stages in the reaction between hemoglobin and oxygen. After Perutz (1970). State 1 corresponds to deoxy-hemoglobin with intact salt bridges and with one molecule of 2,3-diphosphoglycerate, located between the β-chains; state 6 is the fully oxygenated state of the hemoglobin. States 2 − 5 are intermediate.

properties of hemoglobin from the following point of view. Let α and β be the conformations of the subunits not bound to the substrate, and let α^S and β^S be the conformations of the subunits which are complexed with O_2; if we further denote by $[\ \ ]^T$ and $[\ \ ]^R$ the nonreactive and the reactive quaternary structures respectively, the scheme (Fig. 62) may be written as follows:

$$[\alpha_2\beta_2]^T \overset{+2S}{\rightleftharpoons} [\alpha_2\beta_2]^T \rightleftharpoons [\alpha_2^3\beta_2]^R \overset{+2S}{\rightleftharpoons} [\alpha_2\beta_2]^R \ . \qquad (6.45)$$

This scheme comprises elements which are characteristic of both the Monod model and the Koshland model. $[\ \ ]^T$ and $[\ \ ]^R$ represent the 'square' and the 'round' conformations of the allosteric protein according to Monod; the equilibrium shift between the two is caused by the conformational subunits after the addition of ligands, as postulated by Koshland.

In this context, H^+ and diphosphoglycerate, which reinforce the contact between the β-subunits of hemoglobin, may be regarded as allosteric effectors of hemoglobin, which shift the equilibrium towards the T-form and reduce its affinity to oxygen. Thus, according to Perutz' model, hemoglobin is a dynamic system, in which the tertiary structure of each subunit and the quaternary structure of the hemoglobin molecule as a whole oscillate between the oxy- and the deoxy-conformations.

Since the method of X-ray diffraction analysis merely yields average images of the system, and does not respond to the dynamics of the changes taking place in it, conclusions based on the results yielded by this method must be independently verified.

The Dynamic Action Model

Käiväräinen (1978c, 1980) proposed a mechanism for the functioning of hemoglobin, based on the dynamic model of behavior of protein in water, which is compatible with all the known facts. He starts out from the fact that hemoglobin contains four heme 'pockets' and a central cavity, which become changes when they bind oxygen. In accordance with the dynamic model each cavity fluctuates at the rate of $10^7 - 10^4$ sec^{-1} between the 'open' and 'closed' conformations, while displacing and sorbing a certain amount of water. The formation of the first complex stabilizes the 'closed' (a) state of one of the active sites, and destabilizes the 'open' state (b). The result is a deformation of the tertiary subunit structure, and a perturbance of the cooperative system of water molecules in the 'open' (B) state of the central cavity, which reduces the stability of this state. The leftward shift of the $A \rightleftharpoons B$ equilibrium according to the principle of back interaction induces a similar shift in the equilibrium between states a and b of the vacant active sites, and reduces the activation energy of the formation of the successive complex. Each successive reaction stage intensifies this effect. The $T \to R$ transition is accompanied by an increase in the association rate constant and a decrease in the dissociation rate constant, in accordance with the derived equations.

Since the sequence of oxygen binding by the α- and β-chains of hemoglobin has not yet been successively established, we shall assume that the reaction may proceed in any arbitrary sequence. We shall also assume that the effect of the heme cavities on one another is much weaker than the interaction between each such cavity and the center cavity, characterized by the coefficients κ_1, κ_2, κ_3 and κ_4. Then, in accordance with the approach adopted at the beginning of this chapter, the rate constants of the association of four oxygen molecules with four subunits — k_1, k_2, k_3 and k_4 — may be represented as

$$k_1 = P_{b_1}^{\text{coll}} \frac{kT}{h} \exp\left\{ -\frac{(G_1^{b \to b^*} + G_1^{b^* \to a^*}) - \varkappa_1[G_0^{A \to B} + \Delta G_1^{A \to B}(1 - e^{-t/\tau_1})]}{RT} \right\},$$
(6.46)

$$k_2 = P_{b_2}^{\text{coll}} \frac{kT}{h} \exp\left\{ -\frac{\begin{array}{l}(G_2^{b \to b^*} + G_2^{b^* \to a^*}) - \varkappa_2[G_0^{A \to B} + \Delta G_1^{A \to B} + \\ + \Delta G_2^{A \to B}(1 - e^{-t/\tau_2})]\end{array}}{RT} \right\},$$
(6.47)

$$k_3 = P_{b_3}^{\text{coll}} \frac{kT}{h} \exp\left\{ -\frac{\begin{array}{l}(G_3^{b \to b^*} + G_3^{b^* \to a^*}) - \varkappa_3[G_0^{A \to B} + \Delta G_1^{A \to B} + \\ + \Delta G_2^{A \to B} + \Delta G_3^{A \to B}(1 - e^{-t/\tau_3})]\end{array}}{RT} \right\},$$
(6.48)

$$k_4 = P_{b_4}^{\text{coll}} \frac{kT}{h} \exp\left\{ -\frac{\begin{array}{l}(G_4^{b \to b^*} + G_4^{b^* \to a^*}) - \varkappa_4[G_0^{A \to B} + \Delta G_1^{A \to B} + \Delta G_2^{A \to B} + \\ + \Delta G_3^{A \to B} + \Delta G_4^{A \to B}(1 - e^{-t/\tau_4})]\end{array}}{RT} \right\},$$
(6.49)

where $P_{b1,2,3,4}^{\text{coll}}$ is the probability of collision of O_2 molecule with the respective active sites in the b-state; $G_{1,2,3,4}^{b \to b^*}$, $G_{1,2,3,4}^{b \to a^*}$ and $G_0^{A \to B}$ are the free activation energies of the corresponding transitions in hemoglobin; $\Delta G_{1,2,3,4}^{A \to B}$ are the maximum changes in the free activation energy of transition of the central cavity from the 'closed' A-state to the 'open' B-state as a result of the relaxation process induced by the binding of the ligand. It is assumed that the binding of each successive O_2 molecule enhances the disequilibrium of the process and strongly accelerates the processes induced by the binding of the preceding ligand, i.e., the relaxational processes induced by the addition of the first, second and third ligands practically never proceed to completion spontaneously. It is only the disequilibrium caused by the last, fourth ligand which may become fully degenerated, during the time τ_4, which is characteristic of this stage.

The hemoglobin-oxygen dissociation rate constants k_{-1}, k_{-2}, k_{-3} and k_{-4} may be represented in a similar manner. It should be remembered in this connection that the decrease in the free energy of the a-state of the active site by ΔG_a and the increase in G_b by ΔG_b result in the following changes in the free energies of activation of the fluctuations of the central cavity, due to the reestablishment of thermodynamic equilibrium:

$$\Delta G_{1,2,3,4}^{A \to B} = \sigma_{1,2,3,4} \Delta G_b^{1,2,3,4},$$
(6.50)

$$\Delta G_{1,2,3,4}^{B \to A} = \sigma_{1,2,3,4} \Delta G_a^{1,2,3,4},$$
(6.51)

where $\sigma_{1,2,3,4}$ are the 'flexibility' coefficients of the central cavity in hemoglobin. Accordingly, the K_{ass} of each of the four stages of binding O_2 by hemoglobin may be represented as follows:

$$K_{diss}^{(1)} = \left(1/P_{b_1}^{coll}\right) K_{b \rightleftharpoons a*}^{(1)} K_{B \rightleftharpoons A}^{x_1} \lambda_1^{x_1\sigma_1}\left(1-e^{-t/\tau_1}\right), \tag{6.52}$$

$$K_{diss}^{(2)} = \left(1/P_{b_2}^{coll}\right) K_{b \rightleftharpoons a*}^{(2)} K_{B \rightleftharpoons A}^{x_2} \lambda_1^{x_2\sigma_1}\lambda_2^{x_2\sigma_2}\left(1-e^{-t/\tau_2}\right), \tag{6.53}$$

$$K_{diss}^{(3)} = \left(1/P_{b_3}^{coll}\right) K_{b \rightleftharpoons a*}^{(3)} K_{B \rightleftharpoons A}^{x_3} \lambda_1^{x_3\sigma_1}\lambda_2^{x_3\sigma_2}\lambda_3^{x_3\sigma_3}\left(1-e^{-t/\tau_3}\right), \tag{6.54}$$

$$K_{diss}^{(4)} = \left(1/P_{b_4}^{coll}\right) K_{b \rightleftharpoons a*}^{(4)} K_{B \rightleftharpoons A}^{x_4} \lambda_1^{x_4\sigma_1}\lambda_2^{x_4\sigma_2}\lambda_3^{x_4\sigma_3}\lambda_4^{x_4\sigma_4}\left(1-e^{-t/\tau_4}\right), \tag{6.55}$$

where $K_{b \rightleftharpoons a*}^{(1)} = K_{b \rightleftharpoons b*}^{(1)} K_{b* \rightleftharpoons a*}^{(1)}$.

These equations have been derived on the assumption that $\Delta G_b - \Delta G_a = (G_{b*} - G_{a*}) - (G_b - G_a)$.

$$\lambda_1 = K_{b* \rightleftharpoons a*}^{(1)}/K_{b \rightleftharpoons a}^{(1)}, \qquad \lambda_2 = K_{b* \rightleftharpoons a*}^{(2)}/K_{b \rightleftharpoons a}^{(2)}, \qquad \lambda_3 = K_{b* \rightleftharpoons a*}^{(3)}/K_{b \rightleftharpoons a}^{(3)},$$

$$\lambda_4 = K_{b* \rightleftharpoons a*}^{(4)}/K_{b \rightleftharpoons a}^{(4)}.$$

Since the effect of the ligand is to shift the equilibrium towards the a-state, all these coefficients are smaller than unity. In hemoglobin, the subunits are equivalent by pairs, so that $\lambda_1 = \lambda_2 \equiv \lambda_1$ and $\lambda_3 = \lambda_4 \equiv \lambda_4$.

It follows from equations (6.46) - (6.49) that, as $G^{A \rightarrow B}$ increases and $G^{B \rightarrow A}$ decreases at each successive stage, i.e., as the equilibrium $A \rightleftharpoons B$ of the central cavity of hemoglobin shifts to the left, the association rate constants increase, while the dissociation rate constants decrease at the same time. This will clearly result in a corresponding stepwise decrease in K_{diss}. We thus obtain the cooperative effect of binding oxygen by hemoglobin, which is demonstrated by experiment.

Since the clusterophilic interactions of the central hemoglobin cavity containing ordered water are the most sensitive to all kinds of perturbance, it may be expected that the decrease of $K_{diss} = k_{-i}/k_i$ will be due not so much to the increase in the association rate constant k_i as by the decrease in the hemoglobin-ligand dissociation rate constant k_{-i}. This follows from the fact that, as is seen from equations (6.46) - 6.49), k_i is a function of $G^{A \rightarrow B}$, while k_{-i} is a function of $G^{B \rightarrow A}$, i.e., of clusterophilic interactions.

The change in the quaternary structure of hemoglobin, denoted by Monod *et al.* 1965) as $T \rightarrow R$, corresponds to a change in the equilibrium constant of the central cavity $K_{A \rightleftharpoons B} \rightarrow K_{A* \rightleftharpoons B*}$; however, whereas Monod's model recognizes the existence of only two equilibrium forms of the tetramer: T and R – the dynamic model recognized the existence of not only the two equilibrium forms of hemoglobin – the free form and the ligand-bound form – but also three quasi-equilibrium forms, which correspond to hemoglobin with one, two and three ligands respectively.

The physical mechanism of the effect of the central cavity on heme-containing centers, which results in an increase of k_i, may consist in the fact that the stepwise shift of the equilibrium $(A \rightleftharpoons B) \rightarrow (A^* \rightleftharpoons B^*)$ deforms the geometry of the vacant cavities of the active sites so as to reduce the activation barrier which prevents the formation of complexes between the ligands and the heme iron. Other factors involved may be the transition of Fe from the high-spin to the low-spin state, in which the iron perturbs the water structure in the active site, and its affinity to oxygen greatly increases. The presence of enthalpy-entropy compensation in many reactions involving hemoglobin is an indication of changes in the properties of the hydrate hull which resemble phase transitions of the first order (Lumry and Rajender, 1970).

Due to a historical accident, the cooperative properties of hemoglobin were interpreted in terms of Hill's equation before it became known that each O_2 molecule is separately bound to its subunit:

$$Hb + nX \overset{K}{\rightleftharpoons} HbXn, \qquad (6.56)$$

where K is the equilibrium constant of bonding n moles of ligand X with one mole of hemoglobin in a single stage.

A consequence of this kinetic equation is that the degree of saturation y of hemoglobin is a function of the concentration of the ligand X:

$$y = KX^n/(1 + KX^n). \qquad (6.57)$$

The value of Hill's constant n, for which the calculated curve corresponds to that obtained experimentally, characterizes the cooperative nature of ligand bonding. If $n = 1$, cooperativeness, i.e., interaction between the bonding centers, is absent.

The curves are plotted in the coordinates $\log [y/(1-y)]$ *versus* $\log X$ or $\log p$, where p is the partial pressure of oxygen. At 50% saturation $X = 1/K$, while the slope of the curve in its middle part is n.

The interaction between the ligand and hemoglobin is correctly described by means of four equations:

$$\left.\begin{aligned} Hb + X &\overset{K_1}{\rightleftharpoons} HbX_1, \\ HbX_1 + X &\overset{K_2}{\rightleftharpoons} HbX_2, \\ HbX_2 + X &\overset{K_3}{\rightleftharpoons} HbX_3, \\ HbX_3 + X &\overset{K_4}{\rightleftharpoons} HbX_4, \end{aligned}\right\} \qquad (6.58)$$

where $K_{1,2,3,4}$ are the constants of association between hemoglobin and the first, second, third and fourth ligand respectively. Accordingly, the following equation was deduced for the saturation of hemoglobin:

$$y = \frac{K_1X + 2K_1K_2X^2 + 3K_1K_2K_3X^3 + 4K_1K_2K_3K_4X^4}{4(1 + K_1X + K_1K_2X^2 + K_1K_2K_3X^3 + K_1K_2K_3K_4X^4)}. \qquad (6.59)$$

Using the experimental data in conjunction with this formula, it is possible to find the values of K_{ass} and K_{diss} for all four reaction stages.

Dynamic Properties

Results of several studies indicate the presence of relaxational processes in hemoglobin, which affect its affinity to oxygen. Alpert *et al.* (1972), who used the method of pulse photolysis, found that the photo-dissociation of oxyhemoglobin takes place in two stages: the 'intermediate' stage is established less than 10^{-9} sec, while the actual dissociation follows within not less than 10^{-7} sec. Fesenko *et al.* (1975) conducted similar studies on a number of heme proteins and came to the conclusion that the second stage is connected with certain conformational changes in the heme zone, which proceed at a rate of 10^8 - 10^5 sec^{-1}, depending on the external conditions. According to the dynamic model, they correspond to the fluctuations of the heme-containing cavity between state a* and state b. This is probably followed by a relaxational transition to the equilibrium, non-ligand conformation of the protein, which is accompanied by a decrease in K_{ass}. The available data confirm that such a transition in fact takes place in hemoglobin at pH 9.0 (Gibson, 1959); a study of the kinetics of recombination of the Hb-4O_2 complex following its dissociation revealed that the binding rate of O_2 decreases during the relaxation of the protein to its deoxy form with a rate constant of about 200 sec^{-1}. A number of relaxational processes, taking place at somewhat different rates, were recorded by Blumenfeld *et al.* (1977a) following the reduction of methemoglobin by the method of pulse radiolysis. The possible reasons for such processes were discussed above for the case of cytochrome and myoglobin. According to the model, the first two or three stages are caused by a change in the electronic structure of the heme, in the dynamics of behavior of the active site, and by the mechanical interaction between the active center and the auxiliary cavities of the protein. The slowest stage of all may reflect the perturbational interactions between the protein cavities.

The existence of non-equilibrium forms of hemoglobin was also noted during the reduction of water-glycerol solutions of the aquo met-form of hemoglobin and its complexes with inositol hexaphosphate, F$^-$, OH$^-$, N$_3^-$, CN$^-$ and imidazole by thermolyzed electrons at 77°C (Magonov *et al.,* 1978a). In the first three cases both low-spin and high-spin ferro forms are obtained. This is compatible with the fact that certain methemoglobin derivatives (aquo and fluoro met-forms of hemoglobin) in solution are in dynamic equilibrium between the T-form and the R-form. Organic phosphates, including inositol hexaphosphate, displace the equilibrium towards the T-form. Perutz *et al.* (1974 b, c) noted that the R $\rightleftharpoons$ T shift affects the equilibrium between the low-spin and the high-spin states of the aquo and the hydroxy met-forms. Other low-spin derivatives of methemoglobin — hemoglobin azide, cyanide and imidazole — have not been identified in the T-form (Magonov *et al.,* 1978a).

Thus, the presence of some high-spin state during the low-temperature reduction of certain methemoglobin derivatives may be explained by postulating that when solutions of these derivatives are frozen, it is not only the R-form, but also the T-form which contains the iron in the high-spin state which becomes stabilized.

The capacity of the quaternary structure of hemoglobin to execute fluctuations was demonstrated by Ogawa and Shulman (1972) and by Ogawa *et al.* (1974) by high-resolution NMR method. The hemoglobin studied in this work was the deoxy form of human HbA cleaved by carboxypeptidase B at the α-chains of the C-terminal arginine. It has a higher affinity to oxygen than hemoglobin A. At pH 7 the NMR spectrum of such a preparation resembles that of the T-form of the intact deoxy-hemoglobin A. However, at pH 9 the NMR spectrum of modified hemoglobin A changes considerably and seems to correspond to that of the R-form. The variation of the spectrum with the pH made it possible to estimate the rate constant of the T $\rightarrow$ R transition of such a protein as being between $3 \cdot 10^2$ and $5 \cdot 10^2$ sec^{-1} at pH 7.6 and between $6 \cdot 10^2$ and $15 \cdot 10^2$ sec^{-1} at pH 7.7 at 25°C. The corresponding rate of the R $\rightarrow$ T transition under these conditions is about 10^4 sec^{-1}. These R $\rightleftharpoons$ T transition rates are considerably slower than the value of 10^6 sec^{-1} reported by McCray (1972) who had irradiated a solution of native hemoglobin A with laser pulses.

It may be expected that the observed fluctuations of the quaternary structure of hemoglobin would be accompanied by A $\rightleftharpoons$ B transitions of the central cavity of hemoglobin. If the frequency of A $\rightleftharpoons$ B transitions determines the rate of signal transmission between the active sites of hemoglobin subunits, it should be comparable to the association rate of hemoglobin with oxygen, which is about 10^6 sec^{-1}. Viewed in this light, McCray's results appear to be realistic.

Useful information was obtained during a study of bonding of the triphosphate-based spin label to hemoglobin A and to the mutant Chesapeake hemoglobin as a function of the saturation of these hemoglobins with CO (Ogata and McConnel, 1972). It was found that the label is capable of 1:1 complex formation only if Hb is present in the T-form. The results can be readily described in terms of the model of Monod *et al.* (1965). The allosteric equilibrium constant L for deoxy-hemoglobin A (L = [T]/[R]) was found to be 3000 for hemoglobin A, while being 0.53 for Chesapeake hemoglobin, which is distinguished by weak allosteric properties and a high affinity to oxygen.

In accordance with the Boltzmann distribution, the significance of the equilibrium constants is that the T-form of deoxy-hemoglobin A is more stable by 20.1 KJ/mole than the R-form: ΔG (deoxy) = $G_T^1 - G_R^1$ = -20.1 KJ/mole; in the case of anomalous hemoglobin, on the contrary, G_T is larger by 1.6 KJ/mole than G_R.

For oxyhemoglobin and CO-hemoglobin L = 10^{-6}, while for methemoglobin L = 10^{-2} (Perutz *et al.*, 1974a), i.e., the equilibrium is shifted towards the R-form. In the former case G_T is higher by 34.8 KJ than G_R; $\Delta G(oxy) = G_T^{(2)} - G_R^{(2)}$ = 34.8 KJ/mole, while in the latter it is higher by 11.3 KJ/mole. It would appear that the leftward shift R $\rightleftharpoons$ T caused by the ligands is due not so much to the stabilization of the R-state as to the destabilization of the T-state by weakening of the salt bridges. Moreover, according to the dynamic model, this also involves a perturbance of the clusterophilic interaction between the central cavity of hemoglobin and water. If we assume that the change in G_R which accompanies the transition of the hemoglobin from the oxy to the deoxy-state is small, then it may be deduced from the above data that G_T increases: ΔG_T (oxy $\rightarrow$ deoxy) = $G_T^{(2)} - G_T^{(1)}$ = 34.8 + 20.1 = 54.9 KJ/mole. This value is in satisfactory agreement with the energy of rupture of six salt bridges during the T $\rightarrow$ R transition, viz., 25.1–50.2 KJ/mole, and with the interaction energy of the subunits of deoxy-hemoglobin of about 50.2 KJ/mole. Since, in accordance with the model, the water in the central cavity undergoes a transition similar to a phase transition of the first kind, the change in its state does not much affect the free energy of T $\rightleftharpoons$ R transitions.

The destabilization of the 'open' state of the central cavity of hemoglobin during its oxygenation may be expected to shorten its lifetime and to enhance the frequency of A ⇌ B transitions. It is possible that the increase in the rate of hydrogen exchange, noted by Englander and Rolfe (1973), is an experimental confirmation of this deduction.

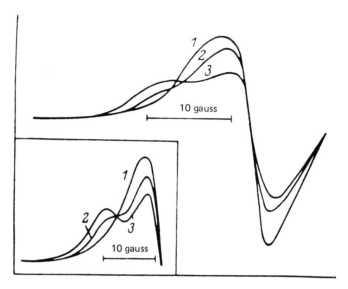

Figure 63. EPR spectra of human hemoglobin labeled with N-(1-oxyl-2,2,6,6-tetra-methyl-4-piperidyl)-iodoacetamide. *Insert:* EPR spectra of spin-labeled Capetown hemoglobin without cooperative properties. After Baldassare *et al.* (1970).
1 — deoxy-hemoglobin; 2 — hemoglobin partly saturated with CO; 3 — CO-hemoglobin.

The spin-label method (2,2,6,6-tetramethyl-4-iodoacetamidepiperidine-1-xyl) (I) revealed that hemoglobin may exist not only in its fully ligand-bound form (R) and in its fully free form (T), but also in intermediate conformational states. This is indicated by the absence of isobestic point of the EPR spectrum of hemoglobin spin-labeled at the β93 SH-group during oxygenation (Fig. 63) (Ogawa *et al.*, 1968). However, this absence may also be due to the decrease in the effective Stokes radius of hemoglobin during the T → R transition, with a corresponding increase in the resulting mobility of the N-O group of the label (Käivaräinen, 1975a). According to the differential Fourier analysis carried out on hemoglobin by Moffat (1971), the A- and B-components of the EPR spectrum correspond to two different orientations of the label with respect to the protein. In the A-state the label becomes immobilized in the cavity of the β-chain, in the so-called tyrosine pocket, the C-end of the β-chain being exposed to the solvent. According to the model of Perutz, this corresponds to the oxy-conformation of the β-subunits. In the B-state, the N-O group of the label rotates freely in the solvent, having been displaced from the TyrHC2 (145)β 'pocket'. Such a condition corresponds to the deoxy-conformation of the β-chains of hemoglobin. Thus, the ratio between the A- and B-components of the spin label reflects the conformational equilibrium of the C-ends of the β-chains of hemoglobin.

Such behavior of the spin label cannot be the result of impairment of the native properties of hemoglobin during its modification. A comparison between the structures of horse CO-hemoglobin and methemoglobin at 0.28 nm resolution (Heidner *et al.,* 1976) led to the conclusion that the SH-group of F9(93)β-cystein, to which the label is bound in methemoglobin, is also in equilibrium between two discrete states, in one of which it is exposed to the solvent, while in the other it is enclosed in a tyrosine 'pocket'. CO-hemoglobin represents the stabilized latter state; as a result, the TyrHC2(145)β side chain is displaced from this 'pocket' in its entirety. When the hemoglobin is in its deoxy state, on the contrary, Tyrβ-145 resides in its 'pocket' most of the time. It may be said that the SH-group of β-93 and Tyrβ-145 in methemoglobin fluctuate in counter-phase, as it were, 'opening' and 'closing' the tyrosine cavity in turn.

The equilibrium β-Tyr (external) $\rightleftharpoons$ βSH(external) is shifted to the left by low-spin ligands and to the right by high-spin ligands. No significant changes were noted in the tyrosine 'pockets' of α-subunits.

The existence of an interaction between the conformational states of α- and β-subunits was confirmed by the spin-label method. This follows from the fact that in a hybrid hemoglobin molecule, in which the β-chains have been inactivated by $\alpha_2(\beta$-ferri-I$)_2$, the oxygenation of the α-chains still results in a change of the EPR spectrum of the (I) label bound to β-93 cysteine (McConnel *et al.,* 1968; Ogawa *et al.,* 1968).

Asakura and Drott (1971) succeeded in preparing mono-spin-labeled and di-spin-labeled proto-hemes, in which the label was bound in the sixth or seventh site in both positions; the EPR spectrum of the hybrid (SL- α-ferricyanide$)_2\beta_2$ was observed to change during the complex formation between the heme β-subunits and oxygen. This is direct evidence of heme-heme interaction. Even more interesting changes were observed during the deoxygenation of a solution of (SL- α-ferrifluoride$)_2\beta O_2$. The changes in the EPR spectrum corresponded not only to a changed mobility of the N-O group of the label, but also to a decrease in the average distance between the magnetic moments of the unpaired electron of the label and the Fe(III) of the heme. The heme groups of α-subunits are more sensitive to the state of the hemes of the β-chains than are the hemes of the β-chains to the state of the hemes of the α-chains.

It was conclusively demonstrated by Perutz *et al.* (1974a, b, c) that the cooperative properties of hemoglobin, just like the Bohr effect, depend on the capacity of its quaternary structure to undergo R $\rightleftharpoons$ T transitions. By virtue of mechanical interactions and of discrete degrees of freedom, changes in the quaternary structure of hemoglobin determine the particular location of the Fe(II) atom with respect to the plane of the porphyrin ring. The 0.075 nm shift of Fe(II) from the center of the molecule into the plane of the heme during the T $\rightleftharpoons$ R transition is accompanied by a change in its electronic state from the high-spin to the low-spin state. In the latter case the radius of the iron atom is 0.055 nm, while in the former it is 0.060 nm. In the low-spin state the ion is sufficiently small to be accommodated in the plane of the porphyrin ring.

In the ferro-hybrids Hb($\alpha_2 NO\beta_2$), in which the α-subunits are always in the low-spin state, the derivatives of the β-subunits βCN^-, βNO^-, βO_2 and βCO are also in the low-spin state, while $\beta H_2 O$, βF^- and β(deoxy) are in the high-spin state.

It has been shown by NMR and optical methods that the aquo-form of methemoglobin in solution is in dynamic equilibrium between R and T. The fluctuation frequency of the spin state of the methemoglobin iron is 10^7 sec^{-1} (Perutz *et al.*, 1978). The equilibrium is shifted towards the R-form at high pH-values, and towards the T-form at low pH values, and under the effect of organic phosphates such as inositol hexaphosphate (Perutz *et al.*, 1974b) which, unlike diphosphoglycerate, forms a 1:1 complex with hemoglobin not only in the deoxy T-form, but also in the oxy R-form. The R-form of aquo-methemoglobin becomes stabilized on crystallization.

The affinity of oxygen to hemoglobin in the R-form is about the same as to the isolated subunits, and is higher by a factor of 70 than in the case of hemoglobin in the T-form. In the view of Perutz, the low affinity to oxygen of the subunits of deoxy-hemoglobin may be explained by postulating that "the globin reduces the affinity of ligands to hemes in some way" rather than by the physical narrowness of the heme cavities in the T-state. This worker previously suggested that the affinity of hemoglobin to oxygen is determined by the spin state of Fe(II), which in turn depends on the orientation of this ion with respect to the plane of the porphyrin ring. However, verification tests revealed that hemoglobins with very different affinities to oxygen had similar magnetic susceptibilities, i.e., the spin state of the iron was the same.

According to the ideas propounded in the present book, the rate constant of the association of the ligand-protein complex is determined by the free energy of activation of the $b \rightarrow b^* \rightarrow a^*$ transitions under the effect of the ligand, the $b \rightarrow b^*$ transition corresponding to the perturbation of the ordered water structure in the active site and to the penetration of the ligand into the heme cavity, while the $b^* \rightarrow a^*$ transition corresponds to a change in the conformation and dynamics of behavior of the heme cavity. It is probable that the interaction between the subunits of hemoglobin in the T-form enhances the activation barriers of these transitions, i.e., stabilizes the non-ligand state of the active site.

Deatherage *et al.* (1976) carried out a crystallographic comparison between the cyano form and the ligand-less form of methemoglobin, and observed conformational changes in the active site, induced by the ligand. According to these workers, structural changes in hemoglobin which are responsible for cooperativeness may be initiated not only by a change in the bond between the proximal histidine and the porphyrin, but also by changes in non-covalent interaction of the globin with the ligand and with the porphyrin.

According to the dynamic model, the action of the allosteric effects which stabilize the T-form favors the destabilization of the 'closed' A-state, a decrease of $G^{A \rightarrow B}$ and the stabilization of the 'open' B-state (increase in $G^{B \rightarrow A}$). It follows from equations (6.46) and (6.49) that a decrease of $G^{A \rightarrow B}$ of the central cavity will result in lower association rate constants, while an increase in $G_0^{B \rightarrow A}$ will result in larger dissociation rate constants. The consequences of this model concerning the association rate constants have been experimentally confirmed (Tauma *et al.*, 1973).

The importance of the role played by the central cavity in cooperative bonding of oxygen by hemoglobin is illustrated by the fact that the interaction of two α- and β-subunits in $\alpha\beta$-dimers alone is not sufficient for it to become manifested. The structural assemblage of α- and β-subunits into a hemoglobin tetramer affects in different manners the dissociation of O_2 from these units; thus the dissociation rate of O_2 from isolated α- and β-chains is 28±6 and 16±5 sec^{-1} respectively (Olson *et al.*, 1971), while the corresponding rates in the tetramer are 13±07 and 21±1 sec^{-1}. Thus, the dissociation rate in isolated α-chains is more than double that observed when these chains form part of the hemoglobin tetramer, while in the case of β-chains it is somewhat slower.

The non-equivalence of heme environments in α- and β-subunits has been conclusively demonstrated by Olson and Gibson (1972) in the case of the reaction between hemoglobin and *n*-butyl isocyanide. At pH 7 the association rate constants and the dissociation rate constants of the β-chains with this ligand are about 10 times as high as those of the α-chains. On the whole, the bonding of *n*-butyl isocyanide by hemoglobin is a two-phase process. It has been shown that it is the β-subunits that are responsible for the fast stage, while the α-subunits are responsible for the slow stage. When the concentration of the ligand increases, the rate of the slow stage increases more than proportionally, while that of the rapid stage increases almost linearly. In the presence of allosteric effectors — diphosphoglycerate and inositol hexaphosphate — the association rates of the α- and β-chains with the ligand decrease, while the dissociation rates increase.

At pH 9.1, with the $R \rightleftharpoons T$ equilibrium shifted to the left, the slow stage becomes less evident, and Hill's coefficient decreases from 2.3 (at pH 7.0) to 1.8. The initial reaction rates at pH 9.1 are almost directly proportional to the ligand concentration, even under conditions far removed from saturation. All this indicates that the differences between the α- and β-chains decrease.

At pH 7 the clusterophilic interactions of the center and of the heme-containing cavities of the α-chains with the ordered water are more marked than with β-chains. This raises the activation energy of the $b \rightarrow b^*$ transition and suppresses the association of the ligand with the α-chain. If the ligand is regarded as an agent perturbing the structure of water, an increase in its concentration may be expected to result in a decrease of the activation barrier of the $b \rightarrow b^*$ transition, i.e., to a non-linear increase in the association rate — which is what happens in actual fact.

Mobility of the Constituent Subunits of Hemoglobin

We used the method of separate determination of the correlation times of spin-labeled proteins and the labels to which they are bound (para. 3.6) to analyze the structural flexibility of human oxyhemoglobin and methemoglobin when these are acted upon in different ways. The hemoglobin was modified in the sulfhydryl groups of β-chains in position 93 by labels I, II and III (Fig. 64).

The EPR specta of oxyhemoglobin and methemoglobin with labels I, II and III differ from one another in their component A-to-component B ratios (Fig. 65). The value of τ_{R+M} was calculated from the shift of the A-components of EPR spectra in solutions of spin-labeled hemoglobins containing saccharose in various concentrations.

In this way we were the first to show that at $t \leqslant 25°C$ the mobility of oxyhemoglobin subunits is higher than those of methemoglobin. At 30°C and above, the correlation times of these two compounds become equal (τ_M = 40 nsec) (Fig. 66), and they behave as rigid particles or else fluctuate at $\leqslant 10^7$ sec^{-1}. These effects cannot be the result of a shift in

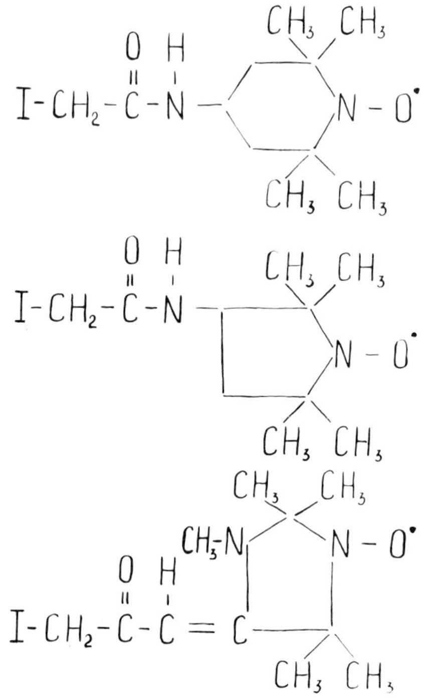

Figure 64. Spin labels used in experiments on hemoglobin.

the equilibrium tetramer ⇌ dimer, since the concentration of oxyhemoglobin dimers is much lower ($< 1\%$) than that of tetramers, and their contribution to the experimental value of τ_M is negligible.

The value of τ_M, calculated from the known size dimensions of the tetrameric hemoglobin and from its degree of hydration on the assumption that its constituent subunits are at rest relative to each other, is at least 35 nsec. τ_R decreases with increasing temperature owing to the usual processes of thermal activation of the mobility of spin labels with respect to the protein (Fig. 66).

An increase in τ_M, i.e., an intensified interaction between hemoglobin subunits at higher temperatures, indicates that hydrophobic forces make a major contribution to the stabilization of its quaternary structure. It is known that hydrophobic interactions are intensified at elevated temperatures (Brands, 1967).

The rather more independent behavior of the individual subunits of hemoglobin at lower ($5 - 25°C$) temperatures affords an explanation for the variation of the oxygenation of hemoglobin with the temperature. It is known that when the temperature is decreased, the curves representing the saturation of hemoglobin with oxygen are shifted towards smaller partial oxygen pressures (Irzhak, 1975). This is in agreement with our own results, since the isolated hemoglobin subunits have a higher affinity to oxygen than those forming part of the hemoglobin tetramer (Perutz, 1979). If the fluctuations of the

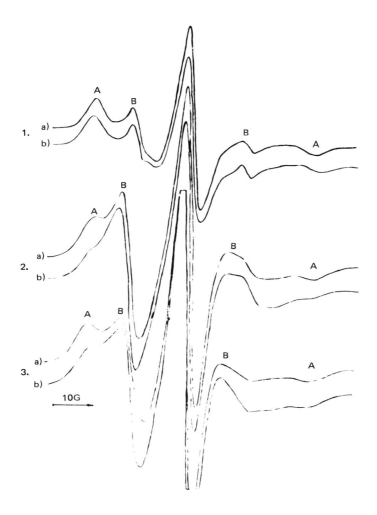

Figure 65. EPR spectra of spin-labeled human hemoglobin: a) oxyhemoglobin; b) methemoglobin. Spin labels: 1 — label III; 2 — label I; 3 — label II. Phosphate buffer, 0.01 M + 0.15 M NaCl (pH 7.3). 24°C.

tertiary and quaternary structures of methemoglobin are not rigidly conjugated, then both quaternary forms of hemoglobin (R and T) may correspond to each of the states of r or t subunits. Accordingly, both conformers (R and T) may contribute to the magnitude of the correlation time τ_M as determined from the shift of the A-components in EPR spectra.

The experimental value of τ_M for hemoglobin may be expressed in terms of the correlation times τ_M^R and τ_M^T of the R and T forms of hemoglobin respectively as follows:

$$\frac{1}{\tau_M} = \frac{f_R}{\tau_M^R} + \frac{f_T}{\tau_M^T} \tag{6.60}$$

where f_R and f_T are the relative times spent by the hemoglobin in the R- and T-forms respectively; these times are interconnected by the relationship $f_R = 1 - f_T$.

If $\tau_M^T > \tau_M^R$, it follows from equation (6.60) that the value of τ_M depends not only on the flexibilities of the R- and T-forms of hemoglobin (τ_M^R and τ_M^T), but also on the equilibrium constant:

$$K_{R \rightleftharpoons T} = \frac{f_R}{f_T} = \frac{f_R}{1 - f_R} \tag{6.61}$$

The shift of the $R \rightleftharpoons T$ equilibrium of methemoglobin to the right may be produced by increasing the hydrogen ion concentration (decreasing the pH-value) (Perutz, 1970). As could have been expected a decrease in pH from 7.3 to 6.4 at 5°C increases the value of τ_M for methemoglobin-II from 25 to 33 nsec, owing to the stabilization of the salt bridges between the hemoglobin subunits. The addition of 0.15 M NaCl (pH 7.3; 5°C) and an increase in the temperature intensifies the interaction between methemoglobin-II subunits, and τ_M increases from 15 to 25 nsec. This is in agreement with the fact that an increase in the concentration of inorganic salts (NaCl, KCl, $Na_2 SO_4$, etc.) also suppresses the oxygenation of hemoglobin (Sidwell *et al.*, 1973).

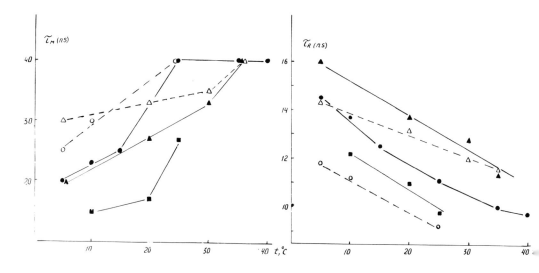

Figure 66. τ_M and τ_R as a function of temperature for labeled methemoglobin and oxyhemoglobin, bound to labels I, II and III: (■) oxyhemoglobin-I; (●) oxyhemoglobin-II; (○) methemoglobin-II; (▲) oxyhemoglobin-III; (△) methemoglobin-III. 0.01 M phosphate buffer (pH 7.3) + 0.15 M NaCl. The values of τ_M have been reduced to standard conditions (water, 25°C). After Käiväräinen and Rozhkov (in press).

The decrease in the flexibility of oxyhemoglobin-I at 17°C is the same in the presence of 4% D_2O as in the presence of 20% D_2O, and τ_M increases from 16 to 21 nsec. This is in agreement with our own results obtained on antibodies and on serum albumin (paras. 5.2 and 6.3) and may be interpreted as the displacement of ordinary water from the hydrate shell of the protein by D_2O or by HOD. The solvation of the segments responsible for the manifestation of structural flexibility of proteins is characterized by a larger association constant. In our view these are the cavities which are found between individual domains and subunits. DOD and DOH form stronger hydrogen bonds, and for this reason the fluctuation frequency of the protein, which is related to its molecular rearrangement, decreases, due to the increasing of clasterophiling interactions.

The disappearance of the isotopic effect at $> 25°C$ shows that this effect is not connected with hydrogen exchange. These results confirm the important role played by certain fractions of the hydrate shell in the structural flexibility of proteins.

According to the dynamic model, the functional activity of proteins is directly connected with their structural flexibility. The above experiments, which were carried out on hemoglobin, demonstrate the correlation between the ligand state of the protein, its $R \rightleftharpoons T$ equilibrium, and the mobility of the subunits which constitute the tetramer $\alpha_2\beta_2$. Nonspecific factors such as temperature, pH or ionic strength, also have a major effect on the interaction between hemoglobin subunits, i.e., on its functional properties.

Changes in the Hydration of Hemoglobin and Thermodynamics of T → R Transition

We studied the hydration of human oxyhemoglobin and methemoglobin by a modified method of Kuntz *et al.* (1969), which was described on p.   (Para 4.1). We compared the integral intensities of a 60 MHz proton magnetic resonance signal of frozen solutions of oxyhemoglobin and methemoglobin between –10 and –40°C, the solutions containing between 0% and 30% D_2O, at pH 7.3 in 0.01 M phosphate buffer. It was found that the hydration of oxyhemoglobin as determined at –10°C increased by 16% as a result of conversion to the met-form by potassium ferricyanide; the hydration determined at –30°C increased by 8% under these conditions. The presence of 5% D_2O in protein solutions at –10°C reduced the signal of the frozen oxyhemoglobin by 10%, and that of methemoglobin by 14%. When the D_2O concentration was increased to 30% (i.e., a six-fold increase), its effect on both samples was enhanced by a factor of only slightly more than three. These results show that the binding constants of D_2O or HDO to proteins are higher than that of H_2O, and that the increase in the concentration of D_2O resulted in a saturation of the centers of preferential D_2O sorption. Our data indicate that the nonpolar surface regions and cavities in proteins are solvated by D_2O and HDO to a greater extent than the polar and the charged groups.

It would appear that the enhanced hydration of the hemoglobin on its conversion to the met-form is due to the increased accessibility to the solvent of its central cavity and of the heme-containing nonpolar cavities, when the proportion of the centers which sorb D_2O and HDO in preference to H_2O increases. The difference between the magnitude and the nature of hydration between oxyhemoglobin and deoxyhemoglobin is even larger. As measured, at –5°C hydration of deoxyhemoglobin is 25% more than that of the oxyhemoglobin.

According to the dynamic model, the resulting heat effect of the transition from deoxy- to oxyhemoglobin is the sum of the contributions made by conformational re-arrangements and by the corresponding change in the state of the water. Nakamura *et al.* (1976) compared the heat effects of CO binding by anomalous Milwaukee hemoglobin M and by Sasketon, containing ferric iron atoms in their β-chains, with isolated α- and β-chains of the normal hemoglobin A. In both anomalous hemoglobins the α-chains alone retained the capacity for adding CO or O_2. It was shown by X-ray diffraction techniques that hemoglobin M in its deoxy-form is isomorphous with hemoglobin A. Absorption spectra indicate that the bonding of CO ligands by the α-chains of anomalous hemo-globins results in a modification of the quaternary structure of these hemoglobins, anal-ogous to the T $\rightarrow$ R transition of hemoglobin A. This involves heme-heme interaction, which is manifested as a stagewise increase in the affinity to the ligand.

The binding of the first ligand with α_1 (Hb M) is an exothermal process, whose thermodynamic parameters at 25°C are as follows:

$$\Delta G = -35.6 \pm 4.4 \text{ KJ/mole}; \Delta H = -71.2 \pm 4.4 \text{ KJ/mole}; \Delta S = -126 \pm 37 \text{ J/mole} \cdot °K$$

The binding of the second ligand with α_2, on the contrary, is accompanied by positive enthalpy and entropy changes:

$$\Delta G = -39.8 \pm 11.3 \text{ KJ/mole}; \Delta H = 24.7 \pm 7.5 \text{ KJ/mole}; \Delta S = 213 \pm 46 \text{ J/mole} \cdot °K$$

Complex formation between CO and isolated α-chains is characterized by the follow-ing parameters:

$$\Delta G = -46.5 \pm 4.2 \text{ KJ/mole}; \Delta H = -62.8 \pm 4.2 \text{ KJ/mole}; \Delta S = -46 \pm 29 \text{ J/mole} \cdot °K$$

On comparing the thermodynamic parameter values just quoted it is seen that the addition of the first ligand to α_1 resembles much more the reaction between CO and an isolated chain than the addition of the second CO molecule to the α_2 of hemoglobin M. It is probable that the first reaction stage between anomalous hemoglobin and CO does not involve any significant rearrangement of its quaternary structure, while the second stage involves a major shift in the equilibrium between the A and B stages of the internal cavity in favor of the A-state. This results in the displacement of a certain amount of ordered water which it contains into the environment and in an increase of the entropy of the solution.

A similar picture is obtained in the case of the thermodynamic parameters of T $\rightarrow$ R transitions in hemoglobin A as a result of binding CO, as a function of pH (Gand *et al.*, 1975).

T $\rightarrow$ R	pH 7.4	pH 9
ΔG, KJ/mole	19.7	16.7
ΔH, KJ/mole	38±10	−50±10
ΔS, J/mole·°K	60	−222

It is known that at high pH values, the R$\rightleftharpoons$T equilibrium is shifted to the left. This means that the number of the water molecules which pass from the ordered to the ordin-ary state in the central cavity as a result of T $\rightarrow$ R transition is smaller at pH 9 than at pH

7.4 Thus, the decrease in the enthalpy and entropy during complex formation between hemoglobin A and CO, and the corresponding conformational rearrangements in the subunits are compensated for to a lesser extent by 'fusion' and water displacement at pH 9 than at pH 7.4. This effect of the changes in the quaternary structure on thermodynamic parameters is confirmed by the fact that the heats ΔH of the reaction with CO for the isolated α- and β-chains, and for the mutant Hb $\alpha_2^{met}\beta_2$)-Iwate, which invariably retains its R-form, are -73.3 and -75.4 KJ respectively. These values are close to the observed value of ΔH of hemoglobin A at pH 9, when the R $\rightleftharpoons$ T equilibrium is shifted leftwards as compared with the neutral pH, and the transition of the quaternary structure of hemoglobin A during the binding of the ligand is not very strongly expressed.

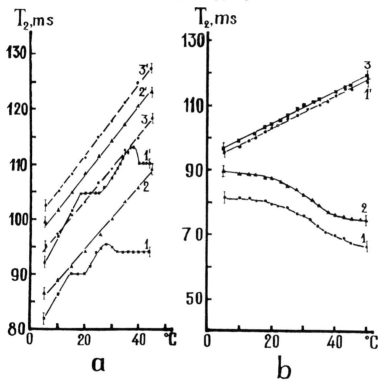

Figure 67. T_2 of water protons as a function of the temperature in solutions of a) methemoglobin (1 and 1') and oxyhemoglobin (2 and 2') and in control buffer solutions not containing D_2O (3) and containing 10% D_2O (3'). 1 — methemoglobin, 0% D_2O; 1' — methemoglobin + 10% D_2O; 2 — oxyhemoglobin, 0% D_2O; 2' — oxyhemoglobin + 10% D_2O. Each point represents the average value of a quintuplicate determination of T_2, carried out in 0.01 M phosphate buffer (pH 7.3) + 0.15 M NaCl. Hemoglobin concentration 5%. b) temperature dependencies of spin-spin relaxation times (T_2) of water protons in solutions of: 1 — methmioglobin in 0.01M phosphate buffer, pH 7.3 + 0.15 M NaCl; 1' — control solution of 0.01M phosphate buffer, pH 7.3 + 0.15 M NaCl; 2 — methmioglobin in Henks' solution (NaCl — 0.14M, KCl — 0.005M, MgSO$_4$ — 0.0003M, MgCl$_2$ — $5\cdot10^{-4}$M, CaCl$_2$ — $1.3\cdot10^{-3}$M, Na$_2$HPO$_4$ — $2.2\cdot10^{-2}$M, KH$_2$PO$_4$ — $8\cdot10^{-3}$M, pH 7.3); 3 — complex of methmioglobin with CN in Henks' solution. After Käiväräinen *et al.* (in press).

The Dynamic Model of Hemoglobin and the Bohr Effect

The T → R transition is connected not only with the cooperative properties of hemoglobin, but also with the Bohr effect. During the transition from the oxy- to the deoxy-form, protons are bound to protein in neutral and in alkaline environment, while in acid environment their concentration in solution increases. The alkaline Bohr effect may be explained by changes in the pK of the imidazole groups, caused by the formation and the rupture of salt bridges in hemoglobin (Perutz, 1970). The acidic Bohr effect may be produced by groups whose pK-values vary between 5.7 in the oxy-form and 4.9 in the deoxy-form. Perutz failed to detect such groups by his X-ray diffraction model; however, at pH 5.7 and below, changes were noted in the crystals of hemoglobin owing to the deformation of its quaternary structure.

The several hypotheses which have been proposed for the Bohr effect will not be discussed here, since none has as yet been universally accepted. We shall merely propose our own interpretation, which is based on the dynamic model. According to one of the postulates of this model, the charged groups in the 'open' state of the protein cavity may be neutralized by H_3O^+ or OH^-. When the cavity passes into the 'closed' state, the charged groups may mutually neutralize each other with formation of bridges, pass into the neutral form and become solvated by undissociated H_2O molecules.

Consider a situation in which, at neutral and alkaline pH values, most of the ionogenic groups in the central cavity of hemoglobin are negatively charged, while at acid pH-values it is the positively charged groups which are the major species. In the deoxy form of hemoglobin, when the A ⇌ B equilibrium is shifted to the right, H_3O^+ is needed to neutralize negative charges, while OH^- is required to neutralize the positive charges. In the former case the pH will decrease when hemoglobin passes from the oxy- to the deoxy-form (alkaline Bohr effect), while in the latter case it will increase (acidic Bohr effect). This mechanism clearly does not exclude the effects of other factors, such as the varying pK-values of imidazole, carboxyl and NH_2-groups in the R- and T-forms of hemoglobin (Irzhakh, 1975).

It follows from the mechanism just proposed that the smaller the leftward shift of the A ⇌ B equilibrium of the central cavity during the oxygenation, the weaker will the Bohr effect become. It may be expected that disturbing agents such as temperature, salts, etc., which reduce the life-time of the B-state, will weaken the Bohr effect as well. In fact, when the temperature is raised from 15 to 37°C, the alkaline and the basic Bohr effects in man decrease by 30% (Roughton, 1964; Antonini, 1967). A similar decrease in the alkaline Bohr effect in human hemoglobin A is also noted in the presence of 5 M NaCl (Antonini and Brunori, 1970). Isolated subunits of hemoglobin show no Bohr effect.

Changes in the Properties of Modified Hemoglobin

A study of the binding of Cu(II) ions by the β-chains of spin-labeled human oxyhemoglobin indicated that the leftward shift of the R ⇌ T equilibrium, and hence also the shift of the A ⇌ B equilibrium of central cavity may result from a perturbation of the structure of hemoglobin (Käiväräinen *et al.*, 1971, 1972b, c). The β-93 SH groups of the oxyhemoglobin were spin-labeled by 2,2,6,6-tetramethyl-4-iodoacetamidopiperidine-1-oxyl (label I) and 2,2,6,6-tetramethyl-4-p-chloromercuribenzamidopiperidine-1-oxyl (label II'). The specific bonding of two Cu(II) ions, each one by each β-chain of I-labeled and II'-labeled oxyhemoglobin, resulted in major changes in EPR spectra (Fig. 68).

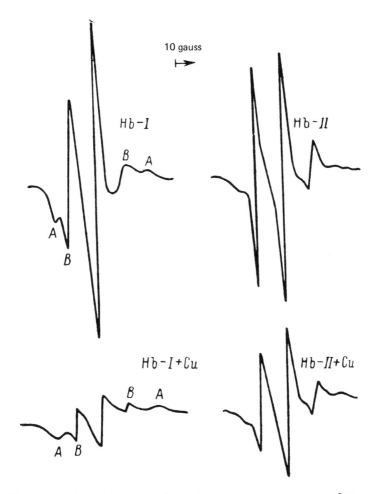

Figure 68. EPR spectra of spin labels I and II' bound to HbO$_2$ cysteins in position β-93, in the absence of Cu(II) ions (Hb-I, Hb-II') and with two Cu(II) ions bound to hemoglobin (Hb-I + Cu; Hb-II' + CU). After Käiväräinen *et al.* (1972c). Carbonate-bicarbonate buffer; pH 9; Hb concentration 5%; 22°C; [Cu] : [Hb] = 2:1.

Special investigations showed that one half of the 80% decrease (i.e., 40% of the decrease) in the amplitude of the EPR spectrum of Hb-I which occurred on the addition of copper ions was caused by the dipole-dipole interaction between the magnetic moments of Cu(II) and of the N-O group of the label; the other half was due to the conformational changes in the hemoglobin. The addition of sodium EDTA, which disrupts the copper-protein complex, fully restored the original forms of the Hb-I and Hb-II spectra. The conformational changes in hemoglobin after the addition of Cu(II), resulted in a shift of the label equilibrium between the A- and B-states towards the former, and label I had become immobilized in the A-position — i.e., the A-component of the spectrum broadened.

Since the equilibrium between the A- and B-states of the label reflects the conformational equilibrium of the C-ends of the β-chains (Moffat, 1971), it follows from the leftward shift in the A $\rightleftharpoons$ B equilibrium that the addition of Cu(II) to Hb-1 reduces the time of residence of TyrHC2(145)β in the 'pocket'. Accordingly, the addition of Cu(II) to HbO$_2$ shifts the equilibrium between the oxy- and the deoxy-conformations of hemoglobin towards the former. This may be the result of the destabilization of the 'open' state of the central cavity.

A more detailed study of conformational changes taking place in hemoglobin as a result of the addition of copper was carried out by determining the titration curves of spin-labeled oxyhemoglobin by a solution of cupric chloride. The titration curves were plotted from the variation of the amplitude of the central component of the EPR spectrum, and from the increase in the correlation time of the label in the A-state as a function of the ratio between the number of Cu(II) ions and that of hemoglobin molecules in the titrated volume ν. By comparing the normalized curves with the calculated curves of variation of the hemoglobin fraction occupied by one and two Cu(II) ions, and of the hemoglobin fraction exclusively occupied by two Cu(II) ions as a function of ν (Fig. 69), we concluded that the decrease in the amplitude of the EPR spectrum of Hb-I may be caused by the addition of both the first and the second Cu(II) ions to hemoglobin. The addition of the first copper ion to the hemoglobin molecule does not produce any conformational changes which would result in an immobilization of the label in the A-state, either in β-1 or in β-2 subunits or in an A $\rightleftharpoons$ B shift. The addition of the second Cu(II) ion, on the contrary, brings about a conformational transition in both β-subunits, so that the mobility of the labels in the A-state decreases and their A $\rightleftharpoons$ B equilibrium becomes shifted to the left.

In this case we must assume a cooperative type of interaction between the β-subunits of hemoglobin during the addition of Cu(II) ions. If we consider the Cu(II)-free hemoglobin and the hemoglobin bound to two Cu(II) ions as two separate conformations, the former one will be stable, since the addition of one ion to hemoglobin does not affect the A $\rightleftharpoons$ B equilibrium.

In the presence of 2M NaCl the cooperative nature of binding the Cu(II) ions is almost completely lost (Käiväräinen *et al.,* 1971). The effect of these NaCl concentrations on the EPR spectrum of spin-labeled Hb-I in the absence of copper was manifested as a decrease in the content of the B-component, and a corresponding leftward shift of the equilibrium A $\rightleftharpoons$ B. Under the experimental conditions employed the hemoglobin was not dissociated in the presence of such NaCl concentrations.

It appears that the interaction between the Cu(II)-binding centers proceeds by a mechanism resembling that of the reaction between hemoglobin and oxygen: a shift in the $(A \rightleftharpoons B) \rightarrow (A^* \rightleftharpoons B^*)$ equilibrium, as a result of which the lifetime of the 'open' state of the

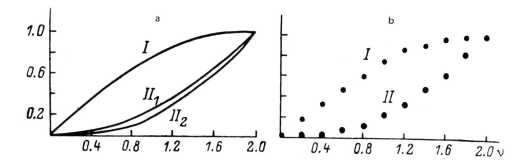

Figure 69. Calculated (a) and normalized (b) experimental titration curves of HbO_2-I by a solution of $CuCl_2$. After Käiväräinen *et al.* (1972). a: I — variation of the fraction of hemoglobin molecules occupied by one and two Cu(II) ions at k_2/k_1 = 1/4; II — variation in the fraction of hemoglobin molecules exclusively occupied by two Cu(II) ions: $II_1 - k_2/k_1$ = 1/4; $II_2 - k_2/k_1$ = 1/6, where k_1 and k_2 are the rate constants of the bonding of the first and second Cu(II) ions to the β-chains of hemoglobin; b: I — variation in the amplitude of the central component of the EPR spectrum; II — variation of the correlation time of label I (τ_{R+M}) in the A-state. Carbonate-bicarbonate buffer; pH 9; HbO_2 concentration 5%, 20°C.

central cavity becomes shorter. In the presence of 2M NaCl, which is an additional perturbing agent, the shift $(A \rightleftharpoons B) \rightarrow (A^* \rightleftharpoons B^*)$ has already taken place to some extent, so that the cooperative properties of the hemoglobin molecule are less conspicuous.

It was deemed of interest to clarify whether or not the cooperative bonding of copper ions by hemoglobin is accompanied by changes in the micro-environment of the spin label at β93 and/or by a change in the mobility of the constituent subunits in the hemoglobin tetramer. We used the relationship between $1/\tau_{R+M}$ and T/η (para 3.6) at 20°C (pH 7.3, 0.01 M phosphate buffer + 0.15 M NaCl) to calculate τ_M and τ_R for oxyhemoglobin-I in the absence of copper, and at $[Cu^{++}]:[Hb]$ molar ratios of 1:1 and 2:1. The τ_M-values found were 17, 20 and 30 nsec respectively; the values of τ_R were 11.6, 11.4 and 10.2 nsec respectively.

Thus, complex formation between the β-chains of hemoglobin and a single Cu^{++} ion has little effect on its flexibility, while the introduction of a second ion strongly enhances the interaction between the subunits of the tetramer and reduces its flexibility (τ_M increases from 17 to 30 nsec).

It is seen that the proposed mechanism of the association and dissociation of proteins and ligands offers a consistent explanation of the experimental data obtained on different objects by different methods. It will be shown below that the dynamic model of protein behavior in water may also be used in a search for a relationship between the kinetic and the conformational properties of enzymes.

Chapter 7

Dynamic Model of Protein Behavior and Mechanism of Enzymatic Function __

The functional model of enzymatic activity which will be presented is really a development of the association and dissociation mechanisms of specific complexes, such as the mechanism described in the preceding chapter. Käiväräinen (1979c) derived expressions interconnecting the rate of enzymatic reaction and the Michaelis constant K_M with the free activation energies of fluctuation of the active site and of the auxiliary cavity of the enzyme between 'closed' and 'open' states during relaxational changes. These expressions yield a simple physical interpretation of the effects of competing and non-competing inhibitors, temperature and other nonspecific agents on the kinetics of the enzymatic process. The corollaries of the model have been confirmed by a large number of experimental results.

7.1 A Possible Connection between the Kinetic and Conformational Properties of Enzymes

In the simplest possible case the kinetic equation of the reaction between an enzyme and a substrate may be written as follows:

$$E + S \underset{k_{-1}}{\overset{k_1}{\rightleftarrows}} [ES] \underset{k_{-2}}{\overset{k_2}{\rightleftarrows}} E + P. \tag{7.1}$$

Assuming that [ES] = const and that k_{-2} is negligibly small, the solution of equation (7.1) will be

$$v = k_2 SE / (K_M + S), \tag{7.2}$$

$$v_{max} = k_2 E \tag{7.3}$$

where $K_M = (k_{-1} + k_2)/k_1 = K_{diss} + k_2/k_1$.

Consider the simplest possible model of behavior of an enzyme containing only two fluctuating cavities, one of which constitutes the active site, while the other fulfils an auxiliary function. The auxiliary cavity may be much smaller-sized than that of the active site. Its function will be determined by the height of the activation barrier separating the two states of the cavity.

The substrate will penetrate into the active site when the latter is in the 'open' state (b); this requires a certain energy of activation, since the substrate must destroy the quasi-crystalline structure of the water and displace it from the active site. The displacement of the water from the active site by the substrate results in a destabilization of the b-state: $\curvearrowright$ (b → b*) and a stabilization of the a-state: $\curvearrowright$ (a → a*) through formation of hydrophobic, hydrogen and other kinds of bonds between the substrate and the groups of the active site (Fig. 70). This is accompanied by a marked increase in the a ⇌ b equilibrium constant (K_0). The rate of this process is usually fast (Almazov and Tverdokhlebov, 1975). Thus, an association of the substrate with the active center is equivalent to the transition b → b* → a*, while a dissociation represents the reverse process a* → b* → b(Käiväräinen, 1979c).

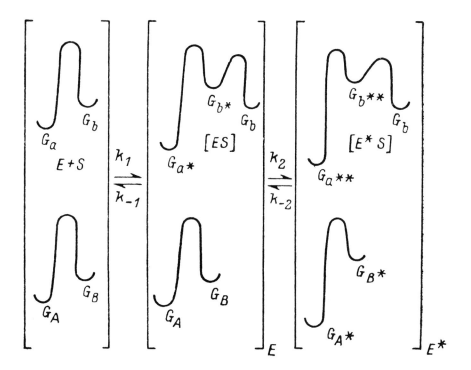

Figure 70. Free energy profiles of the active site and of the auxiliary cavity of the enzyme prior to formation of specific enzyme-substrate complex (E + S) and after: [(ES) and (E*S)] (Käiväräinen), 1979c). E and E* are the states of the enzyme molecule before and after relaxation. Explanation in text.

According to the model, the auxiliary cavity fulfils the following function in this process. When the auxiliary cavity is in the B-state, while the active site is in the a*-state, the result is an increase in G_a^* and a corresponding decrease in the free activation energy of the transition a* → b*. When the auxiliary cavity is in the A-state, the result is a decrease in the activation energy of the transition b* → a*. Since we are assuming that this inter-relationship is not merely forward but also reverse, a similar effect is also exerted by the active site on the behavior of the auxiliary cavity. A cooperation of the opposite (negative) type between the cavities is also possible.

It was shown in the previous chapter that the association and dissociation rate constants k_1 and k_{-1} of specific complexes prior to the beginning of relaxational processes may be represented as

$$k_1 \sim P_b^{\text{coll}} \left(k_{b \to b*} k_{b* \to a*}/k_{A \to B}^x \right) = P_b^{\text{coll}} k_1^0 \exp\left(\varkappa G^{A \to B}/RT \right), \quad (7.4)$$

$$k_{-1} \sim \left(k_{a* \to b*} k_{b* \to b}/k_{B \to A}^x \right) = k_{-1}^0 \exp\left(\varkappa G^{B \to A}/RT \right), \quad (7.5)$$

$$K_{\text{diss}} = k_{-1}/k_1 \sim (1/P_b^{\text{coll}}) K_1 K_2 K_3^x, \quad (7.6)$$

where k are the rate constants of the indexed transitions;

$$k_1^0 = Z \exp\left[-(G^{b \to b*} + G^{b \to a*})/RT \right];$$

$$k_{-1}^0 = Z \exp\left[-(G^{a* \to b*} + G^{b* \to b})/RT \right];$$

$G^{\to}$ are the free activation energies of the respective transitions; $\varkappa$ is the interaction coefficient between the auxiliary cavity and the active site $(0 \leqslant |\varkappa| \leqslant 1)$. In the case of simple enzymes consisting of two subglobules, such as pepsin, lysozyme, papain, α-chromotrypsin — in which changes in the state of the cavities are achieved by a rotation of the subglobules around a common axis — a 'closure' of the active site may be accompanied by an 'opening' of the auxiliary cavity (mechanical interaction between the cavities). This is what is meant by negative cooperation between the cavities $(0 > \varkappa > -1)$.

(B) is the equilibrium constant b- and b*-states of the active site; **(C)** is the equilibrium constant between the b* and a*-states of the active site; **(D)** is the equilibrium constant between the A- and B-states of the auxiliary cavity; and **(E)** are the free energies of the respective states.

The probability of collision between the substrate and the active site in the b-state is

$$P_b^{\text{coll}} \sim [E] [S] \quad K_0 H^b/H^S \tau_{S+E}, \quad (7.7)$$

where [E] and [S] are the respective concentrations of the enzyme and the substrate; H^b/H^S is the ratio of the cross-sectional areas of the active site in the b-state and of the substrate; $1/\tau_{S+E} = 1/\tau_S + 1/\tau_E$ is the resultant correlation time of the relative Brownian motion of the substrate and the enzyme, where **(F)**, and **(G)** is the equilibrium constant between the a- and the b-states of the active site.

We shall assume that the cooperation between protein cavities is of the positive type $(0 < \varkappa < 1)$. Following a rapid stage of the formation of the enzyme-substrate complex $(a \rightleftharpoons b) \to (a* \rightleftharpoons b*)$, there is a slow relaxation of the molecule to a new state of equilibrium, corresponding to the equalization of the equilibrium constants $K_3 \rightleftharpoons K_2$. Since the value of K_2 is determined by the interaction with the substrate, this process takes place mainly due to the change in K_3. The result is the establishment of a new equilibrium A ⇌ B, characterized by the constant K_4.

The relaxation is accompanied by a mutual dynamic adaptation of the active site in the a*-state and the substrate, which is realized owing to the vibration of the active site in the a*-state and the substrate, caused by the vibration of the active site brought about by pulses coming from the auxiliary cavity. Such an adaptation becomes possible owing to a marked increase in the lifetime of the a-state due to the presence of a ligand (a → a*) as compared to the duration of the A-state or of the B-state.

The observed relaxation time t_r of the enzyme is determined by the time required for the dynamic adaptation t_{ad} of the active site and the substrate, i.e., the relaxation need not necessarily proceed to completion.

If these assumptions are correct, t_{ad} should vary inversely with the frequency and the intensity of the pulses induced by the auxiliary cavity, and with the time t_a^* during which these pulses may be received. The frequency of the A ⇌ B transitions may be represented in the form

$$\nu_{A \rightleftharpoons B} \sim 1/(t_A + t_B),\tag{7.8}$$

where t_A and t_B are the lifetimes of the auxiliary cavity in the A- and B-states respectively, while the fluctuation of the free energy of the a*-state may be represented as

$$\Delta G_{a*}(t) = \varkappa G^{B \rightarrow A}(t) \equiv \varkappa G_t^{B \rightarrow A}.\tag{7.9}$$

Thus

$$t_{ad} \sim (t_A + t_B)/t_{a*}\varkappa G_t^{B \rightarrow A}.\tag{7.10}$$

Since equation (7.10) may be represented as
$t_{ad} \sim k_{a* \rightarrow b*}/\varkappa G_t^{B \rightarrow A}$
$$\nu_{A \rightleftharpoons B} = (k_{a* \rightarrow b*}/\varkappa G_t^{B \rightarrow A} \left[\exp(G_t^{A \rightarrow B}/RT) + \right. $$
$$\left. + \exp(G_t^{B \rightarrow A}/RT) \right],\tag{7.11}$$

where

$$G_t^{A \rightarrow B} = G_0^{A \rightarrow B} + P^A(t)\Delta G_p^A,\tag{7.12}$$
$$G_t^{B \rightarrow A} = G_0^{B \rightarrow A} + P^B(t)\Delta G_p^B.\tag{7.13}$$

In equations (7.12) and (7.13), $G_0^{A \rightarrow B}$ and $G_0^{B \rightarrow A}$ are the values of the free energies of activation of the respective transitions of the auxiliary cavity prior to the beginning of the relaxation; ΔG_p^A and ΔG_p^B are the maximum variations in the free energies of A- and B-states as a result of the completion of the relaxation; $P^A(t)$ and $P^B(t)$ are the probabilities of the changes of G^A and G^B, which depend on the relaxation time.

If, during the process of dynamic adaptation, the number of contacts between the substrate and the active site increases, G_a^* may be expected to decrease. On the strength of this finding, t_{ad} may be represented in another manner. If we assume that the necessary preconditions for the adaptation are a decrease in K_2 and the corresponding relaxational change $K_3 \rightarrow K_4$, and that

$$P^A(t) \approx P^B(t) = 1 - \exp(-t_{ad}/\tau), \tag{7.14}$$

where τ is the characteristic relaxation time, we obtain

$$K_4 = \exp\left\{-\frac{G_0^{A \to B} - G_0^{B \to A} + (\Delta G_p^A - \Delta G_p^B)[1 - \exp(-t_{ad}/\tau)]}{RT}\right\}, \tag{7.15}$$

Hence

$$t_{ad} = -\tau \ln\left[1 - RT \ln(K_3/K_4)/(\Delta G_p^A + \Delta G_p^B)\right], \tag{7.16}$$

where K_3 and K_4 are the equilibrium constants of the auxiliary cavity before and after adaptation; ΔG_p^A and ΔG_p^B are the changes in the free activation energies of the respective transitions as a result of the completion of protein relaxation ($t \to \infty$).

If the reaction takes place as a result of adaptation, the transition a**→b**→b corresponds to the dissociation of the enzyme-product complex. It is possible, however, that it is the fluctuation of the active center itself, a** → b**, which is responsible for the instantaneous 'rack' effect (Eyring *et al.*, 1954) and for the consequent diminishing of the activation barrier.

We consider that in the b**-state the interaction of the product with water is thermodynamically more favorable than with the protein environment in the active site; as a result, its desorption and displacement by water are favored. This may be followed by the relaxation of the protein to its initial state (E* → E).

The time required for the reaction [ES] $\xrightarrow{k_2}$ (E + P) comprises the time t_{ad} needed for the dynamic adaptation, and the time required to overcome the energic barriers a** → b** and b** → b(t_{cat}). Accordingly, k_2 may be represented as the reciprocal of the sum of these two periods of time:

$$k_2 \sim 1/(t_{ad} + t_{cat}). \tag{7.17}$$

Obviously, k_2^{cat} may be represented, as may k_{-1}, with allowance for the changes brought about by the relaxation:

$$t_{cat} \sim k_2^{cat} \sim k_{a** \to b**} k_{b* \to b}/k_{B* \to A*}^{\varkappa} = k_2^0 \exp\left(\varkappa G^{B* \to A*}/RT\right), \tag{7.18}$$

where

$$k_2^0 = (kT/h) \exp\left[-(G^{a** \to b**} + G^{b** \to b})/RT\right].$$

Substitution of (7.11) and (7.18) into (7.17) yields

$$k_2 \sim \frac{1}{k_{a* \to b*}/\varkappa G_t^{B \to A} v_{A \rightleftharpoons B} + (1/k_2^0)\exp\left(-\varkappa G^{B* \to A*}/RT\right)}. \tag{7.19}$$

We can now substitute equations (7.4), (7.6), (7.7) and (7.19) into (7.2), and express the rate of the enzymatic reaction in terms of the parameters thus introduced:

$$v \sim \frac{SE\left[k_{a* \to b*}/\varkappa G_t^{B \to A} v_{A \rightleftharpoons B} + (1/k_2^0) \exp\left(-\varkappa G^{B* \to A*}/RT\right)\right]^{-1}}{\left\{K_{diss} + \dfrac{(1/P_b^{coll} k_1^0) \exp\left(-\varkappa G^{A \to B}/RT\right)}{k_{a* \to b*}/\varkappa G_t^{B \to A} v_{A \rightleftharpoons B} + (1/k_2^0) \exp\left(-\varkappa G^{B* \to A*}/RT\right)}\right\}} + \tag{7.20}$$

where the term enclosed in braces equals K_M, and (A) is the dissociation constant of the enzyme-substrate complex.

Substitution of (7.19) in (7.3) yields the expression for the expression for the maximum reaction rate:

$$v_{max} \sim \frac{E}{k_{a* \to b*}/\varkappa G_t^{B \to A} v_{A \rightleftharpoons B} + (1/k_2^0) \exp\left(-\varkappa G^{B* \to A*}/RT\right)}. \tag{7.21}$$

In view of (7.21), and since $k_2/k_1 \ll K_{diss}$, (7.20) may be represented as

$$v \sim P_b^{coll} S v_{max}/(K_1 K_2 K_3^{\varkappa} + S). \tag{7.22}$$

If our initial assumptions are correct, the expressions thus obtained offer a simple physical interpretation for the effect of competing and non-competing inhibitors on the rate of the enzymatic reaction. The competing inhibitor reduces the value of v in equation (7.22) by shifting the equilibrium between the 'open' and the 'closed' states of the active site towards the latter, thus reducing the value of P_b^{coll}.

The non-competing inhibitor, which forms a complex with the auxiliary cavity, shifts the $A \rightleftharpoons B$ equilibrium to the left and stabilizes its fluctuations. This is accompanied by a decrease in $v_{A \rightleftharpoons B}$, and as a result t_{ad} decreases as well (equation (7.11)). Moreover, there is a decrease in $G^{B* \to A*}$, and hence also in the contribution of the auxiliary cavity in overcoming the activation barrier of the reaction. As a result (cf. equation (7.21)), v_{max} decreases.

The effect of the temperature, and of other perturbing agents on the auxiliary cavity is to destroy the clusterophilic interaction and produce a decrease in $G^{B \to A}$ and $G^{B* \to A*}$. Their effect on the former is stronger than on the latter, since the water structure in the B-state is more ordered (more cooperative), and $G^{B \to A} > G^{B* \to A*}$. It follows from equations (7.17) and (7.19) that, as $G^{B \to A}$ decreases, $v_{A \rightleftharpoons B}$ increases and t_{ad} becomes shorter, while t_{cat} increases with decreasing $G^{B* \to A*}$. As long as $(\Delta t_{ad} + \Delta t_{cat}) < 0$, the increase in k_2 and v_{max} should be observed. However, as the temperature and the concentration of the perturbing agents increase, a situation may be produced in which

$$t_{ad} \ll t_{cat}. \tag{7.23}$$

when equation (7.21) may be written as follows:

$$v_{max} \sim E k_2^0 \exp\left(\frac{\varkappa G^{B* \to A*}}{RT}\right). \tag{7.24}$$

Thus, if condition (7.23) is met, we may expect k_2 and ν_{max} to decrease with increasing perturbation of the protein and water structures, i.e., with decreasing $G^{B^* \to A^*}$.

The above is an explanation for the fact that plots of enzymatic activities as a function of the temperature and other perturbing agents are frequently bell-shaped. Numerous workers (Warren *et al.*, 1966; O'Leary and Brummond, 1974; Mockrin *et al.*, 1975; Levin and Zimmerman, 1976) studied the effect of inorganic salts on the kinetics of enzymatic reactions. The activity increased over quite broad ranges of low concentrations (between 0.01 M and 0.1 M), while at higher concentrations (> 0.3 M) the activity decreased, and this decrease was followed, in almost all cases, by a complete inhibition of the enzyme; in other words, the activity curves were bell-shaped. The relative effectiveness of the various salts followed the sequence of the lyotropic Hofmeister series. The specific electrostatic effects attained saturation at concentrations as low as < 0.01 M (Hippel and Schleich, 1973). It is seen that salts act in most cases as non-specific agents, which perturb both the structure of the water and of the proteins.

Equations (7.19) through (7.24) are valid in the case of a positive cooperative interaction between the active site and the auxiliary cavity, when $0 < \kappa < 1$. According to our ideas, such a situation is typical of the interaction between the active sites of the subunits of oligomeric proteins and the central cavity. If alternation of the cavities in counterphase is thermodynamically favored, the equations can be transformed by simply changing the sign of the interaction coefficient (χ).

An increase in NaCl concentration reduces the accessibility of the SH-groups in the active sites of glyceraldehyde-3-phosphate dehydrogenase to iodoacetamide (Mockrin *et al.*, 1975). This corresponds to a decrease of K_0 during the destabilization of the 'open' state of the active center and/or stabilization of the 'closed' state due to intensified hydrophobic interactions.

One of the corollaries of the model is that the observed effects — we may take the effect on v as an example — are independent of the nature of the disturbing agent. In fact, the enzymatic activity varies in the same manner under the effect of both salts and organic solvents (Villanceva *et al.*, 1974), phospholipids, fatty acids and detergents (Gennis and Strominger, 1976). The effect of the perturbing agents, which shift the $A \rightleftharpoons B$ equilibrium to the left, may be expected to cause a similar change not only in the activity of the enzyme, but also in its stability (para 6.1). This conclusion has been experimentally confirmed (Smith and Duerksen, 1975).

If the enzyme molecules in solution are capable of undergoing spontaneous relaxational processes of the type $K_0 \rightleftharpoons K_1$ and $K_3 \rightleftharpoons K_4$, the rates of formation of the enzyme-substrate complex v may be different at different relaxation stages, in accordance with equation (7.22).

The rate of propagation of a sound wave in water at $20°C$ is 1483 meters/sec ($1.483 \cdot 10^{12}$ nm/sec). During the lifetime of the protein cavity in the 'open' state — about 10^{-7} sec — the distance covered by the wave will be $1.5 \cdot 10^5$ nm or 0.15 mm. If the concentration is 10^{-4} M, the average distance between the macromolecules is about 20 nm. Clearly, therefore, if a sonic wave is generated during the cavity fluctuations of one molecule, it may affect a very large number of the surrounding molecules and may stimulate, for example, the transition of their cavities from the 'open' to the 'closed' state.

In the case of an oligomeric protein the activity regulation of each of its subunits may take place over the central cavity of the whole protein, which is common for all subunits. The change in the state of the central cavity, resulting from the interaction between a subunit and the substrate, which resembles that taking place in hemoglobin (para 6.6), may affect the association and dissociation rate constants of the free active sites of remaining subunits with the substrate. Allosteric effectors, which shift the equilibrium A $\rightleftharpoons$ B of the central cavity in either direction, bring about changes not only in the association constants, but also in t_{ad} and t_{cat}, with resulting acceleration or retardation of the reaction.

7.2 Comparison between the Dynamic Model of Enzymatic Activity and Other Models of Enzymatic Catalysis

Most experts in the field of enzyme physics tend to believe that the enzymatic activity involves several basic factors, which are responsible for the acceleration of biochemical reactions. These factors will now be discussed. Since any reaction between an enzyme and a substrate is fully specific, there must exist a perfectly definite complementary correspondence between the groups of the substrate and the active site. In this way a precise mutual orientation of the interacting segments may be attained. According to the theory of the activated complex, this signifies a decrease in the free energy of activation and of the reaction, and a corresponding increase in the reaction rate due to an increase in the entropy of activation. This factor is taken into account in the dynamic model and is realized during the transition of the active site cavity from the 'open' to the 'closed' state under the effect of the ligand and of the subsequent dynamic adaptation. In such a case the change in the conformation of the active center reflects the principle of 'induced fit' (Koshland, 1962). This correspondence is usually attained by structural rearrangements involving the entire globule (Citri, 1973). Koshland (1976) later pointed out the importance of conformational 'flexibility' of the enzymes in mechanisms of non-competitive inhibition and cooperativeness of oligomeric proteins.

The reaction may also be accompanied by a decrease of its activation energy, say, by way of deformation of the substrate molecule under the effect of the active site, delocalization of electron density and weakening of the bonds attacked by the enzymes. This mechanism was proposed by Eyring *et al.* (1954) and received the name of 'rack' effect. Our model allows for a possible instantaneous rack effect accompanying a fluctuation of the active site from the 'closed' to the 'open' state, which is intensified by interaction with auxiliary cavities.

According to the theory of electronic-conformational interactions developed by Volkenstain (1977), the study of the activation of a substrate molecule is essentially a study of the behavior of a substrate molecule in the field of several ligands, which represent the functional groups of the active site. During the formation of the enzyme-substrate complex the molecular orbitals of the ground state and of the excited states of the substrate merge, with the result that the electronic characteristics of the substrate become changed and an acceleration of the reaction becomes possible. It has been shown with the aid of several examples that the activation of the substrate molecule depends on the charges present on the reacting groups and on their mutual locations. If the location pattern of the active groups fails to correspond to the minimum energy of the complex (the 'rack' effect), the activation is much stronger than in a complex at equilibrium state.

In addition to the models known as 'lock-and-key', 'hand-and-glove' and the 'rack', there is also the conformer selection model. It is assumed that one form of the enzyme (E) is fully complementary to the substrate, while the other ($\tilde{E}$) is less so, and may produce the 'rack' effect. According to this model, the substrate is bound to the E-form, which then passes into the $\tilde{E}$-form as a result of thermal fluctuation; the energy of this fluctuation does not become dissipated, but is used in the excitation of the substrate and lowering the activation barrier of the reaction. The fluctuation frequencies of the protein between its two conformers are of the order of 10^3 - 10^6 sec^{-1} and correspond to the number of rotations of the enzyme (Green, 1973; quoted by Volkenstain, 1978).

This model, unlike the 'rack' model, may explain the allosteric effects as a shift in the $E \rightleftharpoons \tilde{E}$ equilibrium to either side. However, difficulties are encountered in describing multi-stage reactions in this way. The dynamic model of enzymatic activity does not suffer from this drawback, since it involves the concept of mutual adaptation between the protein and the ligand, which may take place at any individual intermediate reaction stage, thus facilitating its completion.

Karpeiski (1976) formulated the concept of the active site of the enzyme as a unique organized system, which exists by virtue of numerous contacts between its constituent zones. He identified the catalytic residues in the active site which directly participate in the chemical reaction, the contact segments which ensure the binding of the substrate to the cofactor, and the structural residues responsible for the precise mutual orientation of catalytic and contact segments, and for the connection between the states of the active site and of the globule. Analysis of the experimental data showed that the hydrogen bond network in the active site is mobile, and is readily altered by the action of a ligand or by a change in the external conditions. In the view of Karpeiski, the formation of the enzyme-substrate complex begins with the sorption of the substrate into the active site, whose specificity is low. There follows the highly specific recognition stage, resulting in the formation of the true enzyme-substrate complex, in which the reagents are in a quasi-transitional configuration. It has been shown, for RNA-ase, that this stage is accompanied by a conformational transition of the active site. In terms of the dynamic model, this situation corresponds to the transition b* → a*.

It was suggested by Khurgin *et al.* (1967) that the energy liberated in the course of the reaction is stored by the enzyme in the form of elastic deformation and is again utilized during the subsequent catalytic event. However, these workers later concluded that their model can also function without assuming energy storage in the case of single-stage enzymatic processes. They consider an enzyme as a mechanically deformed system, with discrete degrees of freedom, which are responsible for the specific change in its conformation. The energy is stored as elastic deformations of distinct fragments of the protein. Non-dissipative energy migration is possible if its form is that of coherent low-frequency phonons. This presupposes, on one hand, the absence of any bond ruptures, i.e., the basic structure of the protein must be preserved. On the other hand, the propagation of large amounts of coherent phonons is tantamount to mechanical deformation, i.e., to a conformational change in the protein.

The elastic deformation energy is given by the equation:

$$\varepsilon_0 = SL\Delta_0^2 E, \qquad (7.25)$$

where S is the cross-sectional area, L is the longitudinal dimension, Δ_0 is the relative deformation and E is the rigidity. Approximate calculations showed that, on these assumptions, a sufficient amount of the energy ($\varepsilon_0 \sim 21.9$ KJ/mole) may be stored in a globule about 5 nm in size, after which the energy can be utilized in the enzymatic event.

According to Khurgin *et al.* (1967):

"If the protein molecule includes both rigid and deformable segments, conditions are especially favorable for energy evolution in the mobile segment; during the motion, the energy 'runs off' into sites of the lowest rigidity. Owing to these properties, major deformations in the zone of the active site become possible."

This model is mostly based on the laws of solid body physics and does not take into account the specific properties of biological macromolecules — such as structural flexibility and large thermal fluctuations.

Chernavskii (1978) developed the 'enzyme-machine' model:

"The machine is a system with one discrete degree of freedom and two metastable states, which is capable of transporting energy from one state to another without losses."

'Discrete' degrees of freedom are those related to the collective motion of individual parts of the system over distances much larger than the amplitudes of atomic vibrations. In the system protein-machine the mechanical motions produce a tense state of the system, and the reaction itself takes place due to thermal fluctuations. The machine is so constructed that the mechanical energy stored during the sorption of the substrate, as well as the energy of thermal fluctuations of isolated atoms and groups, are utilized in the most rational manner possible. The substrate 'squeezes' the active site, when the energy is stored as an elastic deformation of a spring. The existence of each metastable state is ensured by 'fixers' — reactive groups which may participate in chemical reactions (Fig. 71). According to Chernavskii, the inorganic ions and water molecules present in an enzyme solution may serve as the constituent parts of the protein-machine. Moreover, if, during a catalytic event, their number changes, they may contribute to the energy storage capacity.

Esipova (1973) considers that the macro motions of individual subunits or parts of enzymes are the cause of their catalytic effect. Such ideas are rather close to ours.

Romanovskii *et al.* (1979), who proposed an electromechanical model of the enzyme-substrate complex, also agree that the active-site-forming domains are capable of mutual translations under the effect of the substrate. In our view this model is valuable since, if suitably modified, it may be employed to describe the interaction between the active site and the auxiliary cavities of the enzyme.

Bukatina *et al.* (1976) adopted a similar approach in proposing a functional scheme of

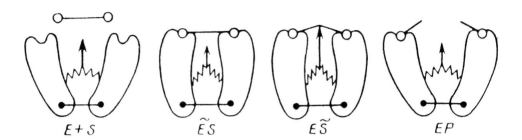

Figure 71. Diagramatic representation of the working mechanism of 'protein–machine' with the four main stages of an enzymatic reaction. After Chernavskii (1978).

an enzymatic mechano-chemical energy converter. In these models the sorption energy of the substrate is stored as the mechanical elastic deformation energy of the enzyme. Our dynamic model of enzymatic activity assumes that the substrate does not enhance the free energy of the protein but, on the contrary, may reduce it through a relaxational process, by forming a complex with the active site. However, both models rest on the assumption that the catalytic process is accompanied by a decrease in the free energy of the enzyme.

One possible explanation for the catalytic effect of an enzyme is that the profiles of chemical free energy and functional free energy of the protein moving along the reaction coordinate are anti-complementary, owing to which the profile of the overall free energy becomes smoother (Fig. 72). It is assumed in this connection that each catalytic stage corresponds to an appropriate conformational state of the protein. The limiting processes seem to be the conformational rearrangements of the protein caused by the change of state of the active site. However, Lamri and Biltonen (1973) fail to propose a definite mechanism of interaction between the active site and the globule, and merely note that 'the disturbance of some contacts and the formation of others' contribute to the thermodynamics of the enzymatic reaction.

According to Sukhorukov and Likhtenshtein (1965), the structure of the hydrate shell plays an important role in the functions of biopolymers. These workers also suggested that, even prior to the formation of the enzyme-substrate complex, the substrate may bring about conformational transformations in the enzyme by altering the hydrate shell. Since the free energy of the overall reaction is constant, it follows that a low activation energy of some stages means that the activation barrier for the other, limiting stages will be high (Likhtenshtein, 1974). According to this worker, the function of the enzyme is to ensure a uniform distribution of the free energy over the individual stages by formation of the appropriate intermediate compounds; the important factors involved in this process are the dynamic adaptation and space- and time-synchronized thermal motion along the reaction coordinate. He further asserts that proteins in their capacity as cooperative systems have large energy and entropy capacities, which may be employed to 'pump over' the energies of individual stages and to equalize the energetic contour.

The dynamic model restates these ideas in a more tangible form. Thus, for instance,

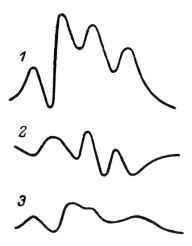

Figure 72. Smoothing the overall free energy profile of the chemical reaction by the enzyme. 1 — profile of chemical free energy; 2 — profile of conformational free energy of the enzyme; 3 — the resulting 'smoothed' profile. After Lamri and Biltonen (1973).

energy may be 'pumped over' from the protein to the active site as a result of the effect of the fluctuations of the auxiliary cavities on the active site and of the formation of 'strained' conformational states; the active center may regulate this process according to the principle of reverse interrelation, using the individual degrees of freedom. Likhtenshtein (1979) subsequently noted the importance of the ability of the protein to change its conformation by relative displacements of domains (blocks), which are interconnected by flexible, hinged 'segments'. In his view, the 'swinging' motion of the domains, resulting from an increase in the temperature, does not make a major contribution to the temperature dependence of the heat capacity of the protein.

Blumenfeld (1977) formulated his conception of the enzymatic process as follows:

"The nature of the conformational change of the enzyme-substrate complex, which follows the addition of the substrate to the active site of the enzyme is that of relaxation, and involves not only the rupture of old and formation of new bonds in the protein macromolecule, but also chemical changes taking place in the substrate by virtue of which the substrate is converted into the reaction product being determined by the rate of this conformational change".

The difference between this approach and the concept of recovery of energy consists in the fact that in the former case the non-equilibrium configuration of the protein is produced directly by the effect of the substrate, and does not involve the energy which has been stored during the preceding act. This approach is more natural and more realistic, since it is unlikely for the fluctuating enzyme molecule to remain in a thermodynamically disfavored state for a long period of time.

According to Blumenfeld, the electron structure of the active site is most strongly affected during the period of vibrational relaxation (10^{-12} to 10^{-13} sec). There follows the transition of the macromolecule to an equilibrium conformation, which proceeds in two stages: at first, changes occur in the protein (10^{-7} to 10^{-8} sec) which do not involve an impairment of secondary bands, but represent a coordinated motion of several atomic groups in the vicinity of the reactive site. This is followed by a slow transition (of the order of a few milli-seconds or even longer) of the entire protein globule to a new state of equilibrium, which is kinetically accessible. Such an approach agrees with our own concepts, if 'fast' processes are understood to mean the penetration of the ligand into the active site, a change in its electronic state, its 'collapse', and the stabilization of the 'closed' state, while a 'slow' process consists of the relaxational shift of the equilibrium between the states of auxiliary cavities, whose rate is determined by the activation energy of the molecular rearrangements involved.

It was noted by Chernavskii (1978) that continuous relaxation, which is postulated in Blumenfeld's model, means that the number of the intermediate states of the protein is very large ($n \sim 10^{10}$ - 10^{13}) and that they are separated by low activation barriers of about kT; in addition, the model does not make use of the energy of random fluctuations, so that the conditions of the catalysis are far from optimal. In our own model the relaxation is not perfectly smooth, but is assumed to be a multi-stage process ($n \ll 10^{10}$, where n is the number of the intermediate stages corresponding, to a first approximation, to the number of values of the equilibrium constants between the 'open' and 'closed' states of the auxiliary cavity of the enzyme ($K_{A \rightleftharpoons B}$)$n$ assumed to exist between the initial, non-equilibrium state of the enzyme-substrate complex and a state which is closer to equilibrium, and during which the chemical reaction takes place. Such a concept involves an optimum utilization of the energy of thermal fluctuations of the macromolecule and a sufficiently large number of intermediate states, by which the most continuous possible profile of the free energy of the reaction is ensured. It has been suggested (Ptitsyn, 1978) in analogy to our own views (Käiväräinen, 1975b, 1979c) that the mobility of the domains is a precondition of an act of enzymatic catalysis, and that the substrate penetrates into the inter-domain cavity of the active site if the cavity is open. The cavity then closes and the enzymatic process begins; during this process large parts of the protein molecule may become displaced with respect to each other without causing any changes in the secondary structure of the protein.

Careri's (1974, 1979) notions, according to which enzymes are proteins oscillating around some state of equilibrium, are very much like our own theories. According to Careri, the key to the mechanism of enzymatic activity is the capacity of the enzyme to execute definite (discrete) conformational fluctuations, which are well correlated in time. This must not be understood as meaning that the fluctuations are rigidly conjugated, but merely that they are statistically correlated, with acceleration of the rate of the enzymatic events. In terms of our own model, this is the interaction between the fluctuating auxiliary cavities and the active site cavity of the enzyme. According to Careri, the coordination of the conformational motions of the protein in space and time is realized through fluctuations of the electric field generated by, say, fluctuations of the pK of the iogenic groups. Correlated shifts are also possible inside the active site of the enzyme.

According to Shnol' (1979), large thermal fluctuations are another important factor in enzymatic activity. He pointed out that the fluctuations in the frequency range (between 1 and 10^4 sec^{-1}) required to overcome the potential barriers of the reaction are of the same order as the common rates of enzymatic reactions.

Thus, the individual elements of the dynamic model of enzymatic activity proposed in this book are typical of several other models. However, our theory has the advantage of operating with more realistic concepts, and is thus susceptible of experimental verification.

7.3 Relaxational Properties of Enzymes

The variation rates of $G^{B \to A}$, $\nu_{A \rightleftharpoons B}$ and $K_{B \rightleftharpoons A}$ are determined by the relaxational properties of the system protein + water. In the case of enzymes these properties may be derived from the variation in the rate of catalysis as a result of a sudden change in the properties of the environment. Such a study was in fact performed by Medvedev *et al.* (1975a, b) on urease and myosin. In both cases, a fast (within 3 seconds) change in the pH from 6.0 to 4.9 and from 4.9 to 6.0 resulted in marked, opposite changes in the activity of the urease, completed within 1 and 8 minutes respectively.

When the pH was reduced from 6.0 to 4.9, the activity decreased in two stages. During the first, fast stage the reaction rate decreased jumpwise, when the time taken was, practically speaking, as long as it took for the pH to become stabilized, after which it slowly decreased by another 16% of the overall activity change. The rapid stage may be explained by a change in the state of ionization of the acidic-basic groups in the protein, with a possible change in the dynamics of the active center. The second stage is the consequence of a slow stabilization of the new thermodynamic equilibrium of auxiliary cavities of the enzyme.

When the pH was suddenly changed from 4.9 to 6.0, i.e., in the opposite sense, the activity of the urease was restored to its former level, but the fast stage was absent. A possible explanation may be that at pH 4.9 the conformation of the protein is such that the rate of reestablishment of the ionized state of the protein groups is limited by the rate of relaxational conformational transitions.

The change in the activity of myosin within 4 minutes after the temperature has jumped from 6°C to 25°C may also be related to the relaxational equalization of the equilibrium constants of auxiliary cavities and the active site. The conformational changes of the myosin around 15°C are confirmed by the temperature variation of the fluorescence of tryptophan.

An analysis of the experimental results of kinetic studies of the transition between the two forms of δ-chymotrypsin (E_A and E_H), carried out by Fersht and Requena (1971) showed that both forms are present in neutral pH, but that the low-active E_H-form constitutes only 15-20% of the total. At alkaline pH values the E_H-form becomes the major species. When a competing inhibitor proflavin (Φ) was added to the enzyme solution at pH 12, the absorption spectrum at 465 nm underwent a two-stage change. The first stage was very fast, while the second one was slow, with a rate constant of 3 sec^{-1} and a large amplitude. This effect may be explained in two different ways. The first interpretation, that of Fersht and Requena (1971) is that the reaction proceeds according to the scheme:

$$E_H \underset{k_{-1}}{\overset{k_1}{\rightleftharpoons}} E_A \overset{\Phi}{\rightleftharpoons} E_A \Phi. \tag{7.26}$$

According to this kinetic equation, the association-limiting process is the rate of spontaneous isomerization $E_H \rightarrow E_A$. From the viewpoint of our dynamic model, proflavin is more likely to react with both E_A and E_H, but its environment and chromophoric properties will be different in the two cases. The first, rapid stage may then be the result of the reaction between Φ and E_H together with the small proportion of the enzyme present in the E_A-form at pH 12, while the second, slow stage corresponds to the relaxation conformational transition $E_H \Phi \rightarrow E_A \Phi$, induced by the binding of the ligand with E_H. According to the model, this transition indicates a simultaneous dynamic adaptation in the zone of the active site, which results in an alteration of the environment of the ligand and of its chromophoric properties.

During the formation of the ternary complex: glutamate dehydrogenase — α-ketoglutarate — nicotinamide-adenine-dinucleotide·H coenzyme, two stages — a fast and a slow one — were also observed by Andree (1978) who utilized the NMR method:

$$E + L \overset{k_1}{\rightleftharpoons} EL \overset{k_2}{\rightleftharpoons} EL'. \tag{7.27}$$

Here, as in the preceding case, the experimental technique employed made it possible to follow the variation in the environment and mobility of the ligand (α-ketoglutarate). The second stage (rate constant k_2) was related by Andree to the isomerization of the complex. According to our own model, this stage reflects dynamic adaptation.

Quantitative analysis showed that the substrate and the coenzyme mutually increase the association constant with the enzyme by more than one order as compared with the binary complex. The dissociation rate constant of the binary complex: enzyme — α-ketoglutarate is about 10^4 sec^{-1}, while the association rate constant is about 10^6 sec^{-1}. Even though the equilibrium association rate constant K_{ass} of such a complex is quite low, the calculated value of the overall correlation time of the complex and the enzyme indicates that the substrate is completely immobilized in the active site.

The allosteric inhibitor guanidine triphosphate and enzyme activator ADP have no significant effect on the binary complex. Nevertheless, judging by the change in the width of the α–ketoglutarate line of the NMR spectra, the effect of guanidine triphosphate on the ternary complex is to increase its K_{ass}.

In another study, Andree and Zantema (1978) succeeded in modifying, with the aid of a specific spin-label, the Lys126 residues located in the active centers of glutamate dihydrogenase. The EPR spectrum reflected the capacity of the label to exist in two states. This may be the result of the $a \rightleftharpoons b$ fluctuations of the active center.

Very interesting dynamic properties were noted during the study of the iso-enzymes of liver hexokinase P_1 and P_2 by the temperature jump method, involving the use of pH-indicators (Jentoft *et al.*, 1977). The rates of the relaxational processes were pH-dependent. At pH 8, the value of $1/\tau$ for both isoenzymes was 6 msec^{-1}, while at pH 6 the values found were $P_1 \sim 50$ msec^{-1} and $P_2 \sim 85$ msec^{-1}. In their interpretation of these data, the authors proposed the following mechanism, which includes a fast protein isomerization, related to ionization processes:

$$\text{EH} \; \underset{k_{-1}}{\overset{k_1}{\rightleftarrows}} \; \text{EH}' \; \overset{k_a}{\rightleftarrows} \; \text{E}' + \text{H}^+, \tag{7.28}$$

where k_1 and k_{-1} are the rate constants of hexakinase isomerization and K_a is the equilibrium constant of proton dissociation. It is believed that this ionization-related stage is faster than the isomerization. The experimentally determined value of the relaxation rate is

$$1/\tau = k_1 + k_{-1} [a_\text{H}/(a_\text{H} + K_a)], \tag{7.29}$$

where a_H is the hydrogen ion activity.

The transition frequency between the states EH and EH' of the isoenzyme P_1 at pH 8 is about 10^4 sec^{-1}, while the rate of rotation of the enzyme under these conditions is 40 sec^{-1}. Generally speaking, at all isomerization levels studied, P_1 and P_2 are at least 50 times higher than an isolated catalytic act of the enzyme. The relaxation rate remained essentially unchanged during the binding of the substrates, which indicates that these processes are largely independent. However, the amplitude of the relaxational process of the enzyme – as evaluated from the change in the absorption of pH indicator (Phenol Red) at 520-580 nm as a result of the temperature jump from 10°C to 15°C – decreased in some cases and increased in others. A decrease in the amplitude was noted in the presence of glucose, Mg, ADP and high concentrations of $Cr(NH_3)_2$-ATP. Mannose and glucose, in the presence of 500 μM $Cr(NH_3)_2$-ATP, enhanced the amplitude of relaxational changes.

Protein denaturation by urea resulted in the disappearance of all kinetic effects related to its pH-dependent fluctuations. This means that they are determined by the properties of the quaternary and tertiary structure of the enzyme. Under normal conditions, both isoenzymes of hexokinase (P_1 and P_2) exist in the dimeric form, with a molecular mass of 100,000, and have an identical functional mechanism, some physicochemical differences notwithstanding. Under physiological conditions they display regulatory properties, and are activated by small dianionic molecules.

According to our own ideas, the interaction with the active sites takes place as a result of the changes in the properties and in the dynamic behavior of the central cavity between the subunits. According to the model, binding of ligands is accompanied by dynamic adaptation in active sites and a change in the equilibrium constant between states A and B of the internal cavity. The corresponding relaxational process $(\text{A} \underset{}{\overset{1/\tau}{\rightleftarrows}} \text{B}) \overset{1/\tau}{\rightarrow}$ $(\text{A}^* \rightleftharpoons \text{B}^*)$ may have the same or the opposite sign to that produced by the temperature jump. In the former case the amplitude of the observed relaxational changes may be expected to decrease, while in the latter it may be expected to increase if the final state of equilibrium is mainly determined by the value of the temperature following the jump. It should be noted that in the presence of perturbing agents – 0.1 M NaCl or KCl – the relaxation amplitude decreases by 25%.

The characteristic relaxation time τ is determined by the properties of the protein component of the system protein + water, which may vary slightly under the effect of ligands. This is probably why τ is independent of the identity of the substrate in the above example. The relaxational process in hexokinase is accompanied by the secretion of protons. According to the dynamic model, the Bohr effect is characteristic not only of hemoglobin but also of other oligomeric proteins (para 6.6, p. 174).

Certain structural features of our dynamic enzyme model were manifested during the analysis of X-ray diffraction data obtained for yeast hexokinase (Steitz *et al.,* 1976; McDonald *et al.,* 1979). When the structures of ligand-free and ligand-bound hexokinases were compared at 0.21 nm resolution, it was noted that during the binding of glucose one of the active-site-forming domains rotates through a 12° angle with respect to the other one, thus diminishing the cavity of the active site. It was shown by the method of small-angle X-ray scattering that this is accompanied by a 1% decrease in the radius of gyration of the enzyme. It is believed that a 'closing' of the active center brings about changes in the interaction between the subunits of the dimeric hexokinase, and that these changes determine the allosteric properties of the enzyme.

7.4 Effect of Perturbing Agents on Enzymatic Activity

In accordance with our model, the effect of nonspecific perturbing agents on the dynamics of protein behavior may be caused by perturbances in the ordered structure of water in the cavities and the geometry of the cavities in the 'open' state.

It follows from equations (7.13) and (7.20) that perturbing agents may affect v by changing the values of $G_0^{B \to A}$ and $G^{B* \to A*}$, i.e., of $K_{B \rightleftharpoons A}$ and of $\nu_{A \rightleftharpoons B}$ and $G^{b \to a*}$, i.e., of $K_{b \rightleftharpoons a}$ and $K_{b* \rightleftharpoons a*}$. They may consequently diminish t_{ad} in equation (7.11), k_{+1} and K_{diss} of equations (7.4) and (7.6) and P_b^{coll} of equation (7.7). If t_{ad}, K_{diss} and k_{+1} decrease the value of v in equation (7.20) should increase, whereas a decrease in P_b^{coll} and $G^{B* \to A*}$ should produce the opposite effect. At high concentrations of the perturbing agents the enzyme may be deactivated by denaturation. The final result will be determined by the sum total of all these contributions.

There are data which indirectly indicate that perturbing agents may in fact influence t_{ad}, which forms part of the complete t_r. The effect of NaCl on the rate of relaxational conformational transition of glutamate decarboxylase following a jumpwise change in pH was studied by the method of stop-flow on the absorption spectrum (O'Leary and Brummond, 1974). When the pH was suddenly changed from 7 to 4, an increase in the experimental first-order rate constant with increasing concentration of NaCl and molarity of the buffer was noted, as well as a tendency to achieve saturation. The jumpwise decrease in pH probably induced a leftward shift of the A $\rightleftharpoons$ B equilibrium of the cavities, while the perturbing agents accelerated this process by decreasing $G_0^{B \to A}$.

On the other hand, the rate constant of conformational changes in this enzyme decreased with increasing salt concentration following an 'upward' jump of pH (from pH 4.5 to pH 7.0) and in the presence of 0.3 M NaCl the transition was practically inhibited altogether. The reason may be that the perturbing agents prevent the pH-induced shift of the A $\rightleftharpoons$ B equilibrium to the right from taking place, just as in the case of a 'downward' pH jump, by decreasing $G_0^{B \to A}$. This effect is believed to be due to the Cl$^-$ anion, which becomes bound to the enzyme in one of its forms and prevents its transition to the other. Such an interpretation presupposes a specific interaction of the anion with the protein, and is not inconsistent with our own views, if it is assumed that the bonding of the atom produces a deformation of protein cavities, as a result of which $G_0^{B \to A}$ decreases.

Low and Somero (1975a, b) made a systematic study of the effect of various inorganic salts on the activity of a number of enzymes including lactate dehydrogenase, malate dehydrogenase and pyruvate kinase. The effect of the ions was related to the change in the volume of the enzyme-substrate complex (V_{ES}) which accompanies its transition to the activated state (E‡S):

$$\Delta V^{\ddagger} = V^{\ddagger}_{E \cdot^{\ddagger} S} - V_{ES}. \tag{7.30}$$

At constant temperature and pressure the absolute reaction rate is:

$$k_2 = (kT/h) \exp\left[-(\Delta E^{\ddagger} + P \Delta V^{\ddagger} - T \Delta S^{\ddagger})/RT\right] = (kT/h) \exp\left(-\Delta G^{\ddagger}/RT\right). \tag{7.31}$$

The salts influenced k_2 in accordance with the Hofmeister series, and the relationship between k_2 and $\Delta V\ddagger$ was in fact exponential. Low and Somero offered the following interpretation for their results. The surface amino acids and peptide bonds may change the extent of their exposure to the solvent in the course of the catalytic process. These transitions of the protein groups are accompanied by large volume changes and energetic changes, produced mainly by the organization of their water environment. Energy and volume changes which accompany the rate-limiting stage of the reaction may contribute to the free energy of activation $\Delta G\ddagger$ of the reaction. The salts may affect the $\Delta V\ddagger$ and hence also the $\Delta G\ddagger$ of the reaction by altering the degree of organization of the water around transferable protein groups.

According to equation (7.31), a change in $\Delta V\ddagger$ by 20 cm^3/mole at 278°K and 1 atm (1.01·10^5 Pa) alters k_2 by less than 1%. However, if a similar increase $\Delta V\ddagger$ takes place under the effect of salts, k_2 decreases by 50%. It is thus seen that a change in the value of $P \Delta V\ddagger$ in equation (7.31) is not the only factor which may affect k_2. Changes in the interaction between the transferable ionogenic groups and the solvent, brought about by the salts, may also contribute to the magnitude of the energetic barrier of the reaction. Thus, for instance, the transfer of a –COO$^-$ group from a nonpolar medium into water is accompanied by an increase in the free energy of 18.85 KJ/mole. the neutralization of the group by a cation will correspondingly reduce the free energy of conformational transition of the enzyme, which must occur for the reaction to take place, and the reaction rate will increase correspondingly. Clearly, this group need not necessarily be located in the active site of the enzyme. At the same time, neutralization of charged proteinic groups by ions may result in an increase in the solution volume, due to decreased electrostriction.

The above interpretation of experimental results, though apparently reasonable, is not the only possibility. Its weak point is the fact that it presupposes a linear relationship between $\ln k_2$ and $\Delta V\ddagger$ up to high salt concentrations (KCl $\sim$ 0.8 M; KSCN $\sim$ 0.5 M; NaI $\sim$ 0.24 M), whereas it is known that the specific electrostatic effects of the interaction between ions and charged protein groups attain saturation even at salt concentrations much below 0.1 M (Hippel and Schleich, 1969). Since in the present case $\Delta V\ddagger$ simply denotes a change of an unknown physical parameter X, which is exponentially related to the reaction rate, equation (7.31) may be written in the form:

$$k_2 = k_2^0 \exp\left(-X/RT\right), \tag{7.32}$$

where k_2^0 is the rate constant in the absence of the salt and X is the physical parameter which depends on the presence of salts.

We shall now inquire into the significance of X in terms of the dynamic model. When the concentration of the disturbing agents (salts) increases, t_{ad} decreases because $v_{A \rightleftharpoons B}$ increases, whereas t_{cat} increases because $G^{B^* \to A^*}$ decreases in accordance with the mechanism described above. If the concentration of the perturbing agents is sufficiently high, $t_{ad} \ll t_{diss}$, and equation (7.19) may be written as

$$k_2 = k_2^0 \exp\left(\varkappa G^{B^* \to A^*}/RT\right). \tag{7.33}$$

Comparison with equation (7.32) yields

$$X = -\varkappa G^{B^* \to A^*}. \tag{7.34}$$

According to our model, the value of $G^{B^* \to A^*}$ is mainly determined by clusterophilic properties, i.e., by the difference between the free energies of the water which is dissociated $(G_{B^*}^{H_2O})$ when the cavity is in the B*-state and a highly organized cluster of water molecules $(G_{B^*}^{Cl})$. Accordingly, the dependence of $G^{B^* \to A^*}$ on the probability P of cluster disruption by the perturbing agent may be represented as follows:

$$G^{B^* \to A^*}(P) = G^{B^* \to A^*}\left[1 - P\left(G_{B^*}^{H_2O} - G_{B^*}^{Cl}\right)\right], \tag{7.35}$$

where P is determined by the concentration of the perturbing agent and its ability to perturb the clusterophilic interactions. The latter property probably determines the sequence of the individual ions in the Hofmeister series. If we assume that the relationship between P and the concentration of the perturbing agent is practically linear, we have an explanation for the exponential dependence of the reaction rate on the salt concentration if the condition $t_{ad} \ll t_{cat}$ is satisfied. Thus, the results of Low and Somero may be regarded as an indirect confirmation of the dynamic model of protein behavior.

Temperature is yet another factor which perturbs the properties of both the hydrate shell and the protein itself. In this connection, the stepwise parallel variation of the activity of lactate dehydrogenase and of the diffusion coefficient of thiourea in solution with increasing temperature (Dreyer, 1971) is of interest. The beginnings of the 'steps' corresponded to 15, 28, 43 and 60°C. They may be caused by the difference in the stabilities of the different fractions of the hydrate shell of the protein. Moreover, lactate dehydrogenase is a multi-globular protein with a quaternary structure, which means that it contains at least two types of cavities: the active site cleft and the central cavity, in which the thermal stabilities of the water clusters may be different. Whatever the interpretation of these results, it is clear that the properties of the water are related to the catalytic activity.

The observed decrease in the activity of carboxypeptidase which accompanied the increase in the concentrations of methanol and glycerol (Gavish and Werber, 1979) led to the conclusion that the overcoming of the activation barrier which limits the reaction rate, and which remains constant at about 56.5 KJ/mole, is accompanied by structural fluctuations of the enzyme, which vary with the viscosity of the environment. Such fluctuations may clearly include the relative thermal translations of the subglobules (domains) of the protein.

7.5 Action of D_2O on Enzymes

X-ray diffraction analysis of protonized and deuterated ribonuclease failed to reveal any differences in the conformations of the two forms of the protein (Bello and Harker, 1961). We may conclude, accordingly, that full deuteration is without effect on the protein structure, but that certain properties of biopolymers — thermal stability or enzymatic activity — may change in D_2O solutions. This is probably due to the stabilization of states A and B of the cavities as a result of enhanced electrostatic, hydrophobic and clusterophilic interactions (para 1.2), and a corresponding decrease in the frequency of transitions A $\rightleftharpoons$ B in D_2O. In accordance with equation (7.11), t_{ad} may be expected to decrease as a result. This conclusion seems to be confirmed by the fact that the rates of non-enzymatic acetylation reaction in ordinary and in heavy water are equal, whereas the corresponding enzymatic reaction proceeds much more slowly in D_2O (Currier and Mautner, 1974).

The activity of simple enzymes in D_2O decreases to not more than one-half of its original value. In the case of allosteric enzymes the isotopic effect is much stronger owing to the larger oligomer association constants and to the strong dependence of the rate of enzymatic reaction on the quaternary structure. D_2O acts much more strongly on the allosteric properties of enzymes than on their catalytic properties (Lobyshev and Kalinichenko, 1978). This is in agreement with the fact that, according to the dynamic model, the subunits of oligomeric allosteric proteins interact by way of phase transitions or quasi-phase transitions of water in the common central cavity. For this reason substitution of ordinary water by D_2O, which forms stronger hydrogen bonds, may be expected to reduce the frequency of fluctuations of this cavity and the sensitivity to allosteric effects. The viscosity of D_2O is almost 20% higher than that of H_2O, which may also contribute to the decrease in the fluctuation frequency of the protein. An enhanced activity of Na, K-ATP-ase was noted by Lobyshev *et al.* (1978) in D_2O concentrations below 1%. Under these conditions, the decrease in t_{cat} due to the intensification of the clusterophilic interactions and enhancement of $G^{B \to A}$ may be larger than the increase in t_{ad} caused by the decrease of $\nu_{A \rightleftharpoons B}$ (cf. eq. (7.21)).

7.6 Glyceraldehyde-3-Phosphate Dehydrogenase

The results obtained by Sloan and Velick (1973), who studied the dehydration of glyceraldehyde-3-phosphate dehydrogenase (GPDH) from yeast bound to nicotinamide-adenine-dinucleotide (NAD) coenzyme, represent an important partial verification of the consequences which follow from the dynamic model. This enzyme is a tetramer, which displays strong positive cooperativeness in binding NAD at pH 8.5, while not being cooperative at pH 7.4

The (apparent) density of the protein and the preferred hydration were determined by density gradient centrifugation in cesium chloride, and also in potassium phosphate solutions of different concentrations (Table 13). 'Preferred' hydration refers to water which can compete with salts for the bonding segments of the protein. The 15.6% decrease in the hydration of GPDH at pH 7.4 during the bonding of NAD is accompanied by a 7% decrease in the effective volume of the macromolecule (Durchslag *et al.*, 1969) which can be observed by the method of small-angle X-ray scattering.

TABLE 14. Apparent density and hydration of GPDH and its complexes with four coenzymes under various conditions

pH	t, °C	Apparent density, gm/cm^3		Preferred Hydration		
		Enzyme	Enzyme + 4 NAD	Enzyme, gm H_2O per gm protein	Enzyme + 4 NAD, gm H_2O per gm protein	Δ, %
7.4	5	1,228	1,234	0.454	0.424	− 6.6
7.4	25	1.213	1.227	0.480	0.405	−15.6
8.5	25	1.224	1.239	0.421	0.348	−17.3

According to the dynamic model, all these effects result from the shift of A $\rightleftharpoons$ B equilibrium in the central cavity of the tetramer to the left, which is accompanied by a corresponding displacement of water and enhancement of the compactness of the molecule. It is difficult to conceive of any other reason for this strong dehydration of the protein.

The possible reasons for the bell-shaped dependence of the enzymatic activity on the shift of the A $\rightleftharpoons$ B equilibrium imposed by external conditions were discussed in a preceding section. One manifestation of such a shift may be a change of the constants of association of the protein with ligands, including coenzymes. The contributions of coenzyme bonding to the change in the equilibrium constant $\Delta K = K_1 - K_2$ at pH 7.4 and 8.5 may be practically equal. In the present case, at 25°C, the magnitude of the dehydration seems to indicate that this is in fact the case, but the K_1 and K_2-values themselves depend on the external conditions. If it is admitted that the hydration is an indication of the K-value, it follows that at pH 7.4 the equilibrium A $\rightleftharpoons$ B is shifted to the right more strongly than at pH 8.5 (Table 14).

It may be expected that at pH 8.5 the water in the B-state of the central cavity is less stable than at pH 7.4. The lower stability and hence the orderliness of the water molecules mean a smaller change in the entropy of water molecules ΔS^{H_2O} during their displacement from the central cavity (B → A transition), and during their exchange against free water when the cavity is in the 'open' state.

We have shown, for the case of the reaction between an antibody and hapten, that the average strength of hydrogen bonds of the water in solution increases during the binding of hapten (sec. 6.3). This may be explained by the decrease of anisotropic fluctuations effectiveness in the water environment of the proteins due to the perturbance of the structure of water in the 'open' state of the cavity between three antibody subunits, and thus a decrease in ΔS^{H_2O}.

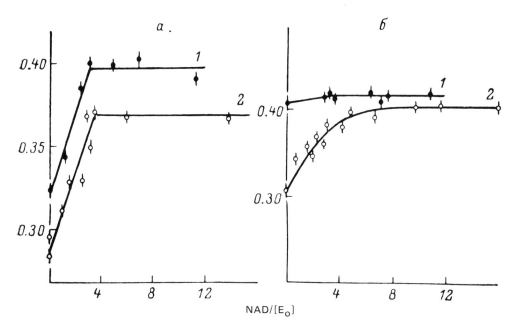

Figure 73. Increase in the relaxation rate of water protons during the titration of a solution of liver GPDH against the NAD coenzyme at 25°C (a) and 40°C (b). After Sloan and Velick (1973). The difference between the rates of spin-lattice relaxation of the protons in the solution under study and of the buffer solution is plotted on the ordinate. 1 − 0.05 M sodium pyrophosphate + 2 mM EDTA, pH 8.5; 2 − 0.05 M calcium phosphate, 0.05 M calcium chloride, 2mM EDTA, pH 7.4.

If the dynamic behavior of GPDH affects the environmental water in a similar manner, we may conclude in accordance with the above that at pH 8.5 the average strength of the hydrogen bonds of the water forming the environment should be higher than at pH 7.4, and should increase as a result of complex formation with NAD. The average strength of hydrogen bonds and correlation times of the rotational motion of water molecules, determined by the NMR method, are interrelated. For this reason an agreement between IR spectroscopic data and NMR data is to be expected, as was seen above for hemoglobin (sec. 6.3). An increase in the rate of spin-lattice relaxation $1/T_1$ (Sloan and Velick, 1973) signifies an increased correlation time of water protons, i.e., a decrease in their mobility (Fig. 73a). These workers thus found themselves in an awkward position: in the first part of their study they demonstrated that at least 120 water molecules become *desorbed* from the protein, so that their mobility must obviously increase; however, NMR results show just the opposite – a *sorption* of 25 water molecules. This result followed from calculations based on the simple model of two states of water – the free state and the bound state. All these contradictions disappear if it is admitted that the fluctuating proteins are capable of perturbing the water environment by way of exchange reactions.

At 40°C and pH 8.5 (Fig. 73b) the structure of the water in the central cavity seems to become disordered to the point where the reaction between GPDH and the coenzyme no longer affects, for all practical purposes, the protein as a generator of thermal fluctuations in the medium. This conclusion seems to be confirmed by the experimental fact that the time of spin-lattice relaxation of protons and the binding of NAD by the enzyme are not interdependent.

Of considerable interest are the calorimetric data obtained by Velick *et al.* (1971) during a study of the reaction between yeast GPDH and NAD at pH 7.3 (Table 15), when the binding of the enzyme is not cooperative, at various temperatures. The association constant between NAD and the enzyme was determined by fluorimetric titration. Two features are noteworthy: firstly, the values of ΔH, ΔS and ΔC_p are negative at 25°C when, according to the results quoted above, the protein is rather extensively dehydrated (the transition of the water from the bound to the free state should be accompanied by an increase in *H*- and *S*-values, so that positive enthalpy and entropy effects could be expected); secondly, that ΔH and ΔS assume very different values at three different temperatures, whereas ΔG° and ΔC_p remain practically unchanged. The latter feature is the so-called enthalpy-entropy compensation effect. The dynamic model explains these phenomena by introducing the concept of intramolecular water-protein compensation.

TABLE 15. Thermodynamic parameters of binding NAD to the enzyme GPDH at pH 7.3.

Fluorometric data		Calorimetric data			
t, °C	K, M^{-1}, x 10^5	ΔG°, KJ/NAD	ΔH, KJ/NAD	ΔC_p, (KJ/°K)NAD	ΔS, (J/°K)/NAD
5	6.55	−31.0	− 7.9	–	+ 83.4
25	2.35	−30.7	−52.0	−2.20	− 70.8
40	1.05	−30.1	−84.2	−2.16	−171.6

According to the model (Chapter 6), the free energy of the association is

$$\Delta G^0 = \Delta G_a + \left(\Delta G^P + \Delta G^{H_2O} \right) \simeq \Delta G_a, \qquad (7.36)$$

where ΔG_a is the change in the free energy due to the transfer of the ligand from the solution into the nonpolar active site cavity and to the changes in the active site itself; ΔG^P and ΔG^{H_2O} are the ligand-induced changes in the free energies of the protein and of the water, both on the protein surface and in its auxiliary cavities.

The condition of water-protein effect is that $\Delta G^{H_2O} = -\Delta G^P$ or, in other words:

$$\Delta H^{H_2O} - T\Delta S^{H_2O} = -\Delta H^P + T\Delta S^P, \tag{7.37}$$

or

$$\Delta\Delta H \equiv \Delta H^{H_2O} + \Delta H^P = T\left(\Delta S^{H_2O} + \Delta S^P\right) \equiv T\Delta\Delta S, \tag{7.38}$$

where, according to the dynamic model, ΔH^{H_2O}, ΔS^{H_2O}, ΔH^P and ΔS^P have opposite signs, and ΔH^{H_2O} and ΔS^{H_2O} are largely determined by the stability of the water clusters in the cavities in the B-state. Therefore, since the clusters are readily fusible, these parameters are more temperature-dependent than are ΔH^P and ΔS^P. A rise in temperature, with its perturbing effect on the structure of the water in the cavities, diminishes the positive values of ΔH^{H_2O} and ΔS^{H_2O} to a greater extent than those of ΔH^B and ΔS^B, so that the experimental values of $\Delta\Delta H$ and $\Delta\Delta S$ decrease (Table 14). Negative $\Delta\Delta H$ and $\Delta\Delta S$-values at 25°C mean that the decrease in H^P and S^P of the system due to enhanced compactness of the system after the bonding of NAD is larger than the increase in H^{H_2O} and S^{H_2O} due to the displacement of water from protein cavities.

Since at pH 7.3 the binding of NAD by yeast GPDH is non-cooperative, the thermodynamic parameters of bonding 4 NAD with 4 GPDH subunits at 25°C usually have the same values: $\Delta G = -31\pm0.5$ KJ/mole, $\Delta H = -52\pm2$ KJ/mole and $\Delta S = -70\pm7$ J(mole·°K) per active site.

In accordance with our notions, these data mean that the binding of each preceding ligand, while accompanied by a leftward shift in the A $\rightleftharpoons$ B equilibrium and by displacement of water from the active site, has no effect on the affinity of the remaining free sites to the ligand. In such a case, the coefficient of the interaction between the active site and the central cavity κ is zero.

7.7 Transaminase

The function of this enzyme (molecular mass of one subunit 45,000), which consists of two identical subunits, is to participate in the transfer of an amino group from an amino acid to a keto acid, with conversion of the latter to another amino acid. The active site of each subunit contains the coenzyme pyridoxal phosphate, which largely acts as a 'shuttle' by detaching the NH_2-group from the amino acid and transferring it to the keto acid (Braunstein and Shemyakin, 1953).

Ivanov and Karpeiskii (1969) suggested that the transition between the two main stages of the enzymatic process is realized by a reorientation of the enzyme, owing to which the reagents come near to the catalytic groups on the protein, which is required for the second stage to begin. After the termination of the reaction and the dissociation of the enzyme with the product, the coenzyme returns to its initial state. The experimental work of Arigoni showed that pyridoxal phosphate is present in different states when in the active site in the absence of the substrate and when in an enzyme-substrate complex, and the reducing agent is faced by two different facets of the coenzyme in the two cases. Data obtained by Ivanov and Karpeiskii indicate that the coenzyme is bound to the protein through all of its functional groups. We may assume, accordingly, that the reorientation of the coenzyme is caused by the appropriate fluctuation of the active site cavity of the apoenzyme during the binding of the substrate.

Judging by the data of X-ray diffraction analysis, the competing inhibitor – hydrazite of isonicotinic acid – strongly affects the structure of transaminase (Borisov *et al.*, 1977). The conformation changes occurring in this protein under the effect of substrates, quasi-substrates and inhibitors were also followed by the spin-label method (Timofeev *et al.*, 1970). The labels were bound to the two accessible SH-groups of each of the two subunits of the enzyme – Cys45 and Cys82. A selective modification of these groups showed that only the Cys45 label responds to changes in the conformation of the enzyme. The distance between this thiol group and the active site is $\geqslant 1.9$ nm. It is important to note that the bindings of ligands by the active site results in qualitatively identical conformational changes in the protein, irrespective of the ligand structure. Thus, all ligands bring about the same type of rearrangement in the active site zone and subsequent processes in the protein.

Makarov *et al.* (1974) studied the effect of the coenzyme on the stability of transaminase. Both native and modified pyridoxal phosphates enhanced the stability of the protein to urea. Irrespective of several modifications, all the coenzymes studied could be classified in two groups as regards their stabilizing effect. This indicates that the bonding site may exist in two conformations, each of which has its own corresponding free energy value of the protein; the coenzyme stabilizes one of these conformations, depending on the structure.

Dissociation of transaminase into subunits results in a loss of activity. Polyanovski *et al.* (1970) showed by the method of fluorescence polarization that its isolated subunits have a high intramolecular mobility in their native state.

X-ray diffraction analysis of transaminase revealed that the formation of an active site by one subunit of the enzyme involves the second subunit as well (Borisov *et al.*, 1977). This means that protein dynamics at the level of its quaternary structure are related to the fluctuations of its active site. It is known that at least two groups, capable of dissociating protons, are present in the active zone of this enzyme; one group belongs to the coenzyme, while the other belongs to the protein.

The temperature jump method was used in a study of relaxational conformational transition of transaminase involving the addition of two protons to the protein (Giannini *et al.*, 1975). The relaxational process was noted in the presence of L-aspartate as substrate, but disappeared with the addition of a competing inhibitor - sodium succinate and 0.5 M NaCl. The characteristic time of this process is about 10^{-3} sec at pH 7. This may reflect the events which accompany the reorientation of the coenzyme. The competing inhibitor and 0.5 M NaCl may be expected to stabilize the a-state of the active site and the corresponding position of the coenzyme. This probably prevents the generation of the conditions required for protons to be bound in the active site zone.

7.8 Myosin

Myosin is one of the contracting proteins (Fig. 74). It should be noted that the sub-fragment TMM C-1 displays all the enzymatic and actin-binding properties of the integral molecule. The most widely accepted model of muscular contraction is based on the assumption that myosin fibers, which are myosin molecules intertwined by their 'tails', are displaced along actin fibers with the aid of protruding 'heads' (TMM C-1 sub-fragments). The displacement takes place only in the presence of Mg^{++} and Ca^{++} in the course of enzymatic hydrolysis of ATP to ADP. Numerous data indicate that the presence of Ca^{++} is required to realize the contact between TMM C-1 and actin and for the manifestation of its ATP-ase activity.

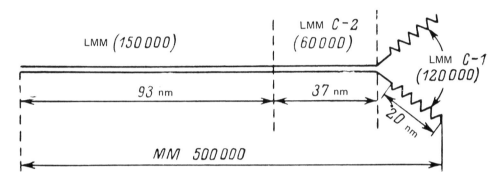

Figure 74. Diagram of myosin molecule based on electron micrographs. After Slayter and Lowey (1967). The long stretched part of the molecule, represented by *two parallel straight lines,* consists of two fibers, intertwined into a super-spiral. LMM and TMM are light and heavy meromyosins respectively. *Dotted lines* denote possible segments of myosin scission by proteolytic enzymes.

It would appear that the myosin heads 'pull up' the actin fibers by successive acts of small periodic contractions (Bendoll, 1970), thus realizing a slippage of myosin and actin fibers against each other. The fundamental question is, accordingly, the nature of the elementary act of contraction of the myosin 'head'. According to Davies (1963), the active site of the myosin contains short segments of polypeptide chain, which may assume two different configurations. In the muscle at rest, in the absence of Ca^{++} ions, there is the stretched form of the polypeptide chain, stabilized by the forces of electrostatic repulsion between Mg-ATP^- and the negative charges on the protein. With the beginning of the activation, the Ca^{++} ions arriving from the triads add onto the bound ATP and form a bridge between it and the ADP attached to the actin. In this way the negative charge on Mg-ATP^{--} is neutralized, as a result of which the polypeptide chain of the myosin becomes shortened. The actin fiber is pulled up, while the ATP shifts to the active site zone of TMM C-1, where it is hydrolyzed to ADP. The ADP is then converted into a new ATP molecule and the system returns to its initial state.

Deshcherevskii (1977) proposed a mathematical formulation of the hypothesis of 'slippage' of myosin against actin fibers.

Volkenstein (1969) developed a molecular theory of muscle contraction based on the assumption that bridge formation between actin and myosin is followed by a conformational transformation of the contracting protein. The consequences of this theory parallel those of the macroscopic treatment of Huxley (1957), but make it possible to express the parameters in the thermomechanical muscle equations in terms of molecular constants.

A first attempt to explain muscle contraction by phase transition (fusion of protein crystals) was made by Engelhardt more than 30 years ago. Subsequently, several workers took up this idea, according to which phase transitions in muscles are an important factor. It was established by Hoeve *et al.* (1963) that a considerable contraction of glycerized muscle fibers immersed in a mixture of ethylene glycol and water, which is produced by ATP, only takes place after the mixture has attained a certain concentration. This is compatible with the mechanism of the action of perturbing agents on cooperative water clusters in protein cavities, described in Chapter 5 above. If the concentration of the perturbing agent (ethylene glycol) is sufficiently high, a sudden shift from the 'open' to the 'closed' state of the cavities is to be expected, with resulting decrease in the linear dimensions of muscle proteins.

Molecular mechanisms of muscle contraction are based on relaxational conformational changes in myosin (Blumenfeld, 1977). The spin-label method revealed conformational changes in myosin (Ignat'eva *et al.,* 1972; Likhtenshtein, 1974). The EPR spectrum of TMM labeled at the S-1 thiol groups indicates that the label exists in two states, A and B, and is strongly immobilized in the former state (correlation time $> 2 \cdot 10^{-8}$ sec). The label is located at some distance from the active site.

In the presence of excess ADP the correlation time of the A-state decreases to $6 \cdot 10^{-9}$ sec, and a redistribution occurs between the A- and B-states of the label, tending towards the relatively mobile B-state. If ATP is added to spin-labeled myosin in the presence of a sufficient amount of Mg^{++} and Ca^{++}, there begins an enzymatic reaction involving the formation of ADP and inorganic phosphate. EPR spectra indicate that the binding of ATP by the active center of the myosin results in a larger increase in the mobility of the microenvironment of the spin label than does the binding of a similar amount of ADP. After the termination of the hydrolysis of ATP, the EPR spectrum becomes identical with that given by the myosin-ADP complex.

It is suggested by Blumenfeld (1977) that ATP hydrolysis proceeds in the following sequence:

1) ATP + M $\underset{k_{-1}}{\overset{k_1}{\rightleftharpoons}}$ ATPM — fast reversible binding of ATP by myosin (M), with conversion of the protein into a non-equilibrium state;

2) ATPM $\overset{k_2}{\longrightarrow}$ ADPM̃ — relaxational change in the conformation of the myosin M → M̃), accompanied by enzymatic hydrolysis;

3) ADPM̃ $\underset{k_{-3}}{\overset{k_3}{\rightleftharpoons}}$ ADP + M̃ — reversible dissociation of the complex enzyme-product;

4) M̃ $\overset{k_4}{\longrightarrow}$ M — relaxation of the protein to its initial state.

Thus when ATP is hydrolyzed off TMM, the latter executes periodic fluctuations between two different conformations. The transition from one conformation to another may correspond to the elementary act of 'pulling up' the actin fiber. It should be noted that the link between myosin and actin is not realized with the aid of ATP, but through a special site on TMM.

Let us now examine this process from the point of view of our dynamic model. After the ATP has penetrated into the active site of the TMM and has stabilized its 'closed' a*-state, a relaxational process takes place in the protein and is accompanied by changes in the geometry of the auxiliary cavities of the TMM; as a result, their lifetime in the 'open' B-state becomes shorter, and the equilibrium $A \rightleftharpoons B$ is accordingly shifted to the left. This may be expected to enhance the compactness, i.e. to shorten the linear dimensions of TMM. The process is accompanied by dynamic adaptation of ATP and the active site of TMM. Dephosphorylation can very probably begin only after the relaxation has proceeded practically to completion, since it determines the functional efficiency of TMM as a contracting protein. After contraction has taken place, ATP has performed its function, and is accordingly cleaved to yield ADP with a smaller binding constant and a phosphate group. The accompanying evolution of heat may favor the fusion of water clusters. After the dissociation of ADP from the active site, TMM becomes detached from the actin, and both its $K_{A* \rightleftharpoons B*}$ and its dimensions are restored to their former value ($K_{A \rightleftharpoons B}$). The protein molecule is now ready to perform its next cycle.

According to this model, the shortening is caused by clusterophilic and hydrophobic interaction, since it is the disturbance of the ordered water structure in an 'open' auxiliary cavity during its deformation which renders this state thermodynamically disfavored and causes the TMM cavities to contract. This is in agreement with the familiar coefficient of development of mechanical stress in collagen, contracting under the effect of the temperature (Bendoll, 1970), since numerous data are available on the presence of an ordered water structure in the central collagen zone (Chapter 5). The increased mobility of the spin label in the presence of ATP, and the redistribution of the components in the EPR spectrum, described above, may result from a decrease in the lifetime of the B-state and to the displacement of the label from the cavity.

The studies of Akopyan (1972), who examined the effect of ultrasonic waves (frequency $\sim 10^6$ sec^{-1}, intensity 0.1 - 2.5 W/cm^2) on the properties of actomyosin, may be regarded as a verification of the dynamic behavior of myosin. Degassing, which is liable to produce side effects, began under these conditions when the intensity was above 0.5 W/cm^2. It may be expected that ultrasonic waves, which perturb the structures of both water and protein, will destabilize the 'open' state of protein cavities and convert the proteins into a more compact form. That this is possible is indicated by the fact that at ultrasonic radiation intensity of less than 0.5 W/cm^2 the number of SH-groups titratable by mercuric chloride and the intrinsic viscosity of actomyosin both decrease. Thus, the available data are in agreement with the dynamic model of TMM contraction.

7.9 Lysozyme

Lysozyme is responsible for the lysis of bacterial cells, which proceeds by way of cleavage of the polysaccharide component off the cell membrane. It consists of 129 amino acid residues, and is stabilized by four intra-chain disulfide bridges (Fig. 75). It has been shown (Phillips and Levitt, 1974) that the active site contracts (transition to the a*-state) and water is displaced from it during the binding of the competing inhibitor tri-N-acetyl-glucosamine (3-NAG).

The intramolecular mobility of lysozyme was observed by high-resolution NMR at 270 MHz (Campbell *et al.,* 1975). In order to explain the nature of the resonance absorption of three tyrosine residues of the enzyme, these workers had to postulate that the tyrosine rings are capable of fluctuating between equivalent states at rates faster than 10^4 sec^{-1}. This presupposes conformational mobility of the surrounding residue in the protein.

The binding of NAG in solution results not only in the displacement of various amino acid residues, but also in reduced rates of fluctuation of overall and local protein structures (Dobson and Williams, 1975). This is confirmed by the lower rate of hydrogen exchange in lysozyme, the effect of protein stabilization by the inhibitor increasing with the temperature (Wickett *et al.,* 1974).

The two-stage mechanism of binding the inhibitor — NAG dimer — by lysozyme was noted by following the reaction by two methods — NMR and temperature jump methods — in parallel (Baldo *et al.,* 1975). One possible explanation is that the ligand L finds itself first in one environment, after which it shifts into another, in which the shielding effect of the indole ring of Try1-8 is marked:

$$E + L \underset{k_{-1}}{\overset{k_{+1}}{\rightleftharpoons}} EL \underset{k_{-2}}{\overset{k_2}{\rightleftharpoons}} EL'. \tag{7.39}$$

An alternative explanation is that the inhibitor is located in the two complexes EL and EL$'$ in the same way, but the transition EL $\rightarrow$ EL$'$ takes place as a result of a major change in the active site, which includes Try108. The question arises as to whether the transition of the active site of lysozyme to the 'closed' state under the effect of the ligand is the result of the reorientation of individual residues constituting this cavity or of the relative shifts of the large blocks (domains) of the enzyme. The work of Ivanov (1978) gives a fairly conclusive answer to this question. This worker attempted to find out whether or not intra-block rearrangements could be observed against the background of overall structural changes in lysozyme.

General transitions were recorded as changes in the mobility of spin labels bound to His15, Lys96, Lys97 and, preferably, to the N-terminal lysine group. The mobility of all labels increased when the NAG and 3-NAG inhibitors were bound by lysozyme, and there was no correlation between these effects and the distance from the active site. The label in the cleft on the side opposite to the active site proved to be the most sensitive (Likhtenshtein *et al.,* 1974). The reason being (Akhmetov *et al.,* 1972; Ivanov, 1978) that the His15 label is located in that cleft of the protein globule which is 'pried apart' when the active site is contracting. The increased mobility of the N-O groups of the labels bound to the protein in its other segments may be the result of its increased compactness in the

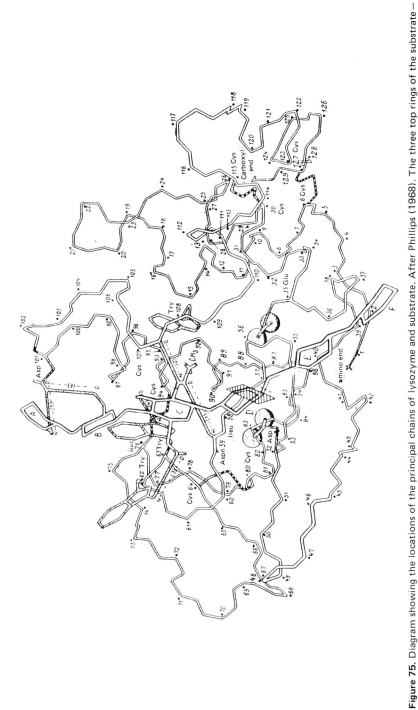

Figure 75. Diagram showing the locations of the principal chains of lysozyme and substrate. After Phillips (1968). The three top rings of the substrate — A, B and C — are retained by six principal hydrogen bonds. The lysozyme molecule fulfils its function by rupturing the substrate chain between the rings D and E.

presence of the ligand, which is equivalent to a decrease in the effective Stokes' radius and to a faster rotation of the protein as a whole. During such a transglobular transition in lysozyme the distance between the amino acids inside the blocks remains unchanged, as shown by the data obtained for the migration of energy between the protein chromophores and the introduced label (luminescence quencher). Thus, conformation transitions in lysozyme are seen to affect entire blocks.

Shashkova *et al.* (1975) employed a direct calorimetric method to study the effect of the pH and of the temperature on the thermodynamics of the reactions of complex formation between lysozyme and the inhibitors NAG (I), its dimer N,N'-diacetylchitobiose (NAG)$_2$ (II) and trimer N, N', N''-triacetylchitotriose (NAG)$_3$ (III), in order to obtain information about the active sites of lysozyme in its different fragments (A, B and C). X-ray diffraction data indicate that inhibitor I, when bound to segment C, forms 4 hydrogen bonds and 30 nonpolar bonds; inhibitor II bound to segments B and C forms 5 hydrogen bonds and 41 nonpolar bonds, while inhibitor III, when sorbed on the segments A, B and C, forms 6 hydrogen bonds and 48 nonpolar bonds (Imoto *et al.,* 1972).

The formation of enzyme-substrate complex may be arbitrarily thought of as proceeding in three stages: formation of new bonds in the complex; changes in the hydration of the interacting species and conformational changes in the enzymes. The conformation of the inhibitors I, II and III does not change on bonding.

The association constants K_{ass} and reaction enthalpies ΔH° were determined from conformational relationships. The changes in the free energy ΔG° and entropy ΔS° were then calculated using the equations $\Delta G^{\circ} = -RT \ln K_{ass}$ and $\Delta G^{\circ} = -\Delta H^{\circ} - T \Delta S^{\circ}$. The experimentally determined enthalpy and entropy contributions to the association as a function of pH indicate that at pH 5.5 and ionic strength of 0.1 the enzyme, and in particular its active site, display the highest conformational mobility. The binding of the sugar residue to the B-segment is accompanied by the largest change of enthalpy and free energy. At all pH values it is only the enthalpy of interaction in segment B which is temperature-dependent, which indicates that segment B has a high structural mobility.

NMR determinations of proton relaxation in solutions of spin-labeled protein made it possible to detect the connection between the conformational mobility of lysozyme macromolecule and the mobility of the nearest layers of its environmental water (Shimanovskii *et al.,* 1977). These workers used the Solomon-Blombergen equations to calculate the effective correlation times τ_c of dipole-dipole interaction between the N-O groups of the spin labels on lysozyme and the water protons. The data thus obtained indicated that a temperature rise enhanced the mobility of water molecules around the N-O group of the labels, as indeed could have been expected, but that an opposite effect – to wit, stabilization of the water molecules forming the environment of the protein – occurs when NAG is bound to spin-labeled lysozyme. This is compatible with the decrease in the conformational mobility of lysozyme produced by NAG, which was observed by NMR and deutrium exchange methods.

The effect of the temperature on the conformational state of lysozyme was studied with the aid of spin labels located in different parts of the globule. All labels displayed breaks in the Arrhenius function at 40 and 50°C, the latter case corresponding to a decrease in enzymatic dependence. However, the variation of the heat capacity with the temperature remains linear until denaturation of the protein has taken place. At the same time the temperature dependence of differential absorption spectra in the pre-denaturation temperature range shows the presence of isobestic points, which indicates either that the equilibrium constants of the two conformers are different or that there is a sequence of transitions between successive similar states of the protein. The fact that conformational changes do not effect the heat capacity of the solution may be the result of intramolecular water-protein compensation processes, related to the stabilization of active site fluctuations and to the displacement of water from the active site.

7.10 α-Chymotrypsin and Pepsin

This enzyme (Fig. 76) is a proteolytic enzyme; however, it is capable of hydrolyzing not only the complex protein molecules, but also low-molecular synthetic substrates such as amides, esters etc. In proteins α–chymotrypsin cleaves the peptide bonds formed by the carbonyl group of phenalalanine, tyrosine and tryptophan.

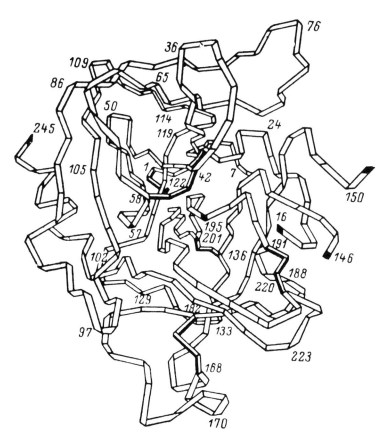

Figure 76. Tridimensional model of an α-chymotrypsin molecule. After Birktoft and Blow (1972).

The hydrolysis of the substrates takes place in several stages:

$$E + S \underset{k_{-1}}{\overset{k_1}{\rightleftarrows}} ES \xrightarrow[\downarrow P_1]{k_2} EA \longrightarrow E + P_2, \qquad (7.40)$$

$$\uparrow$$

$$H_2O$$

where ES is the enzyme-substrate complex, EA is the intermediate acyl enzyme, and P_1 and P_2 are the basic and acidic products of hydrolysis of the substrate.

The active site of α-chymotrypsin comprises a binding segment and a catalytic segment. The binding segment is constituted by the hydrophobic cavity with Met192 residue as the 'lid'. When this residue is modified by a spin label, the enzymatic activity is preserved. This fact makes it possible to follow the conformational states of the active site when the latter is acted upon in different ways.

Ivanov (1978) studied α-chymotrypsin preparations in which various amino acids in the active site and on the periphery of the globule had been modified, and came to interesting conclusions. He found that the 'charge transfer system' Asp102 $\rightarrow$ His57 $\rightarrow$ Ser195, which enhances the nucleophily of the γ-oxygen of the active Ser195 residue, at the same time acts as a connecting link in the active site between two α-chymotrypsin cylinders (blocks) by regulating and producing the optimum conformation of the macromolecule. A break in the 'charge transfer system' moves the polypeptide α-chymotrypsin blocks away from each other, as recorded by the spin label in the contact zone between the blocks. However, the label does not respond to conformational changes when it modifies the amino acids outside the contact zone.

The entry of acridine inhibitor into the hydrophobic active site cavity of the enzyme (binding segment) immobilizes the label on the active Ser195 residue; this may be caused by the collapse of the active site cleft of α-chymotrypsin due to the motion of the enzyme blocks. These results indicate that there is a connection between the catalytic and the binding segments.

Glotov *et al.* (1976a, b) used the method of fluorescent labels to show that the macromolecule of pepsin — another proteolytic enzyme — is conformationally flexible and that its blocks move relative to one another. Binding of a competing inhibitor stabilized the fluctuations of the enzyme.

Koslov *et al.* (1979) made a detailed study of the conformational state of pepsin. A conformational transition, not accompanied by denaturation, takes place in the physiological range; this is indicated by eigen fluorescence polarization data, by the effectiveness of energy transfer from tyrosin to tryptophan, and by the change in dichroic absorption ratio at 292 and 282 nm at pH $\backsim$ 3 when the temperature was raised from 10 tu 80°C. The fluorescence polarization indicates that between 15 and 37°C the intramolecular mobility increases, and decreases again between 37 and 47°C. The nature of the overall variation resembles that obtained for albumin (sec. 5.2). In accordance with the dynamic model, these variations are caused by changes in clusterophilic interactions, in the A $\rightleftharpoons$ B fluctuation frequency of the protein cavities, in the constant of this equilibrium and in the effective Stokes' radius of the macromolecule. It would appear that the absence of any changed in the heat capacity in the zone of conformational transition is due, just as in the cases (described earlier (Fig. 38)), to water-protein enthalpy compensation. The dynamic model predicts a change in the properties of pepsin as a generator of fluctuations in the water environment in the temperature range of the transition.

We shall interpret the variation of the reaction rate constant k_2^{cat} and of K_M as a result of conformational transitions in terms of the dynamic model of the enzyme – cf. equations (7.19) and (7.20). Since in (7.19) $|\kappa| < 1$, while the change in T is small, the principal contribution to the change of k_2 may be expected to originate from the left-hand term of the denominator, which is a characteristic of the dynamic adaptation time between the substrate and the active site. In this term the parameters $k_{a*\to b*}$ and $\kappa G_t^{B\to A}$ vary much less with the temperature than $\nu_{A\rightleftharpoons B}$. If, as a result of thermoinduced transition, $\nu_{A\rightleftharpoons B}$ increases, k_2 should clearly increase as well. This conclusion is confirmed by the results obtained by Koslov *et al.* (1979): at 20.5°C $k_2 = 9.2 \cdot 10^3$ sec^{-1}; at 43°C $k_2 = 31.4 \cdot 10^3$ sec^{-1}.

In the expression for K_M (equation (7.20)) it is $\nu_{A\rightleftharpoons B}$ and P_b^{coll} which vary to the greatest extent with the temperature since, according to the model, the temperature transition is accompanied by a shift in the a $\rightleftharpoons$ b equilibrium to the left. The increase in $\nu_{A\rightleftharpoons B}$ and decrease in P_b^{coll} should result in an increase of K_M. In fact, K_M of pepsin increases from 6.9 at 20°C to 10.2 mM at 43°C. Thus it is seen that the consequences of the dynamic model are convincingly confirmed by experiment.

The available data, taken as a whole, lead to the following conclusion. The functioning of a large number of enzymes with one sub-unit each involves some interaction between their polypeptide blocks, whose relative shifts produce the optimum state of the active site and of the auxiliary cavities in each individual stage. This conclusion is very important for the understanding of the part played by thermal fluctuations of enzyme structure in lowering the activation barriers of the reactions, and is fully compatible with the dynamic model.

An interesting feature of α-chymotrypsin is its ability to retain its catalytic function when in the form of moist powder. It was shown that the higher the salt concentration in α-chymotrypsin powder, the smaller the amount of the water required for the enzymatic activity to become manifest (Khurgin *et al.*, 1978); the correct explanation for this is the tendency of the ions to replace water on the charged and the polar groups of the protein.

In terms of our own theory, the minimum amount of water in the powder which is required for the catalytic act to take place will be that needed to fill the protein cavities in order to ensure the ability of the protein to fluctuate. In the case of α-chymotrypsin this corresponds to 80 water molecules. An analysis of the X-ray diffraction models shows that about 50 molecules may be accommodated in the internal cavities of the protein (Nedev and Khurgin, 1975); it may be assumed that the remaining 30 water molecules become bound by the polar groups on the protein surface, having a high affinity to water.

Khurgin *et al.* (1978) obtained an indirect confirmation of this interpretation by studying the hydration of chymotrypsinogen under dynamic conditions. The dynamic method of plotting the hydration isotherm involves the determination (by Fischer reagent) of the amount of water h bound by the protein in a current of nitrogen (as an inert gas carrier) at a given temperature and water vapor pressure P/P_S.

The water absorption isotherms given by chymotrypsinogen $h = f(P/P_S)$ — were distinctly S-shaped, indicating that the water was coherently bound. Information on the heterogeneity of water sorption sites was obtained by studying the temperature dependence of the size h of the effective single layer on the protein (Fig. 77). The protein retains a smaller amount of water during sorption than during dehydration. The explanation for this difference is that in the sorption process, unlike in the desorption process, the ion-pair-forming groups were not included among the primary hydration centers. The decrease in the hydration with the temperature is probably the result of the exclusion of the sorption sites with a smaller affinity to water. According to the dynamic model, a rise in temperature impairs the clusterophilic interactions, enhances the hydrophobic interactions and thus reduces the effective volume of the protein cavities, with corresponding displacement of water.

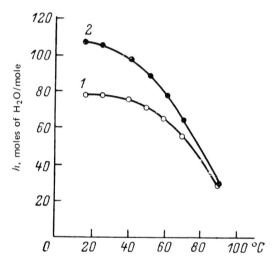

Figure 77. Magnitude h of the water, bound by the protein as a function of the temperature during sorption (1) and desorption (2). After Khurgin *et al.* (1978).

At 90°C, when $h \sim 32$ moles of H_2O per 1 mole of chymotrypsinogen, the effective single layer is formed by the hydration of the water absorption sites which have the highest affinity to water. Since a chymotrypsinogen molecule contains 34 ionogenic groups, it would appear that the remaining water is bound by these groups. The latent molar heat of vaporization of water from the sorbent surface L_0 may be calculated from Clausius-Clapeyron isotherms plotted at different temperatures:

$$d (\ln P)/dT = L_0/RT^2. \tag{7.41}$$

It was found that the largest amount of heat (65 KJ/mole) is evolved when one-third of the effective single layer is filled (about 30 moles of water per mole of chymotrypsinogen), which is roughly equal to the number of ionogenic groups of chymotrypsinogen.

Further sorption of water is accompanied by the evolution of a smaller amount of heat — about 53.2 KJ/mole. As the degree of hydration of the protein increases, it may undergo conformational changes (Khurgin *et al.*, 1978).

Data obtained by Berezin and Martinek (1977) indicate that the water displaced from the nonpolar cavity of the active site of α-chymotrypsin by the substrate undergoes phase transition of the first kind: $\Delta G^{H_2O} = 0$.

If the interaction between water and the nonpolar cavity of the active site is of the conventional, hydrophobic kind, it may be expected that the gain in free energy as a result of its displacement by the ligand will be $2\Delta G_{tr}$, where ΔG_{tr} is the free energy of transfer of the ligand from the water to the nonpolar solvent. However, the experimentally determined gain is only $\sim \Delta G_{tr}$. This may be explained by assuming that the main contribution to the observed change in free energy is made through the shift of the ligand from the water to the nonpolar medium of the active site. It should be noted that the conformational b → a* transition of the active site to the 'closed' state under the effect of the ligand does not produce major changes in the free energy of the system if $G_b = G_a$ and $\Delta G^{H_2O} = 0$, i.e., the resulting transition is close to a transition of the first kind.

7.11 Study of Reactions Involving Papain at Low Temperature

At room temperature and, *a fortiori*, at physiological temperatures many stages of the enzymatic reaction and the corresponding conformational rearrangements of the protein proceed so swiftly that they cannot be recorded. Accordingly, the experimental approach involving studies of enzymatic activities at temperatures much below 0°C in the presence of cryoprotectors, is very promising. This method was used in the study of papain hydrolysis of the methyl ester of N^α-carbobenzoxy-L-lysine in the presence of dimethylsulfoxide as cryoprotector (Fink and Angelides, 1976). At 0°C that value of k^{cat} decreased with increasing concentration of dimethylsulfoxide. Since the latter may be considered as a structure-perturbing agent of both water and protein, its effect on k^{cat} is explained in terms of the dynamic model as resulting from the decrease in $G^{B^* \to A^*}$ (equation (7.18)). On the contrary, the value of K_M increased, probably owing to combined dielectric and competing inhibition. In fact, a decrease in the polarity of the solvent reduces the free energy of substrate transfer from the environment into the nonpolar active site and that of the association constant. The pH-profile of enzymatic activity remains substantially unchanged in the presence of 7.65 M dimethylsulfoxide. Dimethylsulfoxide was also without effect on UV, fluorescence and circular dichroism spectra of papain.

The reaction kinetics were followed by following the variation of the absorption by Try177, localized in the active site, at 276 nm. Three reaction stages were noted prior to the rate-limiting deacetylation stage. In the first stage the pseudo-first-order rate constant was more than 1 sec^{-1} at −65°C, which corresponds to the binding of the substrate. The second stage was pH-dependent, with pK ≤ 3.4, rate constant $k = 8 \cdot 10^{-2}$ sec^{-1} at −40°C and activation energy $E = 31.0$ KJ/mole; this was interpreted as a shift of the imidazole in the active site. The third stage was characterized by $E = 42$ KJ/mole, and was also pH-dependent, with pK* = 4.8, $k = 8.9 \cdot 10^{-4}$ sec^{-1} at −16.5°C, and was attributed to the formation of acyl-enzyme. On the basis of these data, Angelides and Fink (1978) proposed a reaction mechanism postulating the existence of two conformational states of the active site: in one state the His159 imidazole forms a hydrogen bond with Asp175 ($pK_{im} \sim 4; pK_{SH} \sim 8$), while in the other the imidazole is protonized and enters an electrostatic interaction with Asp158-carboxylate and Cys25-thiolate ($pK_{im} \sim 8$, $pK_{SH} \sim 4$). The former state of the cavity, which corresponds to that established by an appropriate crystallographic method, is highly nonpolar and contains ordered water (Drenth *et al.*, 1971). The substrate displaces the water and shifts the equilibrium between the first and second states towards the latter. The shift may be realized by rotation of $C^{\alpha} - C^{\beta}$ bond of His159 through $79°$ and by small shifts of Try177, Asp158 and Cis25. Considerable conformational changes of the active site were also observed by X-ray diffraction method during the binding of the inhibitor by papain (Drenth *et al.*, 1976).

7.12 Allosteric Enzymes

Allosteric proteins are of special interest in the context of signal transmission in macromolecules. It is known that allosteric effects are of two different kinds: heterotropic (interaction between unlike ligands) and homotropic (interaction between like ligands). Two teams of workers attempted to give an interpretation of homotropic allosteric effects: Monod *et al.* (1965) and Koshland *et al.* (1966). The former authors believe that the oligomeric protomers in allosteric proteins have a symmetric organization, and suggest that each protomer (subunit) has only one specific binding site for the ligand of a given type. This model is based on the assumption that oligomers may exist in at least two different states differing from each other by the distribution and/or energy of the bond between the protomers and by the affinity to the ligands. Thus, the ligands are responsible for the equilibrium shift between the oligomers. During a transition from the 'tensioned' T-state to the 'relaxed' R-state with a lower energy of interaction between the protomers all the protomers in the oligomer undergo a change at the same time.

Monod *et al.* further assume that the binding of the ligand with the protein, whatever the state of the latter, is independent of the binding of the other ligands. As an example, let us consider a tetrameric protein in tautomeric equilibrium between two states: the 'round' R-state and the 'square' T-state, with an equilibrium constant $L = [T]/[R]$ (Fig. 78).

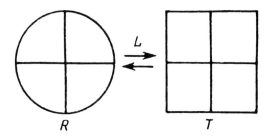

Figure 78. Explanation in text.

The affinity of each subunit to the ligand S is expressed by one of the microscopic dissociation constants: K_R or K_T. These constants are given by the expressions:

$$K_R = \frac{4 - (n-1)}{n} \frac{[RS_{n-1}][S]}{[RS_n]} ; \quad K_T = \frac{4 - (n-1)}{n} \frac{[TS_{n-1}][S]}{[TS_n]} . \quad (7.42)$$

The allosteric properties of the system are described by the tautomeric equilibrium constant L and by the ratio $c = K_R/K_T$, which reflects the relative affinities of the tauto-mers to the ligand S. If L is large and c is small, the proportion of the sites occupied by the ligands in either form as a function of $[S]/K_R$, which describes the ligand concentra-tion, will be S-shaped.

Thus, according to this theory, allosteric interactions between the sites take place by way of an R $\rightleftharpoons$ T equilibrium shift towards the left. In the generalized model of Monod, the R- and T-forms are differentiated from one another not only by their affinities to the substrate, but also by the magnitude of the rate constant of catalytic decomposition of the enzyme-substrate complex (Kurganov, 1978).

The authors of the second theory assume that each of the several protomers forming the allosteric oligomer may exist in two or more tautomeric conformations, but that the ligand can become bound in only one of these conformations. Hybrid conformational states of the oligomeric protein, which contain their subunits in various conformations, are possible. A change in the conformation of one of these subunits may affect the stabi-lity of the conformations of the neighboring subunits and hence also their affinity to the ligand.

This theory takes account of the equilibrium constants of conformational transitions of each subunit before and after the binding of the ligand by one of the tautomeric forms. Parameters describing the forces which are operative between the individual sub-units are also taken into account. According to Koshland's model, such interactions are possible only between subunits located adjacent, but not diagonally, to each other.

The analysis of allosteric properties of a system is based on microscopic equilibrium and on interaction constants. The essential difference between the two theories of allo-sterism is that according to Koshland each one of the bound ligand molecules performs one conformational transition, whereas according to Monod the number of conformation-al transitions is not usually equal to the number of the captured substrate molecules.

Kinetic and structural properties of allosteric enzymes were described in detail in a monograph by Kurganov (1978). The molecules of allosteric enzymes contain not only an active site, but also regulatory sites binding allosteric effectors or modifiers (Fig. 79). Some allosteric enzymes contain effector sites for the substrate. Allosteric interactions are possible in the case of monomeric enzymes as well; they were observed, in particular, for chymotrypsin (Erlanger *et al.,* 1973, 1976) and for ribonuclease (Walker *et al.,* 1975).

Allosteric interactions in oligomeric proteins are highly sensitive to the various distur-bances of the protein structure. They are lost when the proteins are heated in the pre-denaturation temperature range (Gerhardt and Pardee, 1962; Martin, 1963), or as a result of a chemical modification (Rosen *et al.,* 1967) and when enzyme solutions are irradiated by X-rays (Kuzin and Ermekova, 1969). From the viewpoint of the dynamic model all

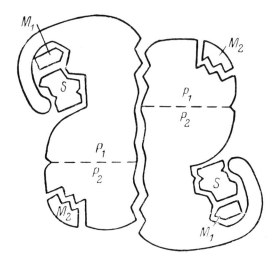

Figure 79. Schematic representation of a symmetrical allosteric enzyme consisting of two protomers. After Koshland and Neet (1968). P_1 and P_2 — two different polypeptide chains or different zones of the same polypeptide chain; S — substrate; M_1 — modifier bound to active site; M_2 — modifier bound to the allosteric site.

these effects impair to varying extents the clusterophilic interactions of the central site with water.

In his discussion of the conformational aspects of regulation of enzymatic activity, Markovich (1977) points out certain aspects of the problem such as the extent of conformational rearrangements and their nature (small shifts of large zones or large shifts of individual segments); whether such rearrangements are induced by ligands or whether certain definite protein pre-conformations are 'selected'; the principal differences between the structural rearrangements of the enzyme as related to catalysis and to regulation respectively.

Fluorescent labels (Markovich *et al.,* 1974) and spin labels localized in the active sites of dehydrogenases, lactate dehydrogenase and glyceraldehyde-3-phosphate dehydrogenase become firmly immobilized. This indicates that the modified active sites are rigid. However, it may be concluded from the results of kinetic analysis and a special modification of Cys153 in the active site that the active site fluctuates between two states, in one of which the Cys153 is accessible to the reagent. It was noted at the same time that the addition of nicotinamide-adenine-dinucleotide enzyme to the active site not only altered the nature of its fluctuations but those of the other sites as well (Vas and Boross, 1974). The most extensive changes were noted after the first molecule of the coenzyme had been bound. Binding glyceraldehyde-3-phosphate dehydrogenase coenzymes enhanced the thermal stability of the protein (Furfine and Velick, 1965) and its stability to proteases (Fenselau, 1972).

In the view of Markovich (1977), dehydrogenases contain three triggering segments: an adenine-bonding, a phosphate-bonding and a nicotinamide-bonding segment. Their specificity is relatively low, since their interaction with effectors of different structures produced identical conformational rearrangements. The spiral blocks, which move as one whole, realize the interactions between mutually distant parts of the macromolecule. Markovich concluded that the structural rearrangements are ordered and resemble the motions of mechanism. This worker showed, by using bifunctional reagents, that the effectors induced new conformational states in dehydrogenases or else alter the mobilities of the corresponding functional groups.

There is no principal difference between the structures of the active and the allosteric sites. However, effectors produce non-dissociative conformational modifications of the quaternary structure of the enzyme which are much slower than those involved in the catalytic act itself. These may be expected to induce relaxational equilibrium shifts between the A and B states of the central cavity of oligomeric proteins, in accordance with the dynamic enzyme model.

A detailed discussion of the mechanism of manifestation of allosteric properties of proteins in terms of the dynamic model will be found in para. 6.6, as exemplified by hemoglobin.

Chapter 8

Interactions between Macromolecules of Different Types and Proteins and Cells in Aqueous Media

It is known that when synthetic polymers of sufficient molecular mass are added to protein solutions in high concentrations, the proteins may form aggregates or may even crystallize out. This is usually explained as an effect of the excluded volume produced by the hydrated polymers, as a result of which the effective protein concentration may increase considerably (Polson, 1977), while their solvation may decrease.

However, not all polymer-produced effects can be explained as simply as that. It was recently shown that nonprecipitating in usual conditions porcine anti–DNPs antibodies form a precipitate when reacted with an antigen-dinitrophenylated serum albumin in the presence of polyethylene glycol (PEG) and dextrans (D) in concentrations of polymers at which there is no nonprecific aggregation of proteins (Franek *et al.*, 1979). This may be considered as indirect evidence of increased flexibility of the antibodies that allow each of them to form the complex with two different antigens. In order to verify this conclusion, we studied the effect of polymers on the flexibility of complexes of porcine *anti-DNP* antibodies with spin-labeled hapten (Ab$_1$ + DNP-SL) by the method described in para. 3.6. We also studied the effect of two polyethylene glycols — PEG-2, MM = 2000, and PEG-20, MM = 20,000 — on the relative mobility of the structural units of spin-labeled oxyhemoglobin.

In order to verify the extent to which the interaction of proteins with polymers is an accurate model of the interaction between proteins of different types *in vivo*, we studied the effect on the flexibility of spin-labeled *anti-dansyl* porcine antibodies (Ab$_2$-SL) of proteins such as human serum albumin, chicken lysozyme (L) and Ab$_1$, as well as of their complexes with the following ligands: human serum albumin with methyl orange, lysozyme with raffinose and Ab$_1$ with DNP-lysine.

We also studied the effect of the proteins and complexes listed above, as well as of oxyhemoglobin, methemoglobin and lysozyme + 3N-acetylglucosamine (3NAG) on the intramolecular mobility of spin-labeled oxyhemoglobin.

We found a correlation between the effects of various polymers and proteins on the intramolecular mobilities of Ab$_1$, Ab$_2$ and oxyhemoglobin on the one hand and, on the other, the effect of these macromolecules on the properties of the solvent. The methods used were IR-spectroscopy, NMR and cryoscopy.

The molar ratio [DNP-SL] : [active site] = 0.8 in Ab$_1$ + DNP-SL complexes was the same in all cases.

The covalent bonding of the spin label SL:(2,2,6,6-tetramethyl-N-1-hydroxypiperidine-4-amino-(N-dichlorotriazine)) by the histidine residues of antidansyl Ab$_2$ and lysozyme was realized in 0.01 M phosphate buffer (pH 7.3) + 0.15 M NaCl. Human oxyhemoglobin was labeled with I (Fig. 64) at the SH-groups of the β–chains as described in para. 6.6.

238

The changes in the activity of water in solutions of polymers of various molecular weights, present in various concentrations, were evaluated by determining the freezing point of their solutions in 0.01 M phosphate buffer (pH 7.3) + 0.15 M NaCl. The closer the freezing point to $0°C$, the greater is the activity of the water in the solution.

8.1. Effect of Polymers on the Flexibility of Antibodies in Complex with Spin-labeled Hapten at Different Temperatures

This study was carried out in collaboration with S.P. Rozhkov and F. Franek.

Determination of the correlation times τ_R and τ_M of Ab_1 + DNP-SL complexes by the method described in para. 3.6 shows that as the polymer concentration increases, τ_M of the protein markedly decreases due to the elevated flexibility of Ab_1 while τ_R remains unchanged (Table 16). PEG-40; D-500; D-70; D-40 correspond to polyethylene glycol and dextrans with molecular mass: 40,000; 500,000; 70,000 and 40,000 respectively. The experimental errors involved in the determination of τ_M and τ_R are ± 2 nsec and ± 0.2 nsec respectively. The concentration of Ab_1 + DNP-SL was kept constant at 1%.

It was deemed of interest to carry out a more-detailed study of the intramolecular mobility of Ab_1 as a function of the polymer concentration. Figure 80 shows such functions of τ_M at $4°C$, $25°C$ and $37°C$ for PEG-6, PEG-40 and D-40 with molecular masses of 6,000 and 40,000 respectively, which indicate that τ_M decreases as the polymer concentration in Ab_1-DNP-SL solution increases. These results cannot be produced by a simple aggregation of the polymer with the antibody, since in such a case τ_M would have increased rather than decreased. The aggregation of Ab_1-DNP-SL, which took place at sufficiently high polymer concentrations ($\geqslant 60$ mg/ml) was accompanied by a sudden immobilization of BPR spectrum, which then ceased to be affected by the presence of saccharose.

The relative mobility of the structural subunits of spin-labeled oxyhemoglobin also increases in the presence of 1% PEG-40 (τ_M decreases from 40 to 23 nsec at $25°C$). However, the low–molecular (MM = 2,000) polyethylene glycol, in the same concentration, had no significant effect on the τ_M of oxyhemoglobin.

TABLE 16. Effect of polymers in two concentrations (1 and 2% w/vol) on the τ_M of Ab_1 + DNP-SL complexes at various temperatures.

°C	In the absence of polymers		PEG-40		D-500		D-70		D-40	
	τ_M nsec :	τ_R nsec	τ_M, nsec 1%	2%	τ_M, nsec 1%	2%	τ_M, nsec 1%	2%	τ_M, nsec 1%	2%
4	30	11.1	15	12	13	12	20	15	26	20
20	40	8.3	20	16	25	15	30	26	38	30
30	50	7.7	30	20	30	20	30	30	46	30
37	60	7.3	35	25	40	30	50	40	58	48

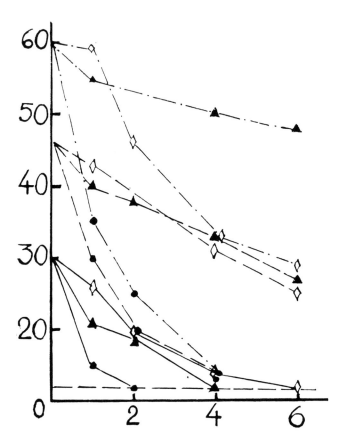

Figure 80. τ_M of AT_1 + DNP-**S**L solutions as a function of the concentrations of: PEG-40 (•);
PEG-6 (▲) and D-40 at $5°C$ (——); $25°C$ (- - -) and $37°C$ (•••). 0.01 M phosphate buffer
(pH 7.3) + 0.15 M NaCl. After Käiväräinen *et al.* (in press).

8.2. Effect of Proteins on the Flexibility of Antibodies and Oxyhemoglobin

The results shown in Table 17 indicate that the flexibility of antibodies is affected not
only by polymers but also by the proteins present in the solution.

Concentration of Ab_2-SL in the solution was in all cases $1 \cdot 10^{-4}$ M. Concentrations
of HSA = $1.5 \cdot 10^{-4}$ M; [Lysozyme] = $4.3 \cdot 10^{-4}$ M; [Ab_1] = $2 \cdot 10^{-5}$ M; [DNP-lysine] =
$4 \cdot 10^{-5}$ M.

TABLE 17. Values of τ_M and τ_R at $20°C$ calculated for the control solution of antidansyl Ab_2-SL and in the presence of human serum albumin (HSA), lysozyme (L) and anti-DNP Ab_1, in the intact and ligand-bound state with methyl orange (MO), raffinose (R) and DNP-Lysine.

	Ab_2-SL (control)	Ab_2-SL + HSA	Ab_2-SL + (HSA-3MO)	Ab_2-SL + L	Ab_2-SL + (L-R)	Ab_2-SL + Ab_1	Ab_2-SL + (Ab_1-DNP)
τ_M	33	25	21	23	30	25	32
τ_M	8.3	9.1	9.1	9.1	9.1	10.5	9.3

It follows from Table 17 that a change in the ligand state of "inductor" proteins (human serum albumin, lysozyme (L) and Ab_1) alters the effect exerted by these proteins on the "receiver" protein (Ab_2-SL). These proteins also have a similar effect on the mobility of the subunits of oxyhemoglobin conjugated with spin-label I (Fig. 64) in position β-93 (Table 18).

The concentrations of all proteins in oxyhemoglobin-I solutions was 5 mg/ml; the concentration of hemoglobin-I was 20 mg/ml in all cases. The values of τ_M and τ_R were determined to within ± 2 nsec and ± 0.2 nsec respectively.

Lysozyme + 3NAG and human serum albumin + 3MO considerably enhance the flexibility of oxyhemoglobin-I. Intact human serum albumin and lysozyme alter the τ_M of oxyhemoglobin-I to a smaller extent. The introduction of intact methemoglobin, oxyhemoglobin or lysozyme + raffinose complex into a solution of oxyhemoglobin-I is practically without effect on the dynamic properties of oxyhemoglobin.

We showed that the flexibility of human serum albumin increases when this protein is bound to methyl orange (MO). The flexibilities of methemoglobin and oxyhemoglobin at $25°C$ are equal and are very low ($\tau_M = 40$ nsec) (para. 6.6).

A special series of experiments carried out on lysozyme (L), spin-labeled at His15, showed that when lysozyme is bonded to a competing inhibitor (3NAG) and raffinose at $36°C$, the effective Stokes radius of the enzyme decreases from 1.9 nm to 1.7 and 1.8 nm respectively. This is manifested as a decrease of τ_M from 6 ± 1 nsec to 4 ± 1 and 5 ± 1 nsec respectively.

TABLE 18. Values of τ_M and τ_R for spin-labeled oxyhemoglobin-I at $25°$ in the presence of proteins without ligands and in complex with them in the same solution.

	Oxyhemoglobin-I (control)	Met Hb	Oxy Hb	Human serum albumin	Human serum albumin + 3MO	Lysozyme	Lysozyme + 3NAG	Lysozyme + raffinose
τ_M	40	40	40	35	30	35	30	40
τ_M	10.8	10.6	10.6	11.0	11.7	11.0	11.8	11.0

Of special interest is the fact that when the rate of rotary diffusion of spin-labeled lysozyme decreases as the concentration of saccharose increases, the EPR spectrum changes from the triplet form into a form reflecting the fluctuations between state *A* (in cavity) and state *B* (on the surface) as at Fig. 46. The possibility to such SL-fluctuations was confirmed by an analysis of a three-dimensional model of lysozyme (Wien *et al.,* 1972). A similar tendency was noted for metmyoglobin-SL. The mobility of the spin label in lysozyme cavity (A-state) corresponds to τ_R of 8 nsec and of 10 nsec in metmyoglobin cavity in agreement with the mobility of the labels in the A-state for spin-labeled oxyhemoglobin, human serum albumin and IgG. It would appear that times of the order of 10^{-8} sec are typical of the specific mobility of peripheral segments of the tertiary structure, which is common to all proteins.

8.3. Effect of Thermality Induced Transitions of Human Serum Albumin on the Mobility of the Subunits of Spin-labeled Oxyhemoglobin

In para. 5.2 above we described changes in the flexibility of human serum albumin at 20° and 35°C; these changes are intensified in the presence of low concentrations of D_2O and are accompanied by changes in the properties of the solvent (Fig. 35).

Fig. 81 shows the relationship between τ_M and temperature of oxyhemoglobin-I solutions and its solutions in the presence of human serum albumin, in the absence of D_2O and in the presence of 10% D_2O. It is seen that there is a direct correlation

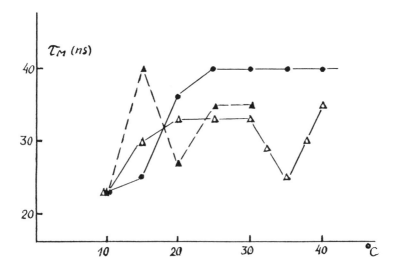

Figure 81. τ_M as a function of the temperature for the control solution of oxyhemoglobin–I (•) and its solutions in the presence of human serum albumin, in the absence of D_2O (△) and in the presence of 10% D_2O (▲). Concentration of oxyhemoglobin–I 20 mg/ml; concentration of human serum albumin 10 mg/ml. 0.01 M phosphate buffer (pH 7.3) + 0.15 M NaCl. After Käivaräinen and Rozhkov (in press).

between the structural transitions in human serum albumin and its effect on the flexibility of oxyhemoglobin.

The presence of 10% D_2O, which stimulates the transition of human serum albumin around $20°C$, is manifested as a large drop of τ_M of oxyhemoglobin-I at this temperature. All these relationships are reversible with temperature decreasing. The formation of associates between human serum albumin and hemoglobin-I would have resulted, on the contrary, in an increase of τ_M. So such explanation of the effect must be excluded. The frequency of collisions between the macromolecules at these concentrations is much less than $1/\tau_M$.

If an interaction between the different types of proteins in solution results in their altered structural flexibility, then in accordance with our models, it must change their functional properties. This statement we supported by the observed change in the activity of alcohol dehydrogenase (ADG) from yeast in the presence of human serum albumin (HSA) in the temperature range of $10°$ to $45°$. ADG concentration was $2.4 \cdot 10^{-6}$ M, HSA $- 2.2 \cdot 10^{-4}$ M, NAD $- 4 \cdot 10^{-4}$ M, ethyl alcohol $- 4 \cdot 10^{-2}$ M. The reaction proceeded in a phosphate buffer (0.01M) + 0.15M NaCl, pH 7.3. The reaction was initiated by adding ethyl alcohol. The initial rate of the reaction was recorded spectrophotometrically by the increase of the concentration of NADH at $\lambda = 360$ nm. The dependence of the initial rate of catalysis on temperature was compared for the ADG solution in the presence and without HSA. In the range of $15-20°$ and $30-40°$ ADG activity was found to rise in the presence of HSA. It is within these ranges that both the flexibility of HSA and its ability to increase the water activity of the solution augment (see Fig. 35 and section 8.5).

It has been shown earlier (Fig. 81) that the relative mobility of the subunits in oxyhemoglobin rises in the same temperature ranges as affected by HSA. A similar phenomenon seems to occur in the ADG tetramer. In keeping with the dynamic model of enzyme action (section 7.1) a rise in the flexibility of proteins must result in their increased activity. The results obtained support this theoretical conclusion.

Because a considerable amount of various enzymes is present together with albumin in serum, the effect revealed may be of practical value.

The next stage in the study was to demonstrate the fact that the effect of polymers and "inductor" protein on the flexibility of antibodies and oxyhemoglobin is directly connected with their effect on the properties of the solvent.

8.4. Effect of Polymers and Proteins in Different Ligand States of Active Sites on the Properties of the Solvent

One possible indication of the effect of macromolecules on water is a shift of the peak attributed to associated band (~ 2130 cm^{-1}), in which specific absorption of polymers and proteins is absent.

The results obtained by IR spectroscopy for PEG-40 and D-40 at two temperatures (Table 19) indicate that the properties of the solvent do in fact change.

It follows from the table that PEG-40 (MM = 40,000) produces a larger low-frequency shift of the water band than does D-40. It is known that the low-frequency shift of

TABLE 19. Location of peak $\nu_2 + \nu_L$ (cm^{-1}) of the water band in solutions containing various concentrations (w/vol) of polyethylene glycol and dextrans.

C°	PEG, MM = 40,000			D, MM = 40,000		
	0%	1%	3%	0%	1%	3%
20	2142	2140.5 (−1.5)	2138.5 (−3.5)	2142	2141 (−1)	2140.5 (2−1.5)
37	2127	2125 (−2)	2122 (−5)	2127	2126 (−1)	2125 (−2)

*   The experimental error in these determinations was ± 0.5 cm^{-1}. Each spectrum was determined five times or more.

** The polymers were dissolved in 0.01 M phosphate buffer (pH 7.3) + 0.15 M NaCl. Figures between brackets indicate the band shift with respect to control in cm^{-1}.

the maximum of the associative water band in solution represents a decrease in the number and energy of hydrogen bonds between the water molecules in solution, while a high-frequency shift reflects an opposite effect.

Cryoscopic data for various polymers of increasing concentrations, confirm the data obtained by IR spectroscopy. Relatively low concentrations ($\leqslant 50$ mg/ml) of the polymers with MM $\geqslant 6,000$, used in our study, raise the freezing point of the solutions, and *hence increase the activity of the water,* while higher concentrations, at which protein aggregation takes place, depress this temperature. No "anomalous" effect of increased water activity was noted in the presence of low-molecular (MM $\leqslant 2,000$) polymers.

Polymers with a sufficiently large molecular mass are loosely structure random coils, which contain a large amount of water in their internal zones. Thermal fluctuations of these coils, which result in a rapid exchange between the internal and the external water, may bring about perturbations in the solvent.

An increase in the concentration of such polymers is accompanied by an increase in viscosity, which in turn results in a decrease in their fluctuation frequency, and the proportion of bound water with a low activity increases. When this effect becomes stronger than the effect of polymer fluctuations, the solutions display ordinary properties and their freezing point decreases with increasing polymer concentration.

In the presence of 5% D_2O the anomalous effect of increased temperature of freezing (t_f) in solution in PEG with MM = 20,000 is enhanced. It is probable that, as in the case of proteins, D_2O enhances the amount and the degree of ordering of water in the polymer coil; during its fluctuations this water becomes exchanged with the free water.

Polymers with a low ($\leqslant 2,000$) molecular mass do not display anomalous properties, probably because they are unable to capture sufficiently large amounts of water in their interior. For this reason their volume fluctuations bringing about the exchange of water are not very effective. This conclusion is in agreement with the fact that the flexibility of oxyhemoglobin-I is not affected by PEG-2.

It was previously shown for Ab_1 (para. 6:3) that intact antibodies bring about a low-frequency shift of the water band relative to the control buffer, while bonding by hapten (H) produces the opposite effect — the water becomes stabilized. This is in agreement with the absence of the effect of the Ab_1 + H complex on the flexibility of Ab_2-SL in contrary with the effect of the intact Ab_1 (Table 17).

The three molecules of methyl orange which are bound to human serum albumin elevate the freezing point of the protein solution while complex formation between lysozyme and one molecule of raffinose results, on the contrary, in a decline of the freezing point. This finding correlates with the opposite effect of bonding ligands by these proteins on the flexibilities of Ab_2-SL (Table 17) and of oxyhemoglobin-I (Table 18) present in the same solution.

Now we have a good background for the assumption that owing to the fluctuations of the polymer and of the protein molecules, the properties of the water in the buffer solutions approach those of pure water. We shall attempt to explain this effect.

A buffer solution of a protein, which is a three–component system, obeys the Gibbs–Duhem equation:

$$n_w d\mu_w + n_p d\mu_p + n_i d\mu_i = 0 \tag{8.1}$$

where n_w, n_p and n_i are, respectively, the molar concentrations of water, protein and ions which are present, while μ_w, μ_p and μ_i are their partial molar free energies or chemical potentials.

The chemical potential of each component of the system is related to its activity as follows:

$$\mu_p = RT \ln a_p + \mu_p^0 \tag{8.2}$$

where μ_p^0 is constant at a given temperature, while the activity a is proportional to the molar concentration of the component in the solution: $a = \gamma C$.

Clearly, if the addition of the protein or of the polymer to the buffer solution results in $n_p d\mu_p > 0$ and $d\mu_w > 0$ this means that μ_i should decrease.

Structural transitions in protein, caused by ligands or by nonspecific factors which alter its chemical potential, should be accompanied by changes in the properties of the salt solution which is used as solvent.

It may be assumed that the fluctuations generated by the proteins, which are propagated through the solution, are capable of disrupting (or reducing in size) the hydrate hulls around the ions present in the buffer. This enhances the quantity of water in solution with an unperturbed structure, and hence also its activity, whereas the activity of the ions μ_i decreases.

The data quoted below confirm that this is in fact the mechanism by which the properties of the solvent change under the action of fluctuating macromolecules.

Decrease in Ion Hydration in the Presence of Human Serum Albumin and Polyethylene Glycol

The PMR method was used to determine the molar concentration of water in the frozen solutions of albumin and polyethylene glycol, containing 0.3 M NaCl, as a function of the temperature (Fig. 82).

The control curve, showing the change in the hydration of 0.3 M NaCl with the temperature in °C was calculated by subtracting the amount of non-frozen water in a solution of 0.01 M phosphate buffer (pH 7.3) not containing NaCl from the amount of non-frozen water in the same buffer containing 0.3 M NaCl (●); the hydration of 0.3 M

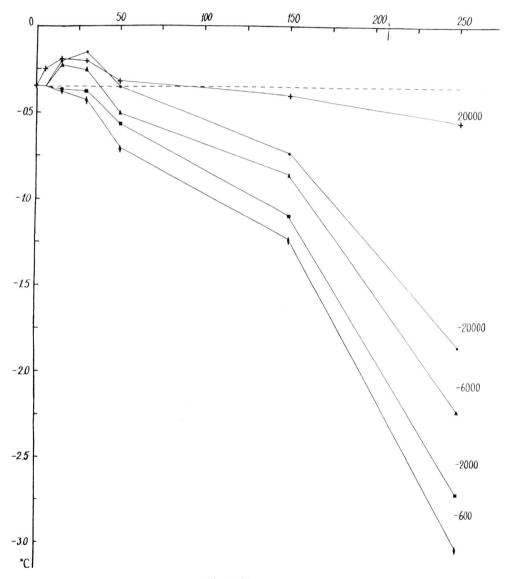

Figure 82.

NaCl in the presence of 1% polyethylene glycol (PEG) was calculated as the difference in the amounts of mobile water in solutions of 0.01 M phosphate buffer (pH 7.3) plus 1% PEG-40 in the presence and in the absence of 0.3 M NaCl (■); the hydration of 0.3 M NaCl in the presence of 2% human serum albumin (+) was calculated as the difference between nonfrozen water in 0.01 M phosphate buffer (pH 7.3) + 2% human serum albumin, in the presence and in the absence of 0.3 M NaCl.

It is readily seen that, throughout the temperature range studied, the presence of PEG and of human serum albumin reduces the amount of water with its structure perturbed by Na^+ and Cl^- ions which, unlike ordinary water, may retain its mobility even at $-20°C$ or below that temperature (Fig. 82). This is a confirmation of the mechanism described above, according to which the presence of macromolecules tends to reduce the number of water molecules involved into the sphere of electrostatic action of ions and as a result the proportion of water with "ordinary" structure increase.

While the properties of the solvent may be affected by a changed dynamic behavior of the properties, a reverse process is also possible.

The general conclusion which may be drawn from the experimental results given above is that an increased activity of the water in solution in the presence of "inductor" macromolecules enhances the flexibility of the spin-labeled "receivers", while shifting the equilibrium towards more highly hydrated conformers. The separation between the "inductor" and the "receiver" molecules is arbitrary, since not only a direct, but *also a reverse relationship* should obtain between the two types. The system [spin-labeled human serum albumin + oxyhemoglobin] was used as an example to show that this relationship in fact exists. At $20°C$ the value of τ_M of spin-labeled human serum albumin increases in the presence of oxyhemoglobin from 29 nsec to 34 nsec. The mechanism by which the activity of water affects the structural flexibility of proteins is described in para. 6.3. It seems that oxyhemoglobin at $20°C$ decreases the activity of water.

The reverse influence of HSA on spin-labeled oxyhemoglobin was described in section 8.3.

There is accordingly reason to believe that we observed an unknown mechanism of interaction between macromolecules (both of the same type and between different types) by way of the solvent, which ensures a mutual alteration of their dynamic and thus also of their functional properties.

This opens a new way in which to model processes taking place in the serum and in the cell protoplasm *in vivo*.

8.5. Interaction between Serum Proteins and Cells*

The tendency of proteins to alter the activity of the water in solutions as a result of thermally induced structural transitions or as a result of ligand bonding, as observed by ourselves, indicates that interaction is possible not only between proteins of different types but also between proteins and cells.

A change in the activity of water in a suspension of cells may be expected to produce

---

* This study was carried out by the author in collaboration with S. D. Kirilyuk.

a shift in the osmotic equilibrium, a change in the content of intracellular water and, in the final result, a change in the cell volume. As a result, the light scattering (turbidity) of cell suspensions may increase or may decrease.

We studied the system: human erythrocytes + human serum albumin between 8°C and 44°C and showed that for this system this in fact is the case (Fig. 83). Complex formation between human serum albumin and ligands (drugs and hormones) strongly affects the ability of human serum albumin to regulate intracellular osmotic pressure.

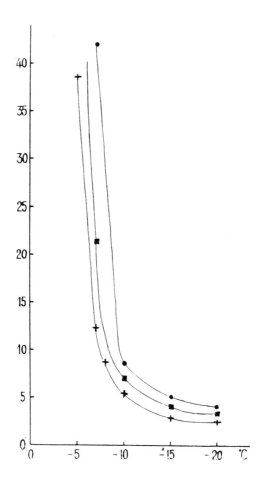

Figure 83. Differential dependence of the molar concentration of mobile water on the temperature in frozen solutions of: 0.3 M NaCl (•); 0.3 M NaCl + 1% PEG–40 (■); 0.3 M NaCl + 2% human serum albumin (+). Explanations in text. All solutions were prepared on the base of 0.01 M phosphate buffer (pH 7.3). After Käivaräinen and Sukhanova (in press).

The variation in the turbidity of a suspension of human erythrocytes in a concentration of $2 \cdot 10^5$ ml^{-1}, containing human serum albumin (15 mg/ml), was determined in a spectrometer equipped with a thermostatted cuvette holder, at a wavelength of 600 nm. All ligands were added in fivefold molar excess with respect to human serum albumin. The erythrocytes were separated out of the blood belonging to Group I, and incubated in 0.01 M phosphate buffer (pH 7.3) + 0.15 M NaCl for at least 24 hours before the measurements were performed for the exhausting of intracellular ATF.

The curves in Fig. 83 reflect the change taking place in the volume of the erythrocytes as a result of thermally induced transitions of intact human serum albumin (Fig. 83a) and its complexes with ascorbic acid (Fig. 83b), nor-adrenaline (Fig. 83c), sodium acetyl-salicylate (Fig. 83d) and adrenaline (Fig. 83e). These experiments also included controls of a possible direct effect of the ligands on the erythrocytes in the absence of human serum albumin.

Ascorbic acid and nor-adrenaline seem to have a similar effect on the structural properties of human serum albumin, which determine its interaction with the solvent. It is manifested as enhanced swelling of erythrocytes in the low- and high-temperature zones, about 20° and 37° respectively, and their contraction in the intermediate zone as compared with the control suspension of erythrocytes. The bonding of human serum albumin by adrenalin seems to result in a steady decrease in the activity of the water in solutions of such complexes when the temperature is decreased from 37°C to 12°C (Fig. 83e). This is manifested as the elimination of water out of the erythrocytes and their contraction, which is accompanied by a decrease in light scattering. The curves shown in Fig. 83 were obtained by subtracting the turbidity of erythrocyte suspensions not containing protein from erythrocyte suspensions containing intact and ligand-bonded human serum albumin. Thus, these differential relationships reflect the contribution of human serum albumin alone and of its complexes to the change in the volume of the erythrocytes when the temperature is increased.

The peaks in Fig. 83 represent the increase in volume (swelling) of the erythrocytes, corrected for the light scattering by human serum albumin alone, when present in the concentration of 15 mg/ml.

It was surprising, that a change in the solvent composition upon the addition of a negligible amount of KCl at a concentration of 0.005 M to the standard 0.01 M phosphate buffer, pH 7.3 + 0.15 M NaCl, results in a quantitative alteration of the curve for the light scattering of erythrocyte suspension (Fig. 84b). HSA has a similar effect on the volume of erythrocytes in the modified Henx's solution (Fig. 84c), which contains KCl at the same concentration as in the previous case. The general composition of the Henx's solution used in our experiments is close to the standard one: 0.14 M NaCl; $5 \cdot 10^{-3}$ M KCl; $3 \cdot 10^{-4}$ M MgSO$_4$; $5 \cdot 10^{-4}$ M MgCl$_2$; $1.3 \cdot 10^{-4}$ M CaCl$_2$; $2.2 \cdot 10^{-2}$ M Na$_2$HPO$_4$; $8 \cdot 10^{-3}$ M KH$_2$PO$_4$; pH = 7.3.

However, in this case glucose was excluded to avoid possible artefacts in osmotic processes due to the active transport of ions.

A sharp change in the curves of light scattering results from the altered character and temperature intervals of thermoinduced transitions in HSA due to the changed composition of the solvent, because these transitions are responsible for the variation

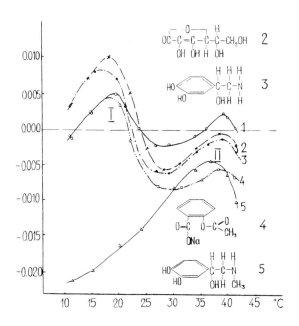

Figure 84. Temperature dependence of turbidity at λ = 600 nm of human erythrocyte suspensions, representing the effect of intact human serum albumin on the volume of the erythrocytes (1) and complexes of human serum albumin with ascorbic acid (2), noradrenaline (3), sodium acetylsalicylate (4) and adrenalin (5). 0.01 M phosphate buffer (pH 7.3) + 0.15 M NaCl. After Käivärainen and Kirilyuk).

of water activity in the erythrocyte suspension. Indeed, in special experiments with spin-labeled HSA in Henx solution we obtain results which point out that the flexibility of HSA in temperature range 20-42° changes in the same way as curve (c) on Fig. 84. For this aim we use the method of separate determination of correlation times of spin-labeled proteins described earlier in Chapter 3 (equation 3.18).

It could be concluded that the interaction between proteins and cells is the result of their sorption (specific or non-specific) on the upper surface of the membranes. However, this conclusion *was rejected* as a result of a number of control experiments.

When the concentration of the protein was halved (from 20 to 10 mg/ml), the effect was attenuated by less than 15%. This makes it unlikely that the effects are caused by *nonspecific* sorption of human serum albumin on human erythrocytes.

The variation of D* with the temperature proved to be qualitatively similar to that represented in Fig. 83 in the case of a suspension of *chicken* erythrocytes containing human serum albumin. This means that there is no *specific* sorption of human serum albumin on human erythrocytes. Qualitatively the same effects we obtained also with immunoglobulins and pool of serum proteins on the erythrocytes volume.

The erythrocytes suspensions were incubated during 1-2 hours in the presence of human serum albumin (5, 15 and 30 mg/ml) at 20°C, centrifuged at 1500 rpm during 10 minutes and the concentration of the protein in the supernatant liquid was determined. This concentration was equal to the initial concentration in all cases, which confirmed that there was no human serum albumin sorbed on erythrocytes under conditions used.

A special osmotic cuvette with an artificial semi-permeable membrane (Amicon UM-20), which did not transmit proteins with a molecular mass of over 20,000, was constructed to exclude completely a direct protein-cell contact as a possible reason for the altered volume of the erythrocytes. The osmotic cuvette consists of two standard spectrophotometric cuvettes, divided with a common wall-semi-permeable membrane. A protein solution is present on the one side of the membrane and a solvent is on the other side. The cuvette is placed in a chamber of the spectrophotometer to be thermostated subsequently.

A decrease of protein concentration in the working part of the osmotic cuvette at a given temperature shows that the water activity of the protein solution is lower than that of the solvent in the other part of the osmotic cuvette. In this case water diffuses through the membrane into the protein solvent and its concentration declines.

If under certain conditions protein increases the water activity of the solvent, then water diffusion is reversed, and the concentration of the protein solution augments. No cells were present in these experiments. Changes of protein concentration were only recorded. Because the HSA concentration in the osmotic cuvette was rather high – over 10-30 mg/ml – the turbidity of the solution at $\lambda = 40$ nm was considered as indicative of changes in its concentration. Results obtained at different temperatures are represented in Fig. 84 by continuous arrows in a standard buffer and by double arrows in the modified Henx's solution. It is obvious that a complete correlation is observed in the osmotic effects produced with biological and artificial membranes. Furthermore, these experiments prove that the ability of proteins to change water activity in the solution is dependent on the salt composition of the solvent.

We can thus conclude that the increase in the volume of the erythrocytes is the result of a shift in osmotic equilibrium, which is accompanied by an increase in the content of intracellular water. We shall inquire into the reason for this effect.

The Boyle van't Hoff law leads for the case when $\pi V = $ const. to the following (Lacke and McCutcheon, 1932):

$$\pi(V - b) = \pi^0(V^0 - b) \tag{8.3}$$

or in another form:

$$V = \frac{\pi^0}{\pi}(V^0 - b) + b \tag{8.4}$$

where V^0 is the volume of the cell in isotonic solution with osmotic pressure π^0; V is the volume of the cell in solution with osmotic pressure π; b is "non-solvent" volume, i.e. the part of cell volume depending on its dry weight and amount of water strongly bound to intracellular structures.

Osmotic pressure in solution of macromolecules may be expressed as follows:

$$\pi \cong \frac{RTC_M}{\overline{V}C_{H_2O}} \tag{8.5}$$

where $\overline{V}$ is volume of one mole of water; C_M and C_{H_2O} are molar concentrations of the macromolecules and water in outer cell medium respectively.

For nonideal water solutions:

$$C_{H_2O} = \gamma a_{H_2O} \tag{8.6}$$

where a_{H_2O} is water activity; γ is coefficient of activity.

After combining the equation (8.4) with equations (8.5) and (8.6) we obtain:

$$V = \frac{a_{H_2O}\pi^0\overline{V}}{\gamma C_M}(V^0 - b) + b \tag{8.7}$$

This equation expresses the variation in the activity of the water in the extra-cellular medium with the variation in the cell volume V. It may obviously be expected that the increase in a_{H_2O} will be accompanied by an increase in V, and by an increase in the turbidity of erythrocyte suspension. Accordingly, peaks I and II in Fig. 83 represent the increase in the activity of water in the extracellular solution as a result of thermally induced structural transitions of human serum albumin. These results are in full agreement with those obtained by the spin-label method (Fig. 35) and NMR (Fig. 37).

The temperature of the blood vessels of warm-blooded animals and of man in the surface layers of the body as a function of the temperature of the environment may vary within a wide range (Barton and Edholm, 1957), including zones I and II. Our results indicate that human serum albumin and other serum proteins may have a definite function in thermal reception and thermal adaptation of organisms, since changes in the concentration and in the activity of intracellular water may be expected to affect the dynamic and the functional properties of enzymes in various blood cells (erythrocytes, lymphocytes, marcophages etc.). Since bonding albumin by hormones or drugs affects its regulatory functions, this question is of major practical importance.

One more interesting effect consisting in redistribution of ions in accordance with water activity changes have been observed by us using the artificial osmotic cuvette. Working part of the osmotic cuvette with lower water activity contained in succession following macromolecules in the modified Henx's solution (without glucose): oxy- and methemoglobin in concentration 15%, dextran (MM = 110,000) in concentrations 10%, 20%, 40% and polyethylene glycol (MM = 40,000) in concentration 2%. The second part of the cuvette connected with the working one by a semipermeable membrane contained pure modified Henx's solution. An osmotic cuvette with the above-noted solutions was incubated for 24 hours at a temperature of 20°. Then we determined the concentration of Na^+, K^+, Cl^{++} and Mg^{++} ions from one and another sides of the semipermeable membrane using the atom-absorption analysis method and a concentration of H^+ using a precision pH-meter. It was found that in every studied case the concentration of ions was higher in Henx's solution without macromolecules, i.e. in solutions with higher initial water activity. It can't be the consequence of the exclusion volume effect in the working part of the cuvette since this effect was excessively compensated by water diffusion and a corresponding rise of the solution level in this part of the cuvette. Dextrans and PEG are the neutral polymers and so Donnans effects are also excluded. Ion concentration difference in the case of oxyhemoglobin was larger than in the case of methemoglobin. In the experiments with dextrane this difference increased almost linearly with the rise of its concentration. The redistribution of Mg^{++} ions was rather more pronounced than that of H^+, K^+ and Na^+. Probably it is due to the different ability of this type of ion to perturb water structure.

Apparently the arrangement of the chemical potentials equilibrium of the solutions divided by a semipermeable membrane may occur not only due to surplus hydrostatic pressure in the part of the cuvette containing the macromolecule solution, but also due to the decrease of water activity in the pure Henx's solution as a result of ionic strength increase in the above solution.

The effect found may play a definite role in the mechanism of passive transport of the ions in cells "in vivo" and must be taken into account in the future.

Conclusion

We attempted to show in the present monograph that the mobility of the protein structure at its various veles is directly related to its hydration, and that it determines to a large extent the functional mechanism of the protein. It is our view that the dynamic model of protein behavior in aqueous media offers a description of the most common properties of proteins. The mechanisms of formation of specific complexes and of enzymatic activities may be derived from the original model.

The large number of examples treated from the viewpoint of the dynamic model indicates that the latter is a satisfactory description of a great variety of experimental data. The basic postulate of the model — fluctuations of protein cavities and relative shifts of polypeptide blocks and subunits in oligomeric proteins — is thus reliably confirmed.

The highly "rational" behavior of biological systems may be considered an additional evidence for the existence of clusterophilic interactions between the ordered water and protein cavities. The phase transitions of the water in the cavities, which proceed practically without any energy losses, enable the active site to regulate the conformation of the protein, its stability and its dynamics as well as to ensure signal transmission over large distances. This mechanism is responsible for the interaction between the active site and the globule, which is indispensable for the protein to fulfill its function.

Antibodies, hemoglobin and glyceraldehyde-3-phosphate dehydrogenase served as examples on which to demonstrate the role played by the central cavity of oligomeric proteins and by the water contained in it in the interaction between the subunits of the protein. An important consequence of the model, which could be experimentally observed, is the capacity of certain proteins with large cavities containing sufficiently highly ordered water to generate anisotropic fluctuations in the environmental medium. Changes in this capacity, produced by the action of specific and nonspecific agents, are in full agreement with the predictions of the theory.

The following "minimum" procedure may be recommended to verify the applicability of our model to any given protein: determination of the "flexibility" of the protein molecule, of its extent of hydration, and of the variation of these parameters under the action of a ligand; detection of two or more stages of conformational transitions of the protein after combination with the ligand or as a result of sudden changes in external conditions such as temperature, pH, ionic strength, etc.; a separate determination of heat effects accompanying conformational changes and hydration changes in the protein (cases in which reliably recorded conformational transitions of the protein are not accompanied by changes in the heat capacity of the solution are of special interest); study of the capacity of proteins possessing sufficiently large cavities to generate anisotropic fluctuations in their aqueous environment as a result of exchange processes, and the dependence of this capacity on the effect of perturbing factors.

It is recommended that this experimental sequence be carried out on an object whose spatial structure is known, and that it be performed under roughly constant conditions (buffer, pH, ionic composition, etc.).

It would be very useful to perform such studies with the aid of some instrument which would yield a simultaneous recording of the kinetics of conformational rearrangements, changes in the state of the water and heat effects. Such an instrument would have the combined performance of a fluorimeter or a dichrograph, an IR spectrometer and a calorimeter, and should work, for example, under stop flow conditions.

The development of a theory of ultrasound absorption by protein solutions and of a suitable experimental technique would be a very promising feature in any study of biopolymer dynamics.

The consequences which follow from the dynamic model of enzymatic activity may be verified by studying the changes in the various kinetic reaction parameters in the presence of specific perturbing agents.

One of the most important results of the present work is that the dissolved macromolecules may interact with each other and cells across the aqueous medium. According to the dynamic model, the intramolecular protein mobility depends on the nature of the exchange process between the fractions of their hydrate hulls and the free water in the solution.

Thermodynamic analysis of the system based on Gibbs-Duhem equation leads to the conclusion that an increase in the activity of water, however produced, should shift the $A \rightleftharpoons B$ equilibrium of protein cavities to the right, i.e., towards the more hydrated conformers, while its decrease should shift this equilibrium to the left. According to the model, B-conformers of proteins have a less rigid structure and a larger number of degrees of freedom than have A-conformers. It would appear that the effect produced by all the macromolecules which increase the flexibility of the immunoglobulins and Hb albumins is due to their ability to enhance the activity of the free water in solution, i.e., its chemical potential, and hence to produce a rightward shift of $A \rightleftharpoons B$ equilibrium.

The reliability of theoretical considerations may be demonstrated not only by the agreement between the expected and the actually obtained experimental results, but also by the fact that the observed effects cannot be interpreted in any other way. The following effects would seem to fall into this category.

1. Effect of low D_2O concentrations on the flexibility of proteins, on the nature of thermally induced transitions in human serum albumin and hemoglobin, and the disappearance of the effect at higher temperatures.

2. The capacity of proteins to affect the properties of their solvent, and their changes induced by specific (ligands) and nonspecific (temperature, salt composition, viscosity, etc.) factors.

3. The water-protein compensation effects, demonstrated by the author in the case of human serum albumin and hemoglobin. They are manifested by the fact that the heat capacity of the solution remains constant despite the major thermally induced changes in the flexibility of the proteins and the consequent changes in the mobility of the solvent molecules.

4. Interactions between macromolecules dissolved in aqueous salt solutions as manifested by the changes in their flexibility.

5. Effect of structural transitions in proteins and the related changes in solvent properties on the volume of the cells present in these solutions.

Results of systematic studies reliably indicate that a solution of a protein in a buffer forms a single cooperative system. If the properties of any one of its components (protein, hydrate shell of the protein, structure of the free solvent, extent of ion hydration) change, the properties of the remaining components are inevitably affected.

The new mechanisms of protein-protein and protein-cell interactions, reported by the author of this book, may be of fundamental importance. They open a new path in the modelling of processes taking place *in vivo* in the blood of animals and in cell cytoplasm.

That the thermally induced changes in the flexibility of human serum albumin affect the activity of water in the range of physiological temperatures is confirmed by the concurrent functional changes in human serum albumin as the regulator of the osmotic processes in blood cells.

This new effect may indicate that human serum albumin fulfils a definite function in the mechanisms of thermal reception and thermal adaptation of organisms. In fact, a changed content of intracellular water should be accompanied by a corresponding change in the concentration of salts and other components of the cytoplasm. This will in turn affect the activity of water and, consequently, will also affect the dynamic and functional properties of cellular enzymes. In addition to erythrocytes, very large amounts of immuno-component cells (lymphocytes, macrophages, etc.) are also carried in the bloodstream, and for this reason a change in the temperature of the environment external to the organism may affect its protective capacity by way of serum proteins.

Since bonding by hormones and drugs may affect the regulatory function of the albumin, this problem is very important in molecular pharmacology.

It can be easily imagined that signal interchange across the medium plays an important part in a large number of individual situations.

Thus, a change in the nature of the interaction between water and enzyme under the effect of a substrate or of an allosteric effector may influence the dynamic and catalytic properties of the enzymes with different specificities in the environment. Such phenomena may perhaps take place in cytoplasmatic polyenzyme systems and are confirmed by our experiments.

Any conformational rearrangements of membranes, induced by mediators or other factors which alter the activity of the intracellular water or of some of its fractions, may also affect the functions of molecular structures dissolved in water or even only partly exposed to it.

A change in the state of a population of serum proteins resulting from a specific response to changes in the organism (e.g., formation of complexes between albumins and adrenalins or other hormones, injected into the bloodstream in conditions of stress) may have a nonspecific influence on the properties of the other proteins in the serum (e.g., on the precipitating and complement-activating properties of the antibody). Whether or not such inter-relationships between serum proteins and intracellular proteins, or between membranes and cytoplasm, in fact exist and if so, what effects they produce – is a problem which can only be solved by further studies.

It is tempting to advance the hypothesis that *inner and outer cell water activity ratio, which affects ion cell homeostasis, may be considered as a factor of cell vital activity, energetics and specialization (differentiation) control.*

There is experimental evidence (mainly NMR) for higher water mobility in cancer cells as compared to that in normal cells. It may be due to the distruction of cytoskeleton in consequence of the malignant transformation of cells and to the correspondent decrease in bound water quantity. Apparently the activity of water in cancer cells is higher as compared to that in normal cells. Thus the higher structural flexibility of enzymes induced by elevated water activity may be considered as one of the factors of accelerated metabolism and division of transformed cells.

The absence of normal intermembrane contacts between malignant tumour cells *leads to the absence of cells division triggering in consequence of insufficient decrease in outer cell water activity.* Therefore the metabolism level of cancer cells remains high and their division continues, i.e. the contact triggering is absent.

It may be assumed on the basis of the proposed scheme that artificially diminished water activity in the blood, and in those parts of the organism where malignant cells are localized, would decrease cancer cell division.

One of the possible ways of water activity decrease in biological liquids is a change in the flexibility of serum proteins by complex formation with specific ligands (drugs and hormones). Such an influence would suppress the processes in normal cells, *including immunocompetent cells,* to a lower degree than in cancer cells, so far as the former contain less free water. Regulation of water activity in the organism may be considered as a new selective approach to the treatment of cancer diseases. The perspectives of this approach are to be investigated.

The possibility of microorganisms growth regulation by means of the change of water activity in the external medium is also a subject of practical interest. We found, for example, that the reproduction rate of E. coli depends on the polyethylene glycol (MM = 40,000) concentration in the culture liquid.

It is not the contention of the author that the interpretations and conclusions given in the present book, which are based on the conception of the dynamic model, are the only ones possible. Nevertheless, the wide range of effects which could be subsequently explained within the framework of the dynamic model and which are not incompatible with it, as well as the experimental confirmation of several nontrivial consequences which follow from it, give grounds for hoping that its fundamental assumptions are not too far removed from the truth.

Should the ideas developed in the present monograph prove useful not only in the interpretation of the available experimental data, but also in pointing the way for further experimental work aimed at a final clarification of the functional principles of biological macromolecules and their interaction with the environment — the task of the author will have been accomplished

References

Abaturov L.V. Hydrogen exchange in proteins. – Usp. nauki i tekhn. Ser. mol. biol., M., 1976, Vol. 8, Pt. 2, pp. 7-126.

Abaturov L.V., Jinoria K.Sh., Varshavsky Yu. M., Yakobashvily N. Effect of ligand and heme on conformational stability (intramolecular conformational mobility) of hemoglobin as revealed by hydrogen exchange. – FEBS Lett., 1977, Vol. 77, pp. 103-106.

Abaturov L.V., Molchanova T.I. Dynamic structure of proteins from hydrogen exchange and proteolytic degradation date. – In book: "Ravnovesnaya dinamika nativnoi struktury belka", Pushchino, 1977, pp. 5-23.

Abaturov L.V., Varshavskii Ya. M. Equilibrium dynamics of tridimensional structure of globular proteins. – Mol. biol., 1978, Vol. 12, pp. 36-46.

Abetsedarskaya L.A., Noftakhutdinova F.G., Fedorov V.D., Mal'tsev N.A. Proton relaxation in solutions and gels of certain proteins. – Mol. biol., 1967, Vol. 1, pp. 451-462.

Akhmetov Yu. D., Likhtenshtein G.I., Ivanov L.V., Kokhanov Yu. V. A Study of transglobular effects in lysozyne by the spin label method. – DAN SSSR, 1972, Vol. 205, pp. 372-376.

Akopyan V.B. The sensitivity of actomyosin of smooth and skeletal muscles to ultrasound. – In book: "Mekhanizmy myshechnogo sokrashcheniya", M., 1972, pp. 45-49.

Aksenov S.I. Study of conformational mobility of proteins in aqueous media by the method of nuclear magnetic resonance. – In book: "Ravnovesnaya dinamika nativnoi struktury belka", Pushchino, 1977, pp. 42-59.

Aksenov S.I. The state of water and its role in the dynamics of biological structures. – Ph.D. summary of thesis, M., 1979.

Aksenov S.I., Filatov A.V. Study of conformational mobility of globular proteins by pulse NMR methods. – Mol. biol., 1978, Vol. 12, pp. 522-532.

Aksenov S.I., Kharchuk O.A. NMR spin-echo study of serum albumin conformational mobility. – Stud. biophys., 1974, Vol. 44, pp. 109-119.

Alanina A.V., Bidnyi S. Yu., Sukharevskii B. Ya. Study of dehydration and of the conformational state of DNA film samples by the method of differential scanning microcalorimetry. – DAN SSSR, 1978, Vol. 241, pp. 703-706.

Aleksandrov V. Ya. Book: "Kletki, makromolekuly i temperature". – M., 1975.

Almazov V.P., Tverdokhlebov E.N. Application of the arrested stream method in molecular biology. – Itogi nauki i tekhn. Ser. mol. biol., M., 1975, Vol. 6, pp. 153-233.

Alpert A., Banerjee R., Lindqvist L. Rapid structural changes in human hemoglobin studied by laser photolysis. – Biochem. Biophys. Res. Comm., 1972, Vol. 46, pp. 913-920.

Amzel L.M., Poljak R.J., Varga J.M., Richards F.F. The three dimensional structure of combining region-ligand complex of immunoglobulin NEW at 3.5 A resolution. – Proc. Nat. Acad. Sci. USA, 1974, Vol. 71, pp. 1427-1430.

Andree P.J. Nuclear magnetic resonance studies on the binding of substrates, coenzymes and effectors to glutamate dehydrogenase. – Biochemistry, 1978, Vol. 17, pp. 772-778.

Andree P.J., Zantema A. Electron spin resonance and nuclear relaxation studies of spin-labeled glutamate dehydrogenase. – Biochemistry, 1978, Vol. 17, pp. 779-783.

Angelides K.J., Fink A.L. Cryoenzymology of papain: reaction mechanism with ester substrate. – Biochemistry, 1978, Vol. 17, pp. 1659-1667.

Antonini E. Hemoglobin and its reaction with ligands. – Science, 1967, Vol. 158, pp. 1417-1425.

Antonini E., Brunori M. On the rate of reaction of an organic phosphate (ATP) with deoxyhemoglobin. – FEBS Lett., 1970, Vol. 7, pp. 351-352.

Antsiferova L.I., Vasserman A.M., Ivanova A.N., Livshits V.A., Nazemets N.S. Atlas of EPR spectra of spin labels and probes. M., 1977.

Artymiuk P.J., Blake C.C.F., Grace D.E.P., Oatley S.J., Phillips D.C. Sternberg J.E. Crystallographic studies of the dynamic properties of lysozyme. – Nature, 1979, Vol. 280, pp. 563-568.

Artyukh R.I., Atanasov B.P., Vol'kenshtein M.V. A study of the conformational properties of whale myoglobin, modified by the spin label method at histidine residues. – Mol. biol., 1977, Vol. 11, pp. 410-417.

Artyukh R.I., Atanasov B.P., Vol'kenshtein M.V., Gerzonde K. Selective carboxymethylation of ferri-hemoglobin of mosquito larvea. – Mol. biol., 1975, Vol. 9, pp. 452-458.

Asakura T., Drott H.R. Evidence of heme-heme interaction in heme-spin-labeled hemoglobin. – Biochem. Biophys. Res. Comm., 1971, Vol. 44, pp. 1199-1204.

Ashman R.F., Kaplan A.P., Metzger H. A search for conformational change on ligand binding in human M macroglobin. I. Circular dichroism and hydrogen exchange. – Immunochemistry, 1971, Vol. 8, pp. 627-641.

Atanasov, B.P. Conformer models of the native state of myoglobin. – Mol. biol., 1970, Vol. 4, pp. 348-356.

Atanasov B. Possible structural and functional differences of the two conformers of myoglobin-like molecules with respect to the heme-heme interaction. – Nature, 1971, Vol. 233, pp. 560-561.

Atanasov B.P., Privalov P.L., Khechinashvili N.N. Calorimetric study of thermal denaturation of cyano-met-myoglobin. – Mol. biol., 1972, Vol. 6, pp. 33-41.

Austin R.H., Beeson K.W., Eisenstein L., Frauenfelder H., Gunsalus I.C. Dynamics of ligand binding to myoglobin. – Biochemistry, 1975, Vol. 14, pp. 5355-5373.

Baldassare J.J., Charache S., Jones R.T., Ho C. EPR studies of spin-labeled hemoglobins. II. Roles of subunit interactions and of intermediate structures in the cooperative oxygenation of hemoglobin and the results on hemoglobin Yakima, hemoglobin J Capetown and carboxypeptidases A and B treated hemoglobin A. – Biochemistry, 1970, Vol. 9, pp. 4707-4713.

Baldo J.H., Halford S.E., Patt S.L., Sykes B.D. The stepwise binding of small molecules to proteins. NMR and temperature jump studies of the binding of 4-(N-acetylaminoglye)-N-acetylglucosamine to lysozyme. – Biochemistry, 1975, Vol. 14, pp. 1893-1899.

Beece D., Bowne S.F., Czégé J., Eisenstein L., Frauenfelder H., Good D., Marden M.C., Marque J., Ormos P., Reinisch L., Yue K.T. The effect of viscosity on the photocycle of bacteriorhodopsin. – Photochem. and Photobiol., 1981, Vol. 33, pp. 517-522.

Beece D., Eisenstein L., Frauenfelder H., Good D., Marden M.C., Reinisch L., Reynolds A.H., Sorensen L.B., Yue K.T. Solvent viscosity and protein dynamics. – Biochemistry, 1980, Vol. 19, pp. 5147-5157.

Bello J., Harker D. Crystallization of deuterated ribonuclease. – Nature, 1961, Vol. 192, pp. 756-765.

Belonogova O.V., Frolov E.N., Illyustrov E.N., Likhtenshtein G.I. Effect of temperature and extent of hydration on the mobility of spin labels in surface protein layers. – Mol. biol., 1979, Vol. pp. 567-575.

Belousov V.P., Konchev I.N., Sidorova A.I. Spectroscopic and thermodynamic properties and structure of aqueous solutions of alcohols. – In book: "Molekulyarnaya fizika i biofizika vodnykh sistem", L., 1974, 2nd Ed., pp. 3-15.

Bendoll J. Book: "Muscles, Molecules and Motion". – (Russian translation).

Bennett W.S., Jr., Steitz T.A. Structure of a complex between yeast hexokinase A and glucose. I. Structure determination and refinement at 3.5 Å resolution. – J. Mol. Biol., 1980, Vol. 140, pp. 183-209.

Bennett W.S., Jr., Steitz T.A. Structure of a complex between yeast hexokinase A and glucose. II. Detailed comparisons of conformation and active site configuration with the native hexokinase B monomer and dimer. – J. Mol. Biol., 1980, Vol. 140, pp. 211-230.

Bennet W.S., Steitz T.A. Structure of a complex between yeast hexokinase A and glucose. – J. Mol. Biol. 1980, V. 140, p. 183-230.

Berendsen H.J.C. Interaction of water and proteins. In: Enzymes: structure and function, 1972, Vol. 29, pp. 19-27.

Berezin I.V., Martinek K. Book: "Osnovy fizicheskoi khimii fermentativnogo kataliza. – M., 1977.

Bernhardt J., Pauly H. Partial specific volume in highly concentrated protein solutions. I. Water-bovine serum albumin and water-bovine hemoglobin. – J. Phys. Chem., 1975, Vol. 79, pp. 584-590.

Birktoft J.J., Blow D.M. Structure of crystalline α-chymotrypsin. V. The atomic structure of tosylα-chymotrypsin at 2 A resolution. – J. Mol. Biol., 1972, Vol. 68, pp. 187-240.

Bishop W.H., Ryan R.J. An investigation of relative subunit mobility in human luteinizing hormone molecules using depolarization of fluorescence. – Biochem. Biophys. Res. Comm., 1975, Vol. 65, pp. 1184-1190.

Bjork I., Karlsson F.A., Berggard I. Independent folding of the variable and constant halves of a lambda-immunoglobulin light chain. – Proc. Nat. Acad. Sci. USA, 1971, Vol. 63, pp.

Blow D.M., Janin J., Sweet R.M. Mode of action of soybean trypsin inhibitor as a model for specific protein-protein interactions. – Nature, 1974, Vol. 149, pp. 54-57.

Blundell T.L., Cutfield J.F. X-ray analysis of globular proteins. – Mol. Struct. Diffract. Meth., 1973, Vol. 1, pp. 385-428.

Blumenfeld L.A., Book: "Problemy biologicheskoi fiziki". – M., 1977.

Blumenfeld L.A., Burbaev D.Sh., Vanin A.F., Vilu R.O., Davydov R.M., Magonov S.N. Non-equilibrium structures of metal-organic complexes functioning as the active sites of the enzymes. – Zhurn. strukt. khim., 1974a, Vol. 15, pp. 1030-1039.

Blumenfeld, L.A., Davydov R.M., Kuprin S.P., Stepanov S.V. Chemical characteristics of conformational non-equilibrium states of metal-containing proteins. – Biofizika, 1977a, Vol. 22, pp. 977-994.

Blumenfeld, L.A., Davydov R.M., Magonov S.N., Vilu R. Studies on the conformational changes of metalloproteins induced by electrons in water-ethylene glycol solutions at low temperatures. Cytochrome C. – Feder. Europ. Biochem. Soc. Letters, 1974, Vol. 45, pp. 256-258.

Blumenfeld, L.A., Davydov R.M., Magonov S.N., Vilu, R. Studies on the conformational changes of metalloproteins induced by electrons in water-ethylene glycol solutions at low temperatures. Hemoglobin. – FEBS Lett., 1974, Vol. 49, pp. 246-248.

Blumenfeld, L.A., Ermakov Yu. A., Pasechnik. Kinetics of the reaction between hemoglobin and carbon monoxide. III. Existence of conformationally non-equilibrated Hb molecules during CO addition and dissociation processes. – Biofizika, 1977b, Vol. 22, pp. 8-14.

Borisov S.V., Borisova S.N., Kachalova G.S., Sosfenov N.I., Voronova A.A., Vainshtein V.K., Torchinskii Yu. M., Volkova G.A., Braunshtein A.E. An X-ray study of aspartate transaminase at 5 A resolution. – DAN SSSR, 1977, Vol. 235, pp. 212-215.

Brands D.F. Conformational transitions of proteins in water and in aqueous solvents. – in book: "Struktura i stabil'nost' biologicheskikh makromolekul". M., 1973, pp. 174-254.

Braunshtein A.E., Shemyakin M.M. Theory of amino acid exchange processes catalyzed by pyridoxal enzymes. – Biokhimiya, 1953, Vol. 18, pp. 393-411.

Brown J.C., Koshland M.B. Activation of antibody Fc function by antigen-induced conformational changes. – Proc. Nat. Acad. Sci. USA, 1975, Vol. 72, pp. 5111-5115.

Brunori M., Antonini E., Fasella P., Wyman J., Rossi-Fanelli A. Reversible thermal denaturation of Aplusia myoglobin. – J. Mol. Biol., 1966, Vol. 34, pp. 497-504.

Buehner M., Ford G.C., Moras D., Olsen K.W., Rossman M.G. Three-dimensional structure of D-glyceraldehyde-3-phosphate dehydrogenase. – J. Mol. Biol., 1974, Vol. 90, pp. 25-49.

Bukatina A.E., Morozov V.N., Shnol' S.E. Mechanochemical transformations in proteins. – In book: "Dvizhenie nemyshechnukh kletok i ikh komponentov", M., 1976.

Bull H.B. Adsorption of water vapor by proteins. – J. Amer. Chem. Soc., 1944, Vol. 66, pp. 1499-1507.

Buontempo U., Careri G., Fasella P. Hydration water of globular proteins: the IR band near 3000^{-1}. – Biopolymers, 1972, Vol. 11, pp. 519-521.

Burkii A. and Cherry R.J. Rotational motion and flexibility of Ca^{2+}, Mg^{2+} - dependent Adenosine 5' -Triphosphatase in sarcoplasmatic reticulum membranes. – Biochemistry, 1981, Vol. 20, pp. 138-145.

Burshtein, E.A. Book: "Sobstvennaya lyuminestsentsiya belka". M., 1977a.

Burshtein, E.A. A study of fast mobility of protein structures by the method of eigen fluoresence. – In book: "Ravnovesnaya dinamika nativnoi struktury belka". Pushchino, 1977b, pp. 60-82.

Burton, A.C. and Edholm, O.G. Book: "Man in a Cold Environment." – (Russian translation).

Butt W.D., Kellin D. Absorption spectra and some other parts of cytochrome C and of its compounds with ligands. – Proc. Roy. Soc., 1962, Vol. 156B, pp. 429-458.

Campbell I.D., Dobson C.M., Williams R.I.P. Proton magnetic resonance studies of the tyrosine residues of hen lysozyme; assignment and detection of conformational mobility. – Proc. Roy. Soc., 1975, Vol. 189B, pp. 503-509.

Careri G. The fluctuating enzyme. – In book: "Quantum statistical mechanics in the natural sciences". Ed. S. Mintz, S. Widmayer. New York-London, 1974, pp. 15-35.

Careri G., Fasella P., Gratton E. Enzyme dynamics: the statistical physics approach. – Ann. Rev. Biophys. Bioeng., 1979, Vol. 8, pp. 69-97.

Carrington A., McLachlan A.D. Book: "Introduction to Magnetic Resonance with Applications to Chemistry". – (Russian translation).

Cathou R.E. Solution conformation and segmental flexibility of immunoglobulins. – In book: "Immunoglobulins". Ed. G.W. Litman, A.A. Good. Plenum Publ. Co., 1978, N4, pp. 37-83.

Cathou R.E., Werner T.C. Hapten stabilization of antibody conformation. – Biochemistry, 1970, Vol. 9, pp. 3149-3155.

Chan S., Feigenson G., Seiter C. Nuclear relaxation studies of lecithin bilayers. – Nature, 1971, Vol. 231, pp. 110-112.

Chernavskii D.C. Principles of biological catalysis. – In book: "Matematicheskaya biologiya i meditsina". M., 1978, Vol. 1, pp. 9-57.

Chernitskii E.A. Book: "Lyuminestsentsiya i strukturnaya labil'nost' belkov v rastvore i kletke". Minsk, 1972.

Chirgadze Yu. N., Ovsepyan A.M. The function of water in the mobility of peptide structures. A study of conformational transitions of polypeptides and oligopeptides during hydration. – Biofizika, 1972a, Vol. 17, pp. 569-575.

Chirgadze Yu.N., Ovsepyan A.M. Hydration mobility in peptide structures. – Biopolymers, 1972, Vol. 11, pp. 2179-2185.

Citri N. Conformational adaptability in enzymes. – Adv. Enzymol., 1973, Vol. 37, pp. 397-648.

Clementi E., Corongiu G., Jonsson B., Romano S. The water structure in the active cleft of human carbonic anhydrase b. A Monte Carlo simulation. – FEBS Lett., 1979, Vol. 100, pp. 28-32.

Cohlberg J.A., Pigiet V.P., Schachman H.K. Structure and arrangement of the regulatory subunits in aspartate transcarbamylase. – Biochemistry, 1972, Vol. 11, pp. 3396-3411.

Coleman J.E. Metallocarbonic anhydrases: optical rotatory dispersion and circular dichroism. – Proc. Nat. Acad. Sci. USA, 1968, Vol. 59, pp. 123-130.

Cone R. Rotational diffusion of rhodopsin in the visual receptor membrane. – Nature New Biol., 1972, Vol. 236, pp. 39-44.

Cooke R., Kuntz I.D. The properties of water in biological systems. – Ann. Rev. Biophys. Bioing., 1974, Vol. 3, pp. 95-126.

Cooper A. Thermodynamic fluctuations in protein molecules. – Proc. Nat. Acad. Sci. USA, 1976, Vol. 73, pp. 2740-2741.

Cser L., Franek F., Gladkikh I.A., Nezlin R.S., Novotny J., Ostanevich Yu. M. Neutron small-angle scattering study on two different precipitin types of pig anti-DNP antibodies. – FEBS Lett., 1977, Vol. 80, pp. 329-331.

Currier S.F., Mautner H.G. On the mechanism of action of choline acetyl transferase. – Proc. Nat. Acad. Sci. USA, 1974, Vol. 71, pp. 3355-3358.

Davies R.E. A molecular theory of muscle contraction. – Nature, 1963, Vol. 199, pp. 1068-1078.

Davies D.R., Padlan E.A., Segal D.M. Three-dimensional structure of immunoglobulins. – Ann. Rev. Biochem., 1975, Vol. 44, pp. 639-695.

Dawson J.W., Gray H.B., Holwerda R.A., Westhead E.W. Kinetics of the reduction of metalloproteins by chromous ions. – Proc. Nat. Acad. Sci. USA, 1972, Vol. 69, pp. 30-33.

Deatherage J.E., Loe R.S., Anderson C.M., Moffat K. Structure of cyanide methemoglobin. – J. Mol. Biol., 1976, Vol. 104, pp. 687-706.

Deisenhofer J., Colman P.M., Epp O., Huber R. Crystallographic structural studies of human Fc-fragment. II. A complete model based on a Fourier map at 3.5 Å resolution. – Hoppe-Seyler's Z. phys. Chem., 1976, Vol. 357, pp. 1420-1434.

Dehl R.E., Noeve C.A. J. Broad-line NMR study of H_2O and D_2O in collagen fibers. – J. Chem. Phys., 1969, Vol. 50, pp. 3245-3251.

Deshcherevskii V.I. Book: "Matematicheskie modeli myshechnogo sokrashcheniya". M., 1977.

Dickerson R.E., Takano T., Kolai O.B. Chain folding reduction mechanism and redox potential in cytochrome C. – In: Fifth Jerusalem Symposium, Jerusalem, 1972.

Dickerson R.E., Timkovich R. Cytochromes C. – Enzymes, 1975, Vol. 11A, pp. 397-448.

Dobson C.M., Williams R.J.P. An NMR study of the dynamics of inhibitor-induced conformational changes in lysozyme. – FEBS Lett., 1975, Vol. 56, pp. 362-367.

Douzou P. La dynamique structurale des proteines et leurs fonctions. – C. r. Soc. biol., 1980, Vol. 174, pp. 574-583.

Dowell L.C., Moline S.W., Rinfret A.P. A low-temperature X-ray diffraction study of ice structures formed in aqueous gelatin gels. – Ciochim. Biophys. Acta, 1962, Vol. 59, pp. 158-167.

Dower S.K., Wain-Hobson S., Gettins P., Givol D., Roland W., Jackson C., Perkins S.J., Suderland Sutton B.J., Wright C.E., Dwek R.A. The combining site of the dinitrophenyl-binding immunoglobulin A myeloma protein MOPS315. – Biochem. J., 1977, Vol. 165, pp. 207-255.

Drenth J., Jansenius J.N., Koekoek R., Walthers B.G. Papain. X-ray structure. – Enzymes, 1971, Vol. 3, pp. 485-500.

Drenth J., Kalk K.H., Swen H:M. Binding of chloromethyl ethyl ketone substrate analogues to crystalline papain. – Biochemistry, 1976, Vol. 15, pp. 3731-3738.

Dreyer G. On the influence of structure anomalies of water at the enzyme function. – Stud. Biophys., 1971, Vol. 27, pp. 145-148.

Drost-Hansen W. Structure and properties of water at biological interface. – In book: "Prepr. cell interface". New York, 1970.

Dudich E.I., Nezlin R.S., Franek F. Fluorescence polarization analysis of various immunoglobulins. – FEBS Lett., 1978, Vol. 89, pp. 89-92.

Durchslag H., Puchwein G., Kratky O., Schuster I., Kirschner K. X-ray small-angle scattering of yeast glyceraldehyde-3-phosphate dehydrogenase as a function of saturation with nicotinamide-adenine-dinucleotide. – FEBS Lett., 1969, Vol. 4, pp. 75-78.

Dwek R.A., Jones R., March D. et al. Antibody hapten interactions in solution. – Phil. Trans. Roy. Soc. London, 1975a, Vol. 272B, pp. 53-73.

Dwek R.A., Knott J.C.A., March D. et al. Structural studies on the combining site of the myeloma protein MOPC315. – Europ. J. Biochem., 1975b, Vol. 53, pp. 25-39.

Edelman G.M. Antibody structure and molecular immunology. – Science, 1973, Vol. 180, pp. 830-840.

Eftink M.R., Chiron C.A. Exposure of tryptophanyl residues in proteins. Quantitative determination by fluorescence quenching studies. – Biochemistry, 1976, Vol. 15, pp. 672-680.

Ehrenstein G. Translational variations in the amino acid sequence of the α-chain of rabbit hemoglobin. Gold Spring Harbor Sympos. Quantit. Biol., 1966, Vol. 31, pp. 705-714.

Eizenberg D., Kautsman V. Book: "Struktura i svoistva vody". L., 1975.

Eklund H., Samama J–P., Wallen L., Bränden C., Akeson A., Jones T.A. Structure of triclinic ternary complex of horse liver alcohol dehydrogenase at 2.9 Å resolution. – J. Mol. Biol. 1981, V. 146, p. 561-587.

El'piner I.E., Zaretskii A.A., Fursov K.P. Absorption of ultrasonic waves by protein solutions. – Biofizika, 1970, Vol. 15, pp. 585-590.

Ely K. Mediation of effector functions by antibodies. – In: Report Workshop, Bethesda, USA, 1978, p. 17.

Emsley J., Feeney J., Sutcliffe L. Book: "High Resolution NMR Spectroscopy". (Russian translation).

Englander S.W., Rolfe A. Hydrogen exchange of respiratory proteins. III. Structural and free energy changes in hemoglobin by use of a difference method. – J. Biol. Chem., 1973, Vol. 248, pp. 4852-4858.

Epp O., Lattman E.E., Schiffer M., Huber R., Palm W. The molecular structure of a dimer composed of the variable portions of the Bence-Jones protein REI refined at 2.0Å resolution. – Biochemistry, 1975, Vol. 14, pp. 4943-4952.

Erlanger B.F., Wasserman N.N., Cooper A.G. Allosteric activation of chymotrypsin-catalyzed hydrolysis of specific substrates. – Biochem. Biophys. Res. Comm., 1973, Vol. 52, pp. 208-215.

Erlanger B.F., Wasserman N.H., Cooper A.G., Monk R.I. Allosteric activation of the hydrolysis of specific substrates by chymotrypsin. – Europ. J. Biochem., 1976, Vol. 61, pp. 287-295.

Esipova N.G. X-ray diffraction analysis of proteins. – Usp. nauki i tekhn. Ser.mol. biol., M., 1973, Vol. 2, pp. 55-131.

Esipova N.G. Makarov A.A. Mgeladze G.N., Madzhageladze G.V., Vol'kenshtein M.V., Monosalidze D.R. Thermal denaturation of pepsin in crystals and in solutions. – Mol. biol., 1978, Vol. 12, pp. 1156-1162.

Esipova N.G., Tumanyan V.G. Factors which determine the formation of the ternary structure of protein globules. – Mol. biol., 1972, Vol. 6, pp. 840-850.

Eyring H., Lumry R., Spikes J. Book: "The mechanism of enzyme action". Baltimore, 1954.

Fabry M.E., Koening S.H., Schillinger W.E. Nuclear magnetic relaxation dispersion in protein solution. IV. Proton relaxation at the active site of carbonic anhydrase. – J. Biol. Chem., 1970, Vol. 245, pp. 4256-4262.

Falk M., Poole A.G., Goymour C.G. Infrared study of the state of water in the hydration shell of Canad. J. Chem., 1970, Vol. 48, pp. 1536-1542.

Fedorov B.A., Timchenko A.A., Denesyuk A.I., Ptitsyn O.V. Comparative analysis of globular protein structure in crystal and in solution in X-ray diffuse scattering. – In: FEBS 12th meet. Dresden, 1978, Vol. 52, pp. 153-158.

Feinstein A., Munn E.A. Conformation of the free and antigen-bound IgM antibody molecules. – Nature, 1969, Vol. 224, pp. 1307-1310.

Fenselau A.P. Structure-function studies on glyceraldehyde-3-phosphate dehydrogenase. – J. Biol. Chem., 1972, Vol. 247, pp. 1074-1079.

Fermi G. Three-dimensional Fourier synthesis of human deoxyhemoglobin at 2.5 Å resolution: refinement of the atomic model. – J. Mol. Biol., 1975, Vol. 97, pp. 237-256.

Fersht A.R., Requena Y. Equilibrium and the rate constant for the interconversion of two conformations of α-chymotrypsin. – J. Mol. Biol., 1971, Vol. 60, pp. 279-290.

Fesenko E.E., Lyubarskii A.L., Kulakov V.N., Vol'kenshtein M.V. The possibility of short-lived intermediate states during the photo-dissociation of ligand-bound derivatives of hemoproteins. – In book: "Konformatsionnye izmeneniya biopolimerov v rastvorakh", Tbilisi, 1975, pp. 187-195.

Filatov A.V. A study of the dynamic properties of biopolymers by pulse NMR. – Candidates's summary of thesis, M., 1978.

Fillips, D. Three-dimensional structure of the enzyme molecule. – In Book: "Molekuly i kletki", M., 1968, pp. 9-28 (Russian translation).

Fink A.L., Angelides K.J. Papain-catalyzed reaction at subzero temperatures. – Biochemistry, 1976, Vol. 15, pp. 5287-5293.

Fisher H.F. A limiting law relating the size and shape of protein molecules to their composition. – Proc. Nat. Acad. Sci. USA, 1964, Vol. 51, pp. 1285-1291.

Franek F., Doskocil J., Simek L. Different types of precipitating antibodies in early and late porcine anti-dinitrophenyl sera. – Immunochemistry, 1974, Vol. 11, pp. 803-809.

Franek F., Olsovska Z., Simek L. Eur. J. Immunol., 1979, Vol. 9, pp. 696-701.

Frank H.S., Evans M.W. Free volume and entropy in condensed systems. – J. Chem. Phys., 1945, Vol. 13, pp. 507-520.

Franks N.P. Structural analysis of hydrated egg lecithin and cholesterol bilayers. I. X-ray diffraction. – J. Mol. Biol., 1976, Vol. 100, pp. 345-358.

Fratiello, A., Schuster R. A NMR solvent exchange study of N, N-dimethylformamide complexes with albumide chloride, bromide and iodide. – J. Phys. Chem., 1967, Vol. 71, pp. 1948-1950.

Frauenfelder H., Petsko G.A., Tsernoglou D. Temperature-dependent X-ray diffraction as a probe of protein structural dynamics. – Nature, 1979, Vol. 280, pp. 558-563.

Frolov E.N., Belonogova O.V. , Likhtenshtein G.I. A study of the mobility of spin and Mössbauer labels bound to macromolecules. In book: "Ravnovesnaya dinamika nativnoi struktury belka. Pushchino, 1977, pp. 99-142.

Frolov E.N., Mokrushin A.D., Likhtenshtein G.I., Trukhanov V.A., Gol'danskii V.I. A study of the dynamic structure of proteins of γ-resonance labels. – DAN SSSR, 1973, Vol. 212, pp. 165-168.

Frushour B.G., Koenig J.L. Raman spectroscopic study of tropomyosin denaturation. – Biopolymers, 1974, Vol. 13, pp. 1809-1819.

Furfine C.S., Velick S.F. The acyl-enzyme intermediate and the kinetic mechanism of the glyceraldehyde-3-phosphate dehydrogenase reaction. – J. Biol. Chem., 1965, Vol. 240, pp. 844-855.

Fursov K.P. Ultrasonic spectrometry of biologically active substances. – Candidate's summary of thesis, M., 1971.

Füst G., Laszlo G., Medgyesi G.A., Czikora K., Rajnavolgi E., Gergely J. Functional properties of human IgA aggregates of different size. In: Abstr. 4th Europ. Immunol. Meet. Budapest, 1978, p. 97.

Gally J.H. Structure of immunoglobulins. – In: The antigens, 1973, Vol. 1, pp. 161-298.

Gamayunov M.I., Vasil'eva L. Yu., Koshkin V.M. Studies of water sorption on biopolymers. – Biofizika, 1975, Vol. 10, pp. 38-40.

Gand H.T., Gills S.J., Barisas B.G., Gersonde K. Heats of carbon monoxide binding by hemoglobin M Iwate. – Biochemistry, 1975, Vol. 14, pp. 4584-4589.

Gavish B., Werber M.M. Viscosity-dependent structural fluctuations in enzyme catalysis. – Biochemistry, 1979, Vol. 18, pp. 1269-1275.

Gelin B.R., Karplus M. Sidechain torsional potentials and motion of amino acids in proteins: bovine pancreatic trypsin inhibitor. – Proc. Nat. Acad. Sci. USA, 1975, Vol. 72, pp. 2002-2006.

Genkin M.V., Davydov R.M., Krylov O.V., Blumenfeld L.A. The effect of salts on the spectroscopic characteristics of immobilized cytochrome C. – Biofizika, 1977, Vol. 22, pp. 162-180.

Gennis R.B., Strominger J.L. Activation of C_{55}-isoprenoid alcohol phosphokinase from *Staphylococcus aureus*. – J. Biol. Chem., 1976, Vol. 25, pp. 1277-1282.

Genzel L., Keilmann F., Martin T.P., Winterling G., Yacoby Y., Frohlich H. Makinen M.W. Low-frequency Raman spectra of lysozyme. – Biopolymers, 1976, Vol. 15, pp. 219-225.

Gerhart J.C., Pardee A.B. The enzymology of control by feedback inhibition. – J. Biol. Chem., 1962, Vol. 237, pp. 891-896.

Giannini I., Barocelli V., Boccalon G., Fasella P. Fast conformational changes at the active site of aspartic aminotransferase. – FEBS Lett., 1975, Vol. 54, pp. 307-310.

Gibson O.H. The photochemical formation of a quickly reacting form of hemoglobin. – Biochem. J., 1959, Vol. 71, pp. 293-297.

Givol D. Mediation of effector functions by antibodies – In: Report Workshop, Bethesda, USA, p. 9.

Glotov B.O., Kozlov L.V., Zavada L.L. Intramolecular mobility of pepsin. – Mol. biol., 1976a, Vol. 10, pp. 161-173.

Glotov B.O., Kozlov L.V., Zavada L.L. A study of the fluorescence of the conjugation products of pepsin with dansyl-lysine chloride. – Mol. biol., 1976b, Vol. 10, pp. 288-293.

Gold V. The D_2O-H_2O solvent system. – In book: "Physicochemical processes in mixed aqueous solutions." Ed. F. Francks, London, 1969.

Goldman S.A., Bruno G.V., Freed J.H. Estimation of slow-motional rotational correlation times for nitroxides by electron spin resonance. – J. Phys. Chem., 1972, Vol. 76, pp. 1858-1860.

Grant E.H., Mittor B.G.R., South G.P., Sheppard R.J. An investigation by dielectric methods of hydration in myoglobin solutions. – Biochem. J., 1974, Vol. 139, pp. 375-380.

Grossberg A.L., Markus G., Pressman D. Change in antibody conformation induced by hapten. – Proc. Nat. Acad. Sci. USA, 1965, Vol. 54, pp. 942-945.

Gurd F.R.N., Rothgeb T.M. Motion in proteins. – Adv. Prot. Chem., 1979, Vol. 33, pp. 74-165.

Hanson D.C., Yguerabide J., Schumaker V.M. Segmental flexibility of immunoglobulin G antibody molecule in solution: a new interpretation. – Biochemistry, 1981, Vol. 20, pp. 6842-6852.

Harbury G.A., Marks R.G. Cytochromes B and C. – In book: "Inorganic Biochemistry". (Russian translation).

Hartman K.D., Ainsworth C.F. Structure of concanavalin A. at 2 Å resolution. – Biochemistry, 1973, Vol. 11, pp. 4910-4919.

Hartsuck J.A., Lipscomb W.W. Carboxypeptidase A. – In: Enzymes. New York, 1971, pp. 11-56.

Harvey S., Hoekstra P. Dielectric relaxation spectra of water adsorbed on lysozyme. – J. Phys. Chem., 1972, Vol. 76, pp. 2987-2994.

Hasted J.B., Elsabeh S.H. The dielectric properties of water in solutions. – Trans. Far. Soc., 1953, Vol. 49, pp. 1003-1013.

Heidner E.J., Ladner R.C., Perutz M.F. Structure of horse carbohemoglobin. – J. Mol. Biol., Vol. 104, pp. 707-722.

Hertz H.G., Zeidler M.D. Measurements of nuclear-magnetic relaxation times in the context of hydration of nonpolar groups in aqueous solutions (in German). – Ber. Bunsenges. Phys. Chem., 1964, Vol. 69, pp. 821-828.

Hill T.L. Effect of rotation on the diffusion-controlled rate of ligand-protein association. – Proc. Nat. Acad. Sci. USA, 1975, Vol. 72, pp. 4918-4922.

Hippel P., Schleich T. Effect of neutral salts on the structure and conformational stability of macromolecules in solution. – In book: "Structure and Stability of Biological Macromolecules." (Russian translation). 1973.

Hodgkin A. Book: "The Nerve Pulse." M., 1965. (Russian translation).

Hoeve C., Willis Y., Martin D. Evidence for a phase transition in muscle contraction. – Biochemistry, 1963, Vol. 2, pp. 282-286.

Hofstee B.H.J. Accessible hydrophobic groups of native proteins. – Biochem. Biophys, Res. Communs., 1975, Vol. 63, pp. 618-624.

Hsi E., Bryant R.G. NMR relaxation in frozen lysozyme solutions. – J. Amer. Chem. Soc., 1975a, Vol. 97, pp. 3220-3221.

Hsi E., Bryant R.G. NMR relaxation in protein crystals. – Amer. Chem. Soc. Polym. Prep., 1975b, Vol. 16, pp. 622-627.

Hsi E., Hilton B.D., Bryant R.G. NMR H relaxation of water in protein powders and frozen solutions. Amer. Chem. Soc. Polym. Prep., 1975, Vol. 16, pp. 642-647.

Hsi E., Jentoft J.E., Bryant R.G. NMR relaxation in lysozyme crystals. – J. Phys. Chem., 1976, Vol. 80, pp. 412-416.

Hsia J.C., Piette L.H. Spin-labeled hapten studies of structure heterogeneity and cross-reactivity of the active site. – Arch. Biochem. Biophysics, 1969, Vol. 132, pp. 446-469.

Hubbell W.L., McConnell H.M. Molecular motion in spin-labeled phospolipids and membranes. – J. Am. Chem. Soc., 1971, Vol. 93, pp. 314-320.

Huber R. Antibody structure. – In: TIBS, 1976, August, pp. 174-178.

Huber R. Conformational flexibility and its functional significance in some protein molecules. – Coll. Ges. Biol. Chem., 1979, N.30, pp. 1-16.

Humphries G.H.K., McConnel H.M. Antibodies against nitroxide spin labels. – Biophys. J., 1976, Vol. 16, pp. 275-277.

Hurbury H.A. and Marks R.H. Cytochromes b and c. In book: "Inorganic Biochemistry", Vol. 2, p. 340, 1975. Ed. G.B. Eichhorn, Elsevier S.P.C.

Huxley A.F. A theory of muscular contraction. – Progr. Biophys. Chem., 1957, Vol. 7, pp. 255-270.

Hvidt A. Book: "Dynamic aspects of conformation changes in biological macromolecules". – D. Reidel Publ. Co., 1973.

Ignat'eva L.G., Seregina T.L., Blumenfeld L.A. Ruuge E.K., Artyukh R.I., Postnikova G.B. A spin-label study of the enzymatic activity of myosin. – Biofizika, 1972, Vol. 17, pp. 533-538.

Imoto T., Johnson L.N., North A.C.T., Phillips D.C., Rupley J.A. Vertebrate lysozymes. – In: Enzymes. New York, 1972, Vol. 7, pp. 665-868.

Irzhak L.I. Book: "Gemoglobiny i ikh svoistva"., M., 1975.

Isenman D.E., Dorrington K.J., Painter R.H. The structure and function of immunoglobulin domains. II. The importance of interchain sulfide bonds and the possible role of molecular flexibility in the interaction between immunoglobulin G and complement. – J. Immunol., 1975, Vol. 114, pp. 1726-1729.

Ivanov L.V. A study of the relationship between the conformational state and the catalytic properties of certain enzymes. – Candidate's summary of thesis, Chernogolovka, 1978.

Ivanov V.I., Karpeiskii M. Ya. Dynamic three-dimensional model for enzymatic transamination. – Enzymol., 1969, Vol. 32, pp. 21-53.

Ivanov V.I., Minchenkova L.E., Minyat E.E., Frank-Kamenetskii M.D., Shchelkina A.K. The B → A transition of DNA in solution. – In book: "Konformatsionnye izmeneniya biopolimerov v rastvorakh. Tbilisi, 1975, pp. 17-22.

Izmailova V.N., Pchelin V.A., Matyukhina L.V. Effect of solubilization on the denaturation of egg albumin. – DAN SSSR, 1963, Vol. 149, pp. 888-890.

Izmailova V.N., Pchelin V.A., Yampol'skaya G.P., Volynskaya A.V. Reaction between benzene and human serum albumin molecules in aqueous solutions. – DAN SSSR, 1966, Vol. 169, pp. 143-145.

Jaton J.C., Huser H., Braun D.G., Givol D., Pecht A., Schlessinger J. Conformational changes induced in a homogeneous anti-type III pneumococcal antibody by oligosaccharides of increasing size. – Biochemistry, 1975, Vol. 14, pp. 5312-5318.

Jaton J.C., Huser H., Riesen W.F., Schlessinger J., Givol D. The binding of complement by complex formed between a rabbit antibody and oligosaccharides of increasing size. – J. Immunol., 1976, Vol. 116, pp. 1363-1366.

Jaton J.C., Wright J.K., Maeda H., Schmidt-Kessen A., Engel J. Interactions of monovalent and multivalent oligosaccharide ligands with homogeneous anti-polysaccharide antibody and its possible effects on the conformation of the IgG molecule. – In: Abstr. 4th Europ. Immunol. Meet. Budapest, 1978, p. 7.

Jentoft J.E., Neet K.E., Struehr, J.E. Relaxation spectra of yeast hexokinases. Isomerization of the enzyme. – Biochemistry, 1977, Vol. 16, pp. 117-121.

Jolicoeur C., Lacroix G. Thermodynamic properties of aqueous organic solutes in relation to their structure. Pt I. Free energy of transfer from H_2O to D_2O for a series of isomeric ketones. – Canad. J. Chem. 1973, Vol. 51, p. 3051-3061.

Kägi J.H., Ulmer D.D. Hydrogen-deuterium exchange of cytochrome C. I. Effect of oxidation state. – Biochemistry, 1968, Vol. 7, pp. 2710-2717.

Käiväräinen A.I. Separate determination of eigen correlation times of spin-labeled proteins and their labels. Mol. biol., 1975a, Vol. 9, pp. 805-811.

Käiväräinen A.I. Dynamic model of behavior of protein in water. – Biofizika, 1975b, Vol. 20, pp. 967-971.

Käiväräinen A.I. Effect of composition of the solvent on the conformational properties of immunoglobulin fragments studied by the spin label method. – In book: "Vsesoyuzn. simpoz. 'Magnitnii rezonans v biologii i meditsine'". Chernogolobka, 1977, pp. 159-160.

Käiväräinen A.I. Effect of disturbing agents on the dynamic properties of immunoglobulin Biofizika, 1978b, Vol. 23, pp. 595-601.

Käiväräinen A.I. Dynamic model of hemoglobin action. – In: Abstr. 12th FEBS Meet., Dresden, 1978, p. 2311.

Käiväräinen A.I. A possible mechanism of immune complex formation. – In: Abstr. 4th Europ. immunol. meet. Budapest, 1978, p. 94.

Käiväräinen A.I. Dynamic model of protein behavior in water. A possible mechanism of the association and dissociation of specific complexes. – Biofizika, 1979a, Vol. 24, pp. 419-425.

Käiväräinen A.I. Dynamic model of protein behavior in water. Relationship between the kinetic and conformational properties of enzymes. – Biofizika, 1979b, Vol. 24, pp. 775-776.

Käiväräinen A.I. Book: "Dynamic behaviour of Proteins in Aqueous Medium and their Functions". L., 1980, Nauka.

Käiväräinen A.I., Käiväräinen E.I. Effect of specific and non-specific agents on the stability of serum albumin. In book: "Tez. dokl. na konf. molodykh uchenykh In-ta biol. Karel. fil. AN SSSR". Petrozavodsk, 1978, pp. 43-47.

Käiväräinen A.I., Käiväräinen E.I., Berestov V.A., Bolotnikov I.A., Franek F. Effect of hapten-antibody complex formation on the capacity to generate anisotropic fluctuations in the surrounding medium. – In book: "Tez. dokl. na IV Vsesoyuz. biokhim. s'ezde". L., 1979, Vol. 2, p. 8.

Käiväräinen E., Käiväräinen E., Franek F., Ol'sovskaya Z. Effect of hapten binding on the interaction between antibody and water. Concept of fluctuating cavities. Immunol. Lett., 1981, Vol. 3, pp. 323-327.

Käiväräinen A.I., Nezlin R.S. Conformational changes in antibodies during formation of specific complexes. – In book: "Konformatsionnye ismeneniya biopolimerov v rastvorakh". Tbilisi, 1975, pp. 253-261.

Käiväräinen A.I., Nezlin R.S. Spin-label approach to conformational properties of immunoglobulins. – Immunochemistry, 1976, Vol. 13, pp. 1001-1008.

Kärväräinen A.I., Nezlin R.S. Evidence for mobility of immunoglobulin domains obtained by spin-label method. – Biochem. Biophys. Res. Comm., 1976, Vol. 63, pp. 270-276.

Käiväräinen A.I., Nezlin R.S. The structure of immunoglobulins. In book: "Immunogenez i kletochnaya differentsirovka". M., 1978, pp. 10-40.

Käiväräinen A.I., Nezlin R.S., Likhtenshtein G.I., Misharin A. Yu., Vol'kenshtein M.V. Conformational changes of spin-labeled antibodies and antigens during the formation of specific complexes. – Mol. biol., 1973b, Vol. 7, pp. 760-768.

Käiväräinen A.I., Nezlin R.S., Volkenstein M.V. A study of the structural changes in antibodies and antigen during the formation of specific complexes by the spin-label method. – In book: "Tez. sekts. dokl. na IV Mezhdunar. biofiz. kongr." M., 1972a, Vol. 4, p. 227.

Käiväräinen A.I., Nezlin R.S., Volkenstein M.V. Spin-spin interaction between iminoxyl radicals located in antibody combining sites. – FEBS Lett., 1973, Vol. 35, pp. 306-311.

Käiväräinen A.I., Nezlin R.S. Volkenstein M.V. The distances between the iminoxyl radicals localized in active sites and the relative freedom of rotation of antibody subunits. – Mol. biol., 1974, Vol. 8, pp. 816-823.

Käiväräinen A.I., Rozhkov S.P., Franek F. Changes of anti-DNP antibodies and their Fab subunits; flexibility under influence of temperature and low concentration of D_2O. – Immunol. Lett. 1980. (In press).

Käiväräinen A.I., Rozhkov S.P., Sykulev Yu. K., Lavrent'ev V.V., Franek F. Change of the antidansyl antibodies flexibility and their Fab and Fc subunits; dynamic properties induced by hapten. – Immunol. Lett., 1981, Vol. 3, pp. 5-11.

Käiväräinen A.I., Timofeev V.P., Volkenstein M.V. Spin-spin interaction between iminoxyl radical and copper as an indicator of conformational changes in hemoglobin. – In book: "Tez. sekts. dokl. na IV Mezhdunar. biofiz. kongr." M., 1972b, Vol. 1, p. 81.

Käiväräinen A.I., Timofeev V.P., Volkenstein M.V. Study of conformational properties of hemoglobin by the method of two paramagnetic labels. – Mol. biol., 1972c, Vol. 6, pp. 875-882.

Käiväräinen A.I., Volkenstein M.V., Timofeev V.P. Applications of the spin-spin interaction between the iminoxyl radical and copper in the study of human hemoglobin. – In book: "Tez. dokl. na Vsesoyuz. konf. po voprosam sozdaniya novykh krovezamenitelei". M., 1971, p. 37.

Kahana L., Shalitin Y. Salt effects on the properties of α-chymotrypsin. I. Effects on the enzymatic activity of chymotrypsin. – Israel J. Chem., 1974, Vol. 12, pp. 573-589.

Kannan K.K., Liljas A., Waara I., Bergsten P. -C., Lövgren S., Strandberg B., Bengtsson U., Carlbam U., Fridborg K., Japur L., Pelef M. Crystal structure of human erythrocyte carbonic anhydrase C. VI. The three-dimensional structure at high resolution in relation to other mammalian carbonic anhydrases. – Cold Spring Harbor Sympos. Quantit. Biol., 1971, Vol. 36, pp. 221-231.

Kannan K.K., Notstrand B., Fridborg K., Lövgren S., Ohlsson A., Peteff M. Crystal structure of human erythrocyte carbonic anhydrase B. Three-dimensional structure at a nominal 2.2 Å resolution. Proc. Nat. Acad. Sci. USA, 1975, Vol. 72, pp. 51-55.

Karpeiskii M. Ya. Active sites of enzymes: stereochemistry and dynamics. – Mol. biol., 1976, Vol. 10, pp. 1197-1210.

Karplus M., MacCammon J.A. Protein structural fluctuations during a period of 100 p.s. – Nature, 1979, Vol. 277, p. 578.

Kauzmann W. Book: "Mechanism of Enzyme Action." – Johns Hopkins Univ. Press, 1954.

Kauzmann W. Some factors in the interpretation of protein denaturation. – Adv. Protein Chem., Vol. 14, pp. 1-63.

Kauzmann W., Moore K., Schultz D. Protein densities from X-ray crystallographic coordinates. – Nature, 1974, Vol. 248, pp. 447-449.

Käiväräinen A.I. Mechanism of formation of specific antibody-antigen complexes. – In book: "Tez. dokl. na nauch. konf. posv. 25-letiyu In-ta biol. Karel. fil. AN SSSR." Petrozavodsk, 1978a, pp. 102-107.

Karplus, M., Gelin, B.R., McCommon J.A. Internal dynamics of proteins. Short time and long time motions of aromatic sidechains in PTI. – Biophys. J., 1980, Vol. 32, p. 603-615, Discuss. 615-618.

Kendrew J.C., Dickerson R.E., Strandberg B.E., Hart R.G., Davies D.R., Phillips D.C., Shore V.C. Structure of myoglobin. – Nature, 1960, Vol. 185, pp. 422-427.

Keyes M.H., Falley M., Lumry R. Studies of heme proteins. II. Preparation and thermodynamic properties of sperm whale myoglobin. – J. Amer. Chem. Soc., 1971, Vol. 93, pp. 2035-2040.

Keyes M.H., Lumry R. Binding of anesthetics of proteins: linkage between the sixth-ligand site of heme iron and the nonpolar binding sites of myoglobin. – Federat. Proc., 1968, Vol. 27, pp. 895-897.

Khaloimov A.I. A spectroscopic study of the effect of the temperature on the nature of the interaction between proteins and water. – In book: "Molekulyarnaya fizika i biofizika vodnykh sistem". L., 1974, 2nd ed., pp. 115-122.

Khaloimov A.I., Zhukovskii A.P., Serova M.N. A spectroscopic study of the effect of dioxane on the denaturation of proteins in aqueous solutions. – In book: "Molekulyarnaya fizika i biofizika vodnykh sistem". L., 1976a, 3rd ed., pp. 55-62.

Khaloimov A.I., Zhukovskii A.P., Shut'ko S.V., Sidorova A.I. Effect of temperature on the nature of the reaction between contracting proteins and water. – In book: "Molekulyarnaya fizika i biofizika vodnykh sistem". L., 1976b, 3rd ed., pp. 46-53.

Khodakovskaya O.A., Yakovlev V.A. A study of blood serum choline esterase with the aid of spin-labeled inhibitor. – Izv. AN SSSR, Ser. biol., Vol. 4, pp. 570-579.

Khurgin Yu. I., Chernavskii D.S., Shnol' S.E. The molecule of proteinic enzyme as a mechanical system. – Mol. biol., 1967, Vol. 1, pp. 419-426.

Khurgin Yu. I., Roslyakov V. Ya., Klyachko-Gurvich A.L., Brueva T.R. Absorption of water vapor by α-chymotrypsin and lysozyme. – Biokhimiya, 1972, Vol. 37, pp. 485-492.

Khurgin Yu. I., Sherman F.B., Tusupkaliev U. Effect of temperature on the hydration of chymotrypsinogen A under dynamic conditions. – Mol. biol., 1978, Vol. 12, pp. 572-579.

Klots I.M. Water. – In book: "Gorizonty biokhimii". M., 1964, pp. 399-410.

Klotz I.M. Comparison of molecular structures of proteins: helix content; distribution of apolar residues. – Arch. Biochem. Biophys., 1970, Vol. 138, pp. 704-706.

Koenig S.H., Halenga K., Shporer M. Protein-water interaction studied by solvent H, ^{2}H and ^{17}O and magnetic relaxation. – Proc. Nat. Acad. Sci. USA, 1975, Vol. 72, pp. 2667-2671.

Koenig S.H., Schillinger W.E. Nuclear magnetic relaxation dispersion in protein solutions. I. Apotransferrin. – J. Biol. Chem. 1969a, Vol. 244, p. 3283-3289.

Koenig S.H., Schillinger W.E. Nuclear magnetic relaxation dispersion in protein solutions. II. Transferrin. – J. Biol. Chem. 1969b, Vol. 244, p. 6522-6526.

Kol'tover V.K., Kutlakhmedov Yu. A., Sukhorukov B.I. Application of paramagnetic probe method to the study of membrances of cell organellae. – DAN SSSR, 1968, Vol. 181, pp. 730-739.

Konev S.V., Lyskova T.I., Chernitskii E.A. Manifestation of the cooperative properties of yeast cells and their reactions to the salt composition of the environment. – Biofizika, 1972, Vol. 17, pp. 833-836.

Koshland D.E. Role of flexibility in the specificity control and evolution of enzymes. – FEBS Lett., 1976, Vol. 62, pp. E47-E52.

Koshland D.E. Comparison and nonenzymic and enzymic reaction velocities. – J. Theor. Biol., 1962, Vol. 2, pp. 75-82.

Koshland D.E., Nemethy J., Filmer D. Comparison of experimental data and theoretical models in proteins containing subunits. – Biochemistry, 1966, Vol. 5, pp. 365-385.

Koshland D.E., Jr., Neet K.E. The catalytic and regulatory properties of enzymes. – Ann. Rev. Biochem., 1968, Vol. 37, pp. 359-410.

Koshland M. Mediation of effector function by antibodies. – In. Report of a Workshop, Bethesda, USA, 1978, p. 26.

Kozlov L.V., Meshcheryakova E.A., Żavada L.L., Efremov E.S., Rashkovetskii L.G. Conformational states of pepsin as a function of the pH and of the temperature. – Biokhimiya, 1979, Vol. 44, pp. 338-349.

Kuntz I.D., Brassfield T.S. Hydration of macromolecules. II. Effects of urea on protein hydration. – Arch. Biochem. Biophys., 1971, Vol. 142, pp. 660-664.

Kuntz I.D., Brassfield T.S., Law G.D., Purcell G.V. Hydration of macromolecules. – Science, 1969, Vol. 163, pp. 1329-1333.

Kuntz I.D., Kauzmann W. Hydration of proteins and polypeptides. – Adv. Protein Chem., 1974, Vol. 28, pp. 239-345.

Kurganov B.I. Book: "Allostericheskie fermenty". M., 1978.

Kurganov B.I., Topchieva I.N., Lisovskaya N.P., Chebotareva N.A., Nataryus O. Ya. Effect of polyethyleneglycol on the association of muscle phosphorylase B. – Biokhimiya, 1979, Vol. 44, pp. 629-633.

Kurzaev A.B., Kvlividze V.I., Kiselev V.F. Special features of phase transition of water in disperse systems. – Biofizika, 1975, Vol. 20, pp. 533-534.

Kuzin A.M., Ermekova V.M. Radiosensitivity of allosteric center of phosphofructokinase. – DAN SSSR, 1969, Vol. 188, pp. 1163-1168.

Kuznetsov A.N. Book: "The spin probe method". M., 1976.

Kuznetsov A.R., Ebert D. The dependence of the character of rotational motion of nitroxyl radicals on their molecular size. – Chem. Phys. Lett., 1974, Vol. 25, p. 342-345.

Kuznetsov A.R., Wasserman A.M., Volkov A., Korst V. Determination of rotational correlation time of nitroxide radicals in a viscous medium. – Chem. Phys. Lett., 1971, Vol. 12, p. 103-106.

Kuznetsov A.N., Ebert B. A study of the conformational changes in a molecule of serum albumin in the pre-denaturation temperature range by the spin probe method. – Mol. biol., 1975, Vol. 9, pp. 407-414.

Kvlividze V.I. Study of the sorbed water by the NMR method. – In book: "Svyazannaya voda v dispersnykh sistemakh. " M., 1970, pp. 41-55.

Ladner J.E., Kitchell J.P., Honzatko K.B., Ke H.K., Wolz K.W., Kalb A.J., Ladner R.C., Lipscomb W.N. Gross quaternary changes in aspartate carbomoiltransferase are induced by the binding of N-(phosphonacetil)-L-aspartate: a 3.5 Å resolution study. Proc. Natl. Acad. Sci. USA 1982, V. 79, p. 125-3128.

Ladner R.S., Heidner E.J., Perutz M.F. The structure of horse methemoglobin at 2.0 Å resolution. – J. Mol. Biol., 1977, Vol. 114, pp. 385-414.

Lakowicz J.R. Fluorescence spectroscopic investigations of the dynamic properties of proteins, membranes and nucleic acids. – J. Biochem. Biophys. Meth., 1980, Vol. 2, p. 91-119.

Lancet D., Pecht I. Kinetic evidence for hapten-induced conformational transition in immunoglobulin MOPS460. – Proc. Nat. Acad. Sci. USA, 1976. Vol. 73, pp. 3549-3554.

Led J.J., Grant D.M., Horton W.J., Sunby F., Wilhelmsen K. Carbon-13 magnetic resonance study of structural and dynamic features in carbamylated insulines. – J. Amer. Chem. Soc., 1975, Vol. 97, pp. 5997-6007.

Leontovich M.A. Book: "Statisticheskaya fizika". M., 1944.

Levin C.J., Zimmerman S.B. Essential arginyl residues in mitochondrial adenosine triphosphatase. – J. Biol. Chem., 1976, Vol. 251, pp. 1775-1780.

Levine D.A., Moore O.R., Ratcliffe R.C., Williams R.J.P. Nuclear magnetic resonance studies of the solution structure of protein. – Intern. Rev. Biochem., 1979, vol. 24, p. 77–141.

Levison S.A., Hicks A.N., Portman A.J., Dandliker W.B. Fluorescence polarization and intensity kinetic studies of antifluorescein antibody obtained at different stages of the immune response. – Biochemistry, 1975, Vol. 14, pp. 3778-3786.

Levitt D.C. The mechanism of the sodium pump. – Biochim. et biophys. acta, 1980, Vol. 604, p. 321-345.

Lewin S. Book: "Displacement of water and its control of biochemical reaction." London-New York, 1974.

Lifshits I.M. Problems in statistical theory of biopolymers. – Zhrn. eksper. teoret. fiz., 1968, Vol. 55, pp. 2408-2420.

Lifshits I.M., Grosberg A. Yu. Phase diagram of the polymeric globule and the problem of auto-organization. – Zhurn. eksper. teoret. fiz., 1973, Vol. 65, pp. 2399-2420.

Likhtenshtein G.I. The role of the effects produced during the hydration of ionic groups in the stabilization of the globular structure of certain proteins. – In book: "Sostoyanie i rol' vody v biologicheshikh ob"ektakh". M., 1967, pp. 101-112.

Likhtenshtein G.I. Book: "Metod spinovykh metok v molekulyarnoi biologii". M., 1974.

Likhtenshtein G.I. Book: "Mnogoyarernye okislitel'no-vosstanovitel'nye metallo-fermenty". M., 1979.

Likhtenshtein G.I., Akhmedov Yu. D., Ivanov L.V., Krinitskaya L.A., Kokhanov Yu. V. A study of the lysozyme macromolecule by the spin-label method. – Mol. biol., 1974, Vol. 8, pp. 48-57.

Likhtenshtein G.I., Avilova T.V. Kinetic features of biological catalysis and the dynamic structure of enzymes. – Usp. sovr. biol., 1973, Vol. 75, pp. 26-45.

Likhtenshtein G.I., Frolov E.N., Belonogova O.V., Kharakhonycheva N.V., Khurgin Yu. I. A study of the dynamic structure of proteins and enzymes by the methods of spin labels and gamma-resonance labels. – In book: "Konformatsionyye izmeneniya biopolimerov v rastvorakh. Tbilisi, 1975, pp. 196-207.

Liljas A., Kanna K.K. Bergstein P.C., Warra I., Fridlorg K., Strandberg B. Crystal structure of human carbonic anhydrase C (2 Å). – Nature New Biol., 1972, Vol. 235, pp. 131-136.

Lim V.I., Ptitsyn O.B. The volume constancy of the hydrophobic nucleus in myoglobin and hemoglobin molecules. – Mol. biol., 1970, Vol. 4, pp. 372-382.

Linderström-Lang K., Schellmann J. Protein structure and enzyme activity. – In: Enzymes, New York, 1959, Vol. 1, pp. 443-510.

Lindskog S., Nyman P.O. Metal-binding properties of human erythrocyte carbonic anhydrases. – Biochem. Biophys. Acta, 1964, Vol. 85, pp. 462-474.

Lindström T.R., Koenig S.H. Magnetic field-dependent water proton spin-lattice relaxation rates of hemoglobin solutions and whole blood. – J. Magn. Res., 1974, Vol. 15, pp. 344-353.

Lipscomb W.N. Relationship of the three-dimensional structure of carboxypeptidase A to catalysis. – Tetrahedron, 1974, Vol. 30, pp. 1725-1732.

Lisyanskii L.I. Fluctuations and sound absorption in aqueous solutions of tert-butanol. – In book: "Molekulyarnaya fizika i biofizika vodnykh sistem"., L., 1974, pp. 63-73.

Lobyshev V.I., Kalinichenko L.P. Book: "Izotopnye effekty D_2O v biologicheshikh sistemakh. M., 1978.

Lobyshev V.I., Shnol' S.E. Factors which stabilize the conformations of protein molecules, as deduced from the isotopic effects of heavy water. – In book: "Konformatsionnye izmeneniya biopolimerov v rastvorakh." Tbilisi, 1975, pp. 323-329.

Lobyshev V.I., Tverdislov V.A., Fogel' Yu., Yakovenko L.V. Activation of K, Na-ATP-ase by small concentration of D_2O and its inhibition by large concentrations. – Biofizika, 1978, Vol. 23, p. 390.

Low P.S., Somero G.N. Activation volumes in enzymatic catalysis, their sources and modification by low molecular weight solutes. – Proc. Nat. Acad. Sci. USA, 1975a, pp. 3014-3018.

Low P.S., Somero G.N. Protein hydration changes during catalysis. A new mechanism of enzymatic enhancement and ion activation inhibition of catalysis. – Proc. Nat. Acad. Sci. USA, 1975b, Vol. 72, pp. 3305-3308.

Lucke, B. and McCutcheon, M. The living cell as an osmotic system and its permeability to water. – Physiological Reviews, 1932, Vol. 12, p. 68-139.

Lumry R., Biltonen R. Thermodynamic and kinetic aspects of protein conformation in physiological functions. – In book: "Structure and Stability of Biological Macromolecules". (Russian translation).

Lumry R., Rajender Sh. Enthalpy-entropy compensation phenomena in water solutions of proteins and small molecules; an ubiquitous property of water. – Biopolymers, 1970, Vol. 9, pp. 1125-1238.

Magonov S.N., Davydov R.M., Blyumenfel'd L.A., Vilu R.O., Arutyunyan A.M., Sharonov Yu. A. Absorption spectra and magnetic circular dichroism spectra of heme-containing proteins in non-equilibrium states. I. Hemoglobin and its derivatives. – Mol. biol., 1978a, Vol. 12, pp. 947-957.

Magonov S.N., Davydov R.M., Blyumenfel'd L.A., Vilu R.O., Arutyunyan A.M., Sharonov Yu. A. Absorption spectra and magnetic circular dichroism spectra of heme-containing proteins in non-equilibrium states. II. Myoglobin and its complexes. – Mol. Biol., 1978b, Vol. 12, pp. 1182-1189.

McCalley R.C., Schimshick E.I., McConnell N.M. The effect of flow rotational notion on paramagnetic resonance spectra. – Chem. Phys. Lett., 1972, Vol. 13, p. 115-120.

McCammon J.A., Gelin B.R., Karplus M. Dynamics of folded proteins. – Nature, 1977, Vol. 267, pp. 585-590.

McCammon J., Karplus M. Internal motions of antibody molecules. – Nature, 1977, Vol. 268, pp. 765-766.

McCammon J.A., Wolunes P.G., Karplus M. Picosecond dynamics of tyrosine side chains in proteins. – Biochemistry, 1979, Vol. 18, pp. 927-942.

McConnel H.M., Ogawa S., Horwitz A.F. Spin-labeled hemoglobin and the heme-heme interaction. – Nature, 1968, Vol. 220, pp. 787-788.

McCray J.A. Oxygen recombination kinetics following laser photolysis of hemoglobin. – Biochem. Biophys. Res. Comm., 1972, Vol. 47, pp. 187-193.

McDonald R.C., Steitz T.A., Engelman D.M. Yeast hexokinase in solution exhibits a large conformational change upon binding glucose or glucose-6-phosphate. – Biochemistry, 1979, Vol. 18, pp. 338-342.

McFarland B., McConnel H. Bent fatty acids in lecithin bilayers. – Proc. Nat. Acad. Sci. USA, 1971, Vol. 68, pp. 1274-1278.

Makarov A.M., Ivanov V.I., Vol'kenshtein M.V. Kinetic of the proteolysis of aspartate amino transferase at various states of the active site. – Mol. biol., 1974, Vol. 8, pp. 433-441.

Malenkov G.G., Minchenkova L.E., Minyat E.E., Shchelkina A.K., Ivanov V.I. The nature of the B→A transition of DNA in solution. – In book: "Konformatsionnye izmeneniya biopolimerov v rastvorakh". Tbilisi, 1975, pp. 10-16.

Maricic S., Ravilly A., Mildvan A.S. Proton relaxation measurements of dissolved and crystalline methemoglobin and metmyoglobin. In book: "Hemes and Hemoproteins". Chance, B., Estabrook, R.W., Yonetate, T. (Eds.). Academic Press, N.Y., 1966.

Markby J.L., Williams M.N., Jardetzky O. NMR studies of the structure and binding sites in enzymes. XII. A conformational equilibrium in staphylococcal nuclease involving a histidine residue. – Proc. Nat. Acad. Sci. USA, 1970, Vol. 65, pp. 645-650.

Markovich D.S. Conformational aspects of regulation of enzymatic activity. – In book: "Molekulyarnaya i kletochnaya biofizika". M., 1977, pp. 7-12.

Markovich D.S., Krapivinskii G.B., Neznaiko N.F. Studies of conformational mobility of the active site of D-glyceraldehyde-3-phosphate dehydrogenase. – Mol. Biol., 1974, Vol. 8, pp. 475-485.

Markus G., McClintock D.K., Castellani I. Ligand-stabilized conformations in serum albumin. – J. Biol. Chem., 1967, Vol. 242, pp. 4402-4408.

Martin R.G. The first enzyme in histidine biosynthesis: the nature of feedback inhibition by histidine. J. Biol. Chem., 1963, Vol. 238, pp. 257-268.

Matthews B.W., Weaver L.H., Kester W.H. The conformation of thermolysin. – J. Biol. Chem., 1974, Vol. 249, pp. 8030-8044.

Mazhul' V.M., Chernitskii E.A., Konev S.V. Conformational transitions of native proteins in solution and in cells. – Biofizika, 1970, Vol. 15, pp. 5-11.

Mazurenko Yu. T. Luminescence polarization of complex molecules during light quenching. – Opt. in spektr., 1973, Vol. 35, pp. 234-245.

Medvedeva N.V., Blyumenfel'd L.A., Ruuge E.K. The relaxational nature of changes in the properties of urease and myosin in solutions during jumpwise changes in the temperature or in the pH. – In book: "Konformatsionnye izmeneniya bio-olimeriv v rastvorakh." Tbilisi, 1975a, pp. 335-342.

Medvedeva N.V., Ruuge E.K., Blyumenfel'd L.A. Temperature dependence of EPR spectra of spin-labeled myosin and its ADP complex. – Biofizika, 1975b, Vol. 20, pp. 26-31.

Metzger H. The effect of antigen on antibodies; recent studies. – Contemp. Topis Mol. Immunol., 1978, Vol. 7, pp. 150-170.

Mockrin S.C., Byers L.D., Koshland D.E. Subunit interaction in yeast glyceraldehyde-3-phosphate dehydrogenase. – Biochemistry, 1975, Vol. 14, pp. 5428-5437.

Moffat J.K. Spin-labeled hemoglobins: a structural interpretation of EPR spectra based on X-ray analysis. – J. Mol. Biol. 1971, Vol. 55, p. 135.

Moffat J.K., Simon S.R., Konnisberg W.H. Structure and functional properties of chemically modified horse hemoglobin. III. Functional consequences of structural alterations for the molecular basis of cooperativity. – J. Mol. Biol., 1971, Vol. 58, pp. 89-101.

Molday R.S., Englander S.W., Kallen E.G. Primary structure effects on peptide group hydrogen exchange. – Biochemistry, 1972, Vol. 11, pp. 150-158.

Monod J., Wyman J., Changeux P. On the nature of allosteric transitions: a plausible model. – J. Mol. Biol., 1965, Vol. 12, pp. 88-118.

Morgan R.S., Peticolas W.L. Frequency-dependent diffraction from enzymatic breathing modes. – Intern. J. Peptide Prot. Res., 1975, Vol. 7, pp. 361-365.

Moult J., Yonath A., Praub W., Smilansky A., Podjarny A., Rabinovich D., Saya A. The structure of triclinic lysozyme at 2.5 Å resolution. – J. Mol. Biol., 1976, Vol. 100, pp. 179-195.

Mrevlishvili G.M., Khutsishvili V.G., Monosalidze D.R., Dzhaparidze G. Sh. A study of the state of water in collagen fibers of native type by calorimetry and NMR. – In book: "Konformatsionnye izmeneniya biopolimerov v rastvorakh". Tbilisi, 1975, pp. 277-305.

Mrevlishvili G.M., Sharimanov Yu. G. Studies of the hydration of intromolecular fusion of collagen by NMR and by calorimetry. – Biofizika, 1978, Vol. 23, pp. 717-719.

Muirhead H., Perutz M.F. Structure of hemoglobin. – Nature, 1963, Vol. 199, pp. 633-638.

Munster A. Theory of fluctuations. In book: "Thermodynamics of Irreversible Processes". – (Russian translation from the German).

Myer Y.P. Conformation of cytochromes. III. Effect of urea temperature, extrinsic ligands and pH variation on the conformation of horse heart ferricytochrome C. – Biochemistry, 1968, Vol. 7, pp. 765-776.

Nakamura T., Sugita Y., Hashimoto K., Yomyama Y., Pisciotta A. Thermodynamics of equilibria of hemoglobin M Milwaukee-1 and Sasketon and isolated chain of hemoglobin A with carbon monoxide. – Biochem. Biophys. Res. Comm., 1976, Vol. 70, pp. 567-572.

Nakanishi M., Tsuboi M. Determination of the number of hydrogen bonds in a protein molecule. – Bull. Chem. Soc. Japan, 1974, Vol. 74, pp. 305-307.

Nedev K.N., Khurgin Yu. I. A study of the surface layer of a protein globule. Hydration of α-chymotrypsin molecule. – Mol. biol., 1975, Vol. 9, pp. 761-767.

Nedev K.N., Volkova R.I., Khurgin Yu. I., Chernavskii D.S. Modeling the structure of the protein globule. – Biofizika, 1974, Vol. 19, pp. 983-986.

Nelson D.J., Opella S.J., Jardetzky O. ^{13}C NMR study of molecular motions and conformational transitions in muscle calcium-binding parvalbumins. – Biochemistry, 1976, Vol. 15, pp. 5552-5560.

Nemethy G., Scheraga H.A. Structure of water and hydrophobic bonding in proteins. IV. The thermodynamic properties of liquid deuterium oxide. – J. Chem. Phys., 1964, Vol. 41, pp. 680-689.

Newcower M.D., Lewis B.A., Quiocho F.A. The radius of gyration of L-arabinose binding protein decreases upon binding of ligand. – J. Biol. Chem. 1981, V. 256, p. 13218–13222.

Nezlin R.S. Book: "Stroenie i biosintez antitel". M., 1972.

Nezlin R.S., Timofeev V.P., Sykulev Yu. K., Zurabyan S.E. Spin labeling in immunoglobulin carbohydrates. – Immunochemistry, 1978, Vol. 15, pp. 143-144.

Nezlin R.S., Zagyanskii Yu. A., Käiväräinen A.I., Stefani D.V. Properties of myeloma immunoglobulin. E. Chemical, fluorescence and spin-labeled studies. – Immunochemistry, 1973, Vol. 10, pp. 681-688.

Nisonoff A., Hopper J.E., Spring S.B. The antibody molecule. New York,

Oakenfull D., Fenwick D.E. Hydrophobic interactions in deuterium oxide. – Austral. J. Chem., 1975, Vol. 152, pp. 849-855.

Ogata R.T., McConnel H.M. Mechanism of cooperative oxygen binding to hemoglobin. – Proc. Nat. Acad. Sci. USA, 1968, Vol. 61, pp. 405-406.

Ogawa S., Patel D.J., Simon S.R. Proton magnetic resonance study of the switch between the two quaternary structures in high-affinity hemoglobins in the deoxy state. – Biochemistry, 1974, Vol. 13, pp. 2001-2006.

Ogawa S., Shulman R.G. High-resolution NMR spectra of hemoglobin. III. The half-ligated state and allosteric interaction. – J. Mol. Biol., 1972, Vol. 70, pp. 315-336.

Ohta Y., Gill T.J., Leung C.S. Volume changes accompanying the antibody-antigen reactions. – Biochemistry, 1970, Vol. 9, pp. 2708-2713.

O'Leary M.N., Brummund W. pH jump studies of glutanate decarboxylase. Evidence for a pH-dependent conformational change. – J. Biol. Chem., 1974, Vol. 249, pp. 3737-3745.

Olson J.S., Andersen M.E., Gibson Q.M. The dissociation of the first oxygen molecule from some mammalian oxyhemoglobins. – J. Biol. Chem., 1971, Vol. 246, pp. 5919-5923.

Olson J.S., Gibson Q.H. The reaction of *n*-butyl isocyanide with human hemoglobin. II. The ligand-binding properties of the α- and β- chains within deoxyhemoglobin. –J. Biol. Chem., 1972, Vol. 247, pp. 1713-1726.

O'Nell J., Adami L. The oxygen isotope partition function ratio of water and the structure of liquid water. – Phys. Chem., 1969, Vol. 73, pp. 1553-1560.

Ovchinnikov Yu. A., Ivanov V.T., Shkrob A.M. Book: "Membranno-aktivnye kompleksony". M., 1974.

Padlan E.A., Segal D.M., Cohen G.H., Davies D.R. The three-dimensional structure of the antigen-binding site of McPO 603 protein. – In: The immune system, genes, receptors, signals. New York, 1974, pp. 7-14.

Patel D.J., Kampa L., Schulman R.G., Yamane T., Fujiwara M. Proton NMR studies of hemoglobin in H_2O. – Biochem. Biophys. Res. Comm., 1970, Vol. 40, pp. 1224-1230.

Pchelin V.A., Izmailova V.N., Bol'shova G.P. Effect of the solubilization of benzene on the catalytic properties of pepsin. – DAN SSSR, 1962, Vol. 142, pp. 950-953.

Pchelin V.A., Izmailova V.N., Ochurova K.T. Certain relationships governing solubilization in protein systems. – DAN SSSR, 1958, Vol. 123, pp. 505-508.

Pchelin V.A., Volynskaya A.V., Izmailova V.N. Internal hydrophobic structure of proteins from the data for the solubilization of carbohydrates. – DAN SSSR, 1969, Vol. 186, pp. 139-141.

Permyakov E.A. Luminescence studies of structural dynamics of proteins at low temperatures. – Candidate's summary of thesis. Pushchino, 1975.

Permyakov E.A. Cooperative freezing of intramolecular mobility of proteins. – In book: "Ravnovesnaya dinamika nativnoi struktury belka". Pushchino, 1977, pp. 83-98.

Permyakov E.A., Burshtein E.A. Relaxation processes in frozen aqueous solutions of proteins; temperature dependence on fluorescence parameters. – Stud. biophys., 1975, Vol. 51, pp. 91-103.

Perutz M.F. Structure and function of hemoglobin. II. Some relation between chain configuration and amino acid sequence. – J. Mol. Biol., 1965, Vol. 13, pp. 669-678.

Perutz M.F. Relation between structure and sequence of hemoglobin. – Nature, 1968, Vol. 194, pp. 914-917.

Perutz M.F. Regulation of oxygen affinity of Lemoglobin: influence of structure of the globin on the heme iron. – Annu. Rev. Biochem., Vol. 46, Palo Alto, Calif., 1979, p. 327–386.

Perutz M.F. Stereochemistry of cooperative effects in hemoglobin. – Nature, 1970, Vol. 228, pp. 726-739.

Perutz M.F. The role of bound water in hemoglobin and myoglobin. – Biosystems, 1977, Vol. 8, pp. 261-263.

Perutz M.F., Fersht A.R., Simon S.R., Roberts G.C. Influence of globin structure on the state of the heme. II. Allosteric transitions in methemoglobin. – Biochemistry, 1974a, Vol. 13, pp. 2174-2186.

Perutz M.F., Heidner E.J., Ladner J.E., Beetlestone J.G., Ho C., Slade E.F. Influence of globin structure on the state of the heme. III. Changes in heme spectra accompanying allosteric transitions in methemoglobin and their implications for heme-heme interaction. – Biochemistry, 1974b, Vol. 13, pp. 2187-2200.

Perutz M.F., Ladner J.E., Simon S.R., Ho C. Influence of globin structure on the state of the heme. I. Human deoxyhemoglobin. – Biochemistry, 1974c, Vol. 13, pp. 2163-2173.

Perutz M.F., Lehmann H. Molecular pathology of human hemoglobin. – Nature, 1968, Vol. 219, pp. 902-909.

Perutz M.F., Rossman M.G., Cullis A.F., Muirhead H., Will G., North A.C.T. Structure of hemoglobin. Nature, 1960, Vol. 185, pp. 416-422.

Perutz M.F., Sanders J.K.M., Chenery D.H., Noble R.W., Pennely R.R., Fung L.W.M., Ho Co., Gianini I., Pörschke D., Winkler H. Interaction between the quaternary structure of the globin and the spin state of the heme in ferric mixed spin derivatives of hemoglobin. – Biochemistry, 1978, Vol. 17, pp. 3640-3652.

Perutz M.F., Ten Eyck L.F. Stereochemistry of cooperative effects in hemoglobin. – Cold Spring Harbor Sympos. Quantit. Biol., 1971, Vol. 36, pp. 295-310.

Phillips D. The three-dimensional structure of enzyme molecules. In book: "Molecules and Cells". (Russian translation).

Phillips D.C., Levitt M. Peptides, polypeptides and proteins. Ed. E.R. Blout et al. New York, 1974.

Pickover C.A., McKey D.B., Engelman D.M., Steitz T.A. Subtrate binding closes the cleit between the domains of yeast phosphoglycerate kinase. – J. Biol. Chem., 1979, Vol. 254, p. 11323-11329.

Pilz I., Kratky O., Karush F. Changes in the conformation of rabbit IgG antibody caused by the specific binding of a hapten. X-ray small-angle studies. Europ. J. Biochem., 1974, Vol. 41, pp. 91-96.

Pilz I., Kratky O., Licht A., Sela M. Shape and volume of anti-poly(d-alanyl) antibodies in the presence and absence of tetra-d-alanine as followed by small-angle X-ray scattering. – Biochemistry, 1973, Vol. 12, pp. 4998-5005.

Poljak R.J. X-ray diffraction studies of immunoglobulins. – Adv. Immunology, 1975, Vol. 21, pp. 1-33.

Polson A.A. A theory for the displacement of proteins and viruses with polyethylene glycol. – Prep. Biochem., 1977, Vol. 7, pp. 129-154.

Polyanovskii O.L., Zagyanskii Yu. A., Tumerman L.A. Molecular parameters and the structure aspartate transaminase and its subunits6– Mol. biol., 1970, Vol. 4, pp. 458-470.

Poo M., Cone R. Lateral diffusion of rhodopsin in the photoreceptor membrane. – Nature, 1974, Vol. 247, pp. 438-441.

Porter R.R. Structural studies of immunoglobulins. – Science, 1973, Vol. 180, pp. 713-720.

Preval C., Fougereau M. Specific interaction between V_H-V_L regions of human monoclonal immunoglobulins. – J. Mol. Biol., 1976, Vol. 102, pp. 657-678.

Privalov P.L. Stability of proteins. Small globular proteins. – Adv. Prot. Chem., 1979, Vol. 33, p. 167-241.

Privalov P.L. Stability of proteins. Proteins which do not present a single cooperative system. – Adv. Prot. Chem. 1982, Vol. 35, p. 1-104.

Privalov P.L. Stability of proteins. Small globular proteins. – Adv. Prot. Chem. 1979, Vol. 33, p. 167-241.

Privalov P.L. Study of thermal transformation of procollagen. I. Enthalpy of denaturation of procollagen with varying amino acid content. – Biofizika, 1968, Vol. 13, pp. 955-963.

Privalov P.L. Calorimetric studies of biopolymer solutions. – Itogi nauki i tekhn. Ser. mol. biol., M., 1975, Vol. 6, pp. 7-33.

Privalov P.L., Khechinashvili N.N. A globular protein structure: a calorimetric study. – J. Mol. Biol., 1974, Vol. 86, pp. 665-684.

Privalov P.L., Serdyuk I.N., Tiktopulo E.I. Thermal conformational transformations of tropocollagen. II. Viscometric and light-scattering studies. – Biopolymers, 1971, Vol. 10, pp. 1777-1794.

Privalov P.L., Tiktopulo E.I. A study of a thermal trans-conformation of procollagen. II. Heat absorption in the pre-denaturation temperature range in a solution of procollagen in the presence of salts. – Biofizika, 1969, Vol. 14, pp. 20-27.

Privalov P.L., Tiktopulo E.I. Thermal conformational transformations of tropocollagen. I. Calorimetric study. – Biopolymers, 1970, Vol. 9, pp. 127-139.

Prusakov V.E., Belonogova O.V., Frolov E.N., Stukan R.A., Gol'danskii V.I., Berg A.I., Mei L. A study of the dynamics of aqueous protein systems by γ-resonance spectroscopy. – Mol. biol., 1979, Vol. 13, pp. 443-449.

Ptitsyn O.B. Interdomain mobility in proteins and its probable functional role. – FEBS Lett., 1978, Vol. 93, pp. 1-4.

Putman F.W. Immunoglobin structure: variability and homology. – Science, 1969, Vol. 163, pp. 633-640.

Radchenko I.V., Shestakovskii F.K. X-ray scattering in water-methanol mixtures. – Zhurn. fiz. khim., 1955, Vol. 29, pp. 1456-1461.

Randall J., Middendorf H.D., Crespi H.L., Taylor A.D. Dynamics of protein hydration by quasi-elastic neutron scattering. – Nature, 1978, Vol. 276, pp. 636-638.

Rao P.B., Bryan W.P. Measurement of strongly held water of lysozyme. – J. Mol. Biol., 1975, Vol. 97, pp. 119-122.

Ratnikova N.V., Ivanov V.V., Moskvich Yu. N. Particular features of nuclear spin relaxation of the protons of adsorbed water in biological matrices. – Biofizika, 1975, Vol. 20, pp. 398-402.

Rifkind D.M. Hemoglobin and myoglobin. – In book: "Neorganicheskaya biokhimiya". M., 1978, Vol. 2, pp. 256-338. (Russian translation).

Romanovskii Yu. M., Tikhomirova N.K., Khurgin Yu. I. An electromechanical model of the enzyme-substrate complex. – Biofizika, 1979, Vol. 24, pp. 442-447.

Rosen O.M., Copeland P.L., Rosen S.M. The dissociation of O-fructose 1,6-diphosphatase by adenosine-5'-phosphate. – J. Biol. Chem., 1967, Vol. 242, pp. 2760-2763.

Roslyakov V. Ya., Khurgin Yu. I. A study of solid-phase hydrolysis of cinnamal-α-chymotrypsin. – Biokhimiya, 1972, Vol. 37, pp. 493-497.

Rossi-Bernardi L., Roughton F.J.W. The specific influence of carbon dioxide and carbonate compounds of buffer power and Bohr effects in human hemoglobin solutions. – J. Physiol., 1967, Vol. 189, pp. 1-29.

Roughton F.J.W. Oxygen in the animal organism. – I.U.B. Sympos. Ser., 1964, Vol. 31, p. 5.

Rupley J.A. Comparison between protein structures in crystal and in solution. – In book: "Structure and Stability of Biological Macromolecules". (Russian translation).

Salemme F.R. Structure and functions of cytochromes C. – Ann. Rev. Biochem., 1977, Vol. 46, pp. 219-229.

Samoilov O. Ya. Book: "Struktura vodnykh rastvorov elektrolitov i gidratatsiya ionov", M., 1957.

Sarvazyan A.N. Book: "Spetsificheskie mekhanizmy biologicheskogo deistva impul'snogo ul'trazvuka, svyazannye s dinamikoi biologicheskikh sistem". M., 1977, pp. 107-113.

Saviotti M.L., Galley W.C. Room temperature phosphorescence and the dynamic aspects of protein structure. – Proc. Nat. Acad. Sci. USA, 1974, Vol. 71, pp. 4154-4158.

Schiffer M., Girling R.L., Ely K.R., Edmindson A.B. Structure of a λ-type Bence-Jones protein at 3.5Å resolution. – Biochemistry, 1973, Vol. 12, pp. 4620-4626.

Schlessinger J., Steinberg I.Z., Givol D., Hochman J. Subunit interaction in antibodies and antibody fragments studies by circular polarization of fluorescence. – FEBS Lett., 1975a, Vol. 52, pp. 231-235.

Schlessinger J., Steinberg I.Z., Givol D., Hochman J., Pecht I. Antigen-induced conformational changes and their Fab fragments studied by circular polarization of fluorescence. – Proc. Nat. Acad. Sci., USA, 1975b, Vol. 72, pp. 2775-2779.

Schoenborn B.P. Structure of alkaline metmyoglobin-xenon complex. – J. Mol. Biol., 1969, Vol. 45, pp. 297-303.

Schoenborn B.P. A neutron diffraction analysis of myoglobin. III. Hydrogen-deuterium bonding in side chains. – Cold Spring Harbor Sympos. Quantit. Biol., 1972, Vol. 36, pp. 569-575.

Schulman R.S., Withrick K., Yamane T., Patel D.J., Blumberg W.E. NMR determination of ligand-induced conformational changes in myoglobin. – J. Mol. Biol., 1970, Vol. 53, pp. 143-157.

Shakhparonov, M.I. Book: "Vvedenie v sovremennuyu teoriyu rastvorov." M., 1976.

Shashkova E.A., Vichutunskii A.A., Zaslavskii B. Yu., Usubalieva L.P., Khorlin A.Ya., Shul'man M.L. A study of the effect of pH and temperature on the thermodynamic parameters of complex formation between lysozyme and chitin oligosaccharides. – Bioorg. khim., 1975, Vol. 1, pp. 965-972.

Shimanouchi T. Structural studies of macromolecules by spectroscopic methods. Ed. K.J. Ivin. New York, 1976.

Shimanovskii N.L., Stepanyants A.U., Lezina V.P., Ivanov L.V., Likhtenshtein G.I. Studies of the microstructure of the water-protein layer of spin-labeled preparations of lysozyme by the method of magnetic relaxation. – Biofizika, 1977, Vol. 22, pp. 811-815.

Shimshick E.J., McConnel H.M. Rotational correlation time of spin-labeled α-chymotrypsin. – Biochem. Biophys. Res. Comm., 1972, Vol. 46, pp. 321-327.

Shnol' S.E. Book: "Fiziko-khimicheskie faktory evolyutsii". M., 1979.

Shnol' S.E., Chetvertikova E.P., Rybina E.P. Macrovolume-synchronous conformational vibrations in preparations of proteins of actomyosin complex and in solutions of creatine kinase. – In book: "Molekulyarnay i kletochnaya biofizika". M., 1977, pp. 79-93.

Sidewell A.M., Munch R.R. Jr., Barron E.S.G., Rogness T.R. – J. Biol. Chem., 1937, Vol. 123, p. 335–

Singler P.B., Blow D.H., Mattheus B.W., Henderson R. Structure of crystalline α-chymotrypsin. II. A preliminary report including a hypothesis for the activation mechanism. – J. Mol. Biol., 1968, Vol. 35, pp. 143-164.

Slayter H.S., Lowey S. The substructure of the myosin molecule as visualized by the electron microscope. – Proc. Nat. Acad. Sci. USA, 1967, Vol. 58, pp. 1611-1617.

Sloan D.L., Velick S.F. Protein hydration ranges in the nicotinamide adenine dinucleotide complexes dehydrogenase of yeast. – J. Biol. Chem., 1973, Vol. 248, pp. 5419-5427.

Smith R.J., Duerksen J.D. Glycerol inhibition of purified and chromatin-associated mouse liver hepatoma RNA polymerase II activity. – Biochem. Biophys. Res. Comm., 1975, Vol. 67, pp. 916-923.

Snyder G.H., Rovan R., Karplus S., Sykes B.D. Complete tyrosine assignments in the high field R NMR spectrum of the bovine pancreatic trypsin inhibitor. – Biochemistry, 1975, Vol. 14, pp. 3765-3776.

Sokolov N.D. Problems in theory of the hydrogen bond. – In book: "Vodorodnaya svyaz' ". M., 1964, pp. 7-39.

Spirin A.S., Gavrilova L.P. Book: "Ribosoma". M., 1971.

Squire P.G., Rimmel M.E. Hydrodynamics and protein hydration. – Arch. Biochem. and Biophys., 1979, Vol. 196, p. 165–177.

Steitz T.A., Bennet W., Anderson C., McDonald R., Pickover C., Engelman D. Closing clefts and control specificity and mechanism of kinases. – In: Abstr. 11th Congr. Biophys. Toronto, 1979, p. 144.

Steitz T.A., Flatterick M., Anderson W.F., Anderson C.M. High-resolution X-ray structure of yeast hexokinase, an allosteric protein exhibiting a non-symmetric arrangement of subunits. – J. Mol. Biol., 1976, Vol. 104, pp. 197-222.

Sukhorukov B.I., Likhenshtein G.I. The kinetics and the mechanism of denaturation of biopolymers. Biofizika, 1965, Vol. 10, pp. 935-942.

Sutton B.J., Gettins P., Givol D., Wain-Hobson S., Willan K.J., Dwek R.A. The cross architecture of an antibody-combining site as determined by spin-label mapping. – Biochem. J., 1977, Vol. 165, pp. 177-197.

Suurkuusk J. Specific heat measurements on lysozyme, chymotrypsinogen and ovalbumin in aqueous solution and in the solid state. – Acta Chem. Scand., 1974, Vol. 28, pp. 409-417.

Swain C.G., Bader R.T.W. The nature of the structure difference between light and heavy water and the origin of the solvent isotope effect. – Tetrahedron, 1960, Vol. 10, pp. 182-190.

Sykulev, Yu. K., Käiväräinen A.I., Rozhkov S.P., Nezlin R.S. IgE dynamical properties changes after disulfide bonds reduction. Spin-label method approach. – Immunol. lett., 1980, Vol. 1, pp. 245-248.

Sykulev Yu. K., Käiväräinen A.I., Rozhkov S.P., Nezlin R.S. A spin-label study of IgE after mild reduction. – In: Abstr. 4th Intern. Congr. Immunol., Paris, 1980.

Syrnikov Yu. P. Certain features of biopolymer-water interaction. – In book: "Molekulyarnaya fizika i biofizika vodnykh sistem. L., 1974, pp. 23-28.

Tabagya M.I. NMR study of the mechanism of electron transfer in cytochrome. – Candidate's summary of thesis. M., 1977.

Takano T. Structure of myoglobin refined at 2.0 Å resolution. – J. Mol. Biol., 1977, Vol. 110, pp. 533-584.

Takano T., Kallai O.B., Swanson R., Dickerson R.E. The structure of ferrocytochrome C at 2.45 Å resolution. – J. Biol. Chem., 1973, Vol. 248, pp. 5234-5255.

Takano T., Swanson R., Kallai O.B., Dickerson R.E. Conformational changes upon reduction of cytochrome C. – Cold Spring Harbor Sympos. Quantit. Biol., 1971, Vol. 36, pp. 397-403.

Takizava T., Hayashi S. Endothermic crystallization of lysozyme to the orthorhombic crystal for the aqueous solution at $35°C$. – J. Phys. Soc. Japan, 1976, Vol. 40, pp. 299-300.

Tanford C. Physical chemistry of marcomolecules. New York - London, 1967.

Tauma I., Imai K., Shimizu K. Analysis of oxygen equilibrium of hemoglobin and control mechanism of organic phosphates. – Biochemistry, 1973, Vol. 12, pp. 1491-1498.

Thompson B.C., Waterman M.R., Cottam G.L. Evaluation of the water environments in deoxygenated side cells by longitudinal and transverse water proton relaxation rates. – Arch. Biochem. Biophys., 1975, Vol. 165, pp. 193-200.

Tiktopulo E.I., Privalov P.L. A study of temperature-produced changes in ribonuclease. – Biofizika, 1975, Vol. 20, pp. 778-782.

Tiktopulo E.I., Privalov P.L., Andreeva A.P., Aleksandrov V. Ya. The mobility of the collagen structure and the temperature adaptation of animals. – Mol. biol., 1979, Vol. 13, pp. 619-624.

Timchenko A.A., Zurupa S.N. Largescale changes in yeast phosphoglyceratekinase upon substrate binding. – Biophizika 1982, V. 27, p. 1017-1021.

Timofeev V.P., Dudin I.V., Sykulev Yu. K., Nezlin R.S. Rotational correlation times of IgG and its fragments spin-labeled at carbohydrate or protein moieties. – FEBS Lett., 1978, Vol. 89, pp. 191-195.

Timofeev V.P., Dudich I.V., Sykulev Yu. K., Nezlin R.S., Franek F. Slow conformational change in anti-dansyl antibody as a consequence of hapten binding. – FEBS Lett., 1979, Vol. 102, pp. 103-106.

Timofeev V.P., Polyanovskii O.L., Vol'kenshtein M.V., Likhtenshtein G.I. Mobility of the paramagnetic label bound by aspartate aminotransferase. – Biochim. Biophys. Acta, 1970, Vol. 220, pp. 357-360.

Troitskii G.V., Zavyalov V.P., Kiryukhin I.F. Temperature dependence of protein conformation using IgG fractions with different isoelectric points and myeloma IgG. – Biochim. Biophys. Acta, 1973, Vol. 223, pp. 53-61.

Tulinsky A., Vandlen R.L., Morimoto C.N., Mani N.W., Wright L.H. Variability in the tertiary structure of α-chymotrypsin at 2.8 Å resolution. – Biochemistry, 1973, Vol. 12, pp. 4185-4192.

Tumerman L.A., Nezlin R.S., Zagyanskii Y.A. Increase of the rotational relaxation time of antibody molecule after complex formation with dansyl hapten. – FEBS Lett., 1972, Vol. 19, pp. 290-292.

Valentine R.C., Green N.M. Electron microscopy of antibody-hapten complex. – J. Mol. Biol., 1967, Vol. 27, pp. 615-619.

Vas M., Boross L. An approach for the determination of equilibrium constant of structural mobility. – Europ. J. Biochem., 1974, Vol. 43, pp. 137-244.

Velick S.F., Baggott J.P., Sturtevant J.M. Thermodynamics of nicotinamide adenine dinucleotide addition to the glyceraldehyde-3-phosphate dehydrogenases of yeast and of rabbit skeletal muscle. An equilibrium and calorimetric analysis over a range of temperatures. – Biochemistry, 1971, Vol. 10, pp. 779-786.

Vengerova T.I., Rokhlin O.V., Nezlin R.S. Retention of complete antigenic activity of L-chains of rat IgG after their splitting into halves. – Immunochemistry, 1972, Vol. 9, pp. 413-420.

Villanceva G.B., Batt C.W., Brunner W. Effects of doxane on thrombin and trypsin activities. – J. Chem. Soc. Chem. Comm., 1974, Vol. 19, pp. 766-767.

Volkenstein M.V. Molecular biology, problems and prospects. – In book: "Sbornik k 70-letiyu akad. V.S. Engel'gardta". M., 1964, pp. 178-188.

Volkenstein M.V. Molecular theory of muscle contraction. – Mol. biol., 1969, Vol. 3, pp. 856-868.

Volkenstein M.V. Book: "Molekulyarnaya biofizika". M., 1975.

Volkenstein M.V. Conformational theory of polymers and biopolymers. – Mol. biol., 1977, Vol. 11, pp. 1269-1273.

Volkenstein M.V. Book: "Obshchaya biofizika". M., 1978.

Volynskaya A.V., Izmailov V.N., Pchelin V.A. The hydrophobic structure of globular proteins. – Biofizika, 1973, Vol. 18, pp. 210-215.

Vuk-Pavlovic S., Blatt Y., Glaudemans C.P.J., Lancet D., Pecht I. Hapten-linked conformational equilibria in immunoglobulins XRPC-24 and J-539 observed by chemical relaxation. – Biophys. J., 1978, Vol. 24, pp. 161-170.

Vuk-Pavlovic S., Isenman D.E., Elgavish G.A., Licht A., Gafni A., Pecht I. Hapten-induced structural changes in rabbit immunoglobulin G with specifically mercurated inter-heavy-chain disulfide. – Biochemistry, 1979, Vol. 18, pp. 1127-1129.

Vuks M.F. Book: "Rasseyanie sveta v gazakh, zhidkostyakh i rastvorakh." L., 1977.

Wagner G., Wütrich R. Structural interpretation of the amide proton exchange in the basic pancreatic trypsin inhibitor and related proteins. – J. Mol. Biol., 1973, Vol. 134, p. 75–94.

Walker E.J., Ralston G.B., Darvey I.G. An allosteric model for ribonuclease. – Biochem. J., 1975, Vol. 147, pp. 425-433.

Wallach D. Effect of internal rotation on angular correlation functions. – J. Chem. Phys., 1967, Vol. 47, p. 5258–5268.

Wang J.H., Robinson C.V., Edelman I.S. Self-diffusion and structure of liquid water. III. Measurement of the self-diffusion of liquid water H^2, H^3 and O^{18} as traces. – J. Amer. Chem. Soc., 1953, Vol. 75, pp. 466-473.

Ward R.L. ^{35}Cl NMR studies of a zinc metalloenzyme carbonic anhydrase. – Biochemistry, 1969, Vol. 8, pp. 1879-1883.

Ward R.L. Zinc environmental differences in carbonic anhydrase isozymes. – Biochemistry, 1970, Vol. 9, pp. 2447-2454.

Warren J.C., Stowring L., Morales M.E. The effect of structure-disrupting ions on the activity of myosin and other enzymes. – J. Biol. Chem., 1966, Vol. 241, pp. 309-316.

Warren S.S., Edwards B.F.P., Evans D.R., Wiley D.C., Lipscomb W.N. Aspartate transcarbomoylase from E. coli: Electron density at 5.5 Å resolution. Proc. Nat. Acad. Sci. USA, 1973, Vol. 70, pp. 1117-1121.

Weber B.H., Storm M.C., Boyer P.D. An assessment of the exchangeability of water molecules in the interior of chymotrypsinogen in solution. – Arch. Biochem. Biophys., 1974, Vol. 153, pp. 1-6.

Wee V.T., Fedlmann R.J., Tanis R.J., Chignell C.F. A comparative study of mammalian erythrocyte carbonic anhydrases employing spin-labeled analogues of inhibitory sulfonamides. – Molec. Pharmacol., 1976, Vol. 12, pp. 832-843.

Wetzel R., Becker M., Behlke J., Billwitz M., Böhm S., Ebert B., Hamman H., Krumbiegel J., Lassman G. Temperature behaviour of human serum albumin. – Eur. J. Biochem., 1980, vol. 108, p. 469-475.

Wickett R.R., Ide G.J., Rosenberg A. A hydrogen-exchange study of lysozyme conformation changes induced by inhibitor binding. – Biochemistry, 1974, Vol. 13, pp. 3273-3277.

Wien R.W., Morriset J.D., McConnel M.M. Spin-label-induced nuclear relaxation. Distances between bound saccharides, histidine- and tryptophan- 123 on lysozyme in solution. II-1990. – Biochemistry, 1972, Vol. 11, p. 3707-3716.

Willan K.J., Golding B., Givol D., Dwek R.A. Specific spin labeling of the Fc region of immunoglobulins. – FEBS Lett., 1977, Vol. 80, pp. 133-136.

Wishnia A. Substrate specificity at alkane-binding sites of hemoglobin and myoglobin. – Biochemistry, 1969, Vol. 8, pp. 5064-5070.

Worcester D.L., Franks N.D. Structural analysis of hydrated egg lecithin cholesterol bilayers. II. Neutron diffraction. – J. Mol. Biol., 1976, Vol. 100, pp. 359-378.

Wu T.T., Kabat E.A. An analysis of the sequence of the variable regions of Bence-Jones proteins and myeloma L. chains and their implication for antibody complementarity. – J. Exp. Med., 1970, Vol. 132, pp. 211-219.

Yamanaka T., Mizushima H., Nozaki M., Horio T., Okunuki K. Studies on cytochrome C. III. Determination of "native" mammalian heart muscle cytochrome C and its physiological properties. – J. Biochem., 1959, Vol. 46, pp. 121-132.

Zagyanskii V.A., Nezlin R.S., Tumerman L.A. Flexibility of IgG molecules as established by fluorescent polarization measurements. – Immunochemistry, 1969, Vol. 6., pp. 787-797.

Zatsepina G.I. Book: "Svoistva i struktura vody". M., 1974.

Zavodszky P., Jaton J.C., Veniaminov S. Vu., Medgyesi G.A. Effect of antigen binding on the conformational stability of antibodies. – In: Abstr. 4th Europ. immunol. meet., Budapest, 1978.

Zavyalov V.P., Demchenko A.P., Sukhomudrenko A.G., Troitskii G.V. Temperature-dependent changes of myelome immunoglobulin G(K)IVA, Bence-Jones protein (K-type)IVA and its fragments. – Biochim. Biophys. Acta, 1977, Vol. 491, pp. 1-15.

Zav'yalov V.P., Tetin S. Yu., Abramov V.M., Troitskii G.V. Effect of salt concentration on immunoglobulin G. structure. – Biochim. Biophys. Acta, 1978, Vol. 533, pp. 496-503.

Zav'yalov V.P., Troitskii G.V., Abramov V.M., Skvortsov V.T. Conformational transitions of antibodies induced by hapten binding. – Biochim. Biophys. Acta, 1977, Vol. 493, pp. 359-366.

Zav'yalov V.P., Troitskii G.V., Demchenko A.P., Generalov I.V. Temperature and pH-dependent changes of immunoglobulin G structure. – Biochim. Biophys. Acta, 1975, Vol. 386, pp. 155-167.

Zav'yalov V.P., Troitskii G.V., Khechinashvili N.N., Privalov P.L. Thermally induced conformational transition of Bence-Jones protein IVA and its proteolytic fragments. – Biochim. Biophys. Acta, 1977, Vol. 492, pp. 102-111.

Zientara G.P., Nagy J.A., Freed J.H. Diffusion-controlled kinetics of protein domain coalescence: effect of orientation, interdomain forces and hydration. – J. Chem. Phys. 1980, Vol. 73, p. 5092-5106.

Index of Authors

281

Index of Subjects

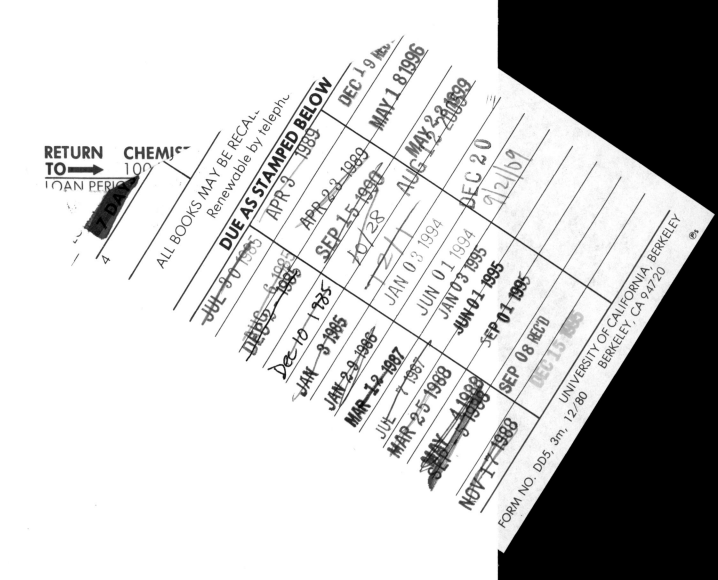

RETURN CHEMIS
TO ➡ 100
LOAN PERIO
7 DAY
4

ALL BOOKS MAY BE RECALL
Renewable by telepho

DUE AS STAMPED BELOW

		DEC 19 REC	
		MAY 1 8 1996	
	APR 3 1989		
	APR 2 1989	MAY 2 8 1999	
	SEP 15 1990	AUG 1 1994	DEC 20
JUL 3 0 1985	10/28		9/21/69
	JAN 03 1994		
	JUN 01 1994		
	JAN 03 1995		
DEC 6 1985	JUN 01 1995		
Dec 10 1985	SEP 01 1995		
JAN 3 1985			
JAN 29 1986		SEP 08 REC'D	
MAR 12 1987	JUL 7 1987	DEC 15 1985	
MAR 25 1988			
MAY 1988	SEP 1988		
NOV 17 1988			

UNIVERSITY OF CALIFORNIA, BERKELEY
BERKELEY, CA 94720

FORM NO. DD5, 3m, 12/80

RETURN **CHEMISTRY LIBRARY** 6427
TO ➡️ 100 Hildebrand Hall 642-3753

LOAN PERIOD 1	2	3
	1 MONTH	
		6